S. D. Bernardi
Associate Professor of Mathematics (Retired)
New York University

BIBLIOGRAPHY OF SCHLICHT FUNCTIONS

PART I (–1965)
PART II (1966–1975)
PART III (1976–1981)

MARINER PUBLISHING COMPANY, INC.
Tampa, Florida

This Bibliography is dedicated to

My wife Helen
My daughter Helen
My daughter Elizabeth

Copyright © by S. D. Bernardi, 1982, New Haven, Connecticut.

All rights reserved

Printed in the United States of America

ISBN: 0-936166-09-6

INTRODUCTION

This Bibliography contains a total of 4263 references to the publications of aproximately 1000 authors which have appeared in approximately 400 mathematical journals and publishing companies in the U.S.A., the Soviet Union, England, Germany, France, Japan, China, and in many other countries. It includes many symposiums, colloquiums, congresses, dissertations, abstracts, technical reports, lecture notes, and books dealing with the theory of analytic univalent (Schlicht) and multivalent mappings of simply and multiply-connected domains. This Bibliography covers the years 1902 (when first results began to appear) through the year 1981, plus several references in the year 1982.

This Bibliography consists of three separate parts. Parts I, II were first published by the Courant Institute of Mathematical Sciences, New York University, Technical Reports, respectively, IMM-35I (May 1966) and IMM-414 (March 1977). I thank New York University for allowing publication of Parts I, II in their present form. Part III has not been published elsewhere. Since much use has been made, since the year 1966, by many authors of the numbered references in Parts I, II it seems reasonable to keep the three parts of the Bibliography separate so as not to disturb the numbered references in the first two Parts. For example, extensive use has been made of the numbered references of Parts I, II, III by Professor A. W. Goodman in his forthcoming two-volume work, *Univalent Functions*, to be published by Mariner Publishing Company.

The principal methods of using this Bibliography and other pertinent information are described in the separate Prefaces to Parts I, II, III.

It is hoped that this extensive Bibliography will be a useful source to researchers in this special field, to students and teachers of function theory, and to engineers and scientists in industry.

The writing of this Bibliography during the past thirty years, consuming literally thousands of hours of trips to mathematics libraries, reading, photocopying, filing, typing, translation of foreign papers, etc., etc. would have been an impossible task without the constant help and encouragement of my wife Helen.

S. D. Bernardi
July 1982

CONTENTS

PART I (–1965) 1
PART II (1966–1975) 133
PART III (1976–1981) 274

BIBLIOGRAPHY OF SCHLICHT FUNCTIONS PART I (–1965)

CONTENTS

Preface vii
Bibliography... 1
Supplementary Bibliography 101
List of Mathematical Journals 113
Expository Papers 116
Topic References 117
Corrections 132

PREFACE

This bibliography contains 1694 references to the publications of more than 570 authors which have appeared in approximately 220 mathematical journals in the U.S.A., the Soviet Union, England, Germany, France, Japan, and in many other countries. It includes many symposiums, colloquiums, congresses, dissertations, abstracts, technical reports, lecture notes, and books dealing with the theory of analytic univalent (schlicht) and multivalent mappings of simply and multiply-connected domains. This survey covers the years 1907 (when the first substantial results were published) through 1965.

The interior of the unit circle is taken as the standard mapping domain for approximately 1400 papers, while the remaining papers deal with half-planes and various other canonical domains.

Of the total number of research papers listed in the bibliography approximately 736 of them contain in their titles the word univalent or multivalent (or their equivalents). However, many results in the class of schlicht functions are contained in papers whose titles do not specifically refer to them, such as typically-real functions, bounded functions, and functions of positive real part; many theorems in these three classes are easily transformed, as is well known, into theorems in the class of schlicht functions.

Included in the bibliography are 219 references which, while not dealing primarily with univalency, contain related results from analytic function theory to which specific references are made by many writers of papers on schlicht functions. These 219 'supporting' references were obtained from the bibliographies of papers on schlicht functions and include treatments of Grunsky coefficients, Faber polynomials, Bieberbach-Eilenberg functions, Cesaro means of power series, Parseval relations, Hadamard compositions of power series, Tchebychef polynomials, Lindelof's principle of subordination, Toeplitz forms, properties of solutions of the differential equation $w''(z) + p(z)w(z) = 0$, Schwarzian derivative, spherical derivative, orthogonal trajectories, level curves and quadratic differential forms. The reader who is familiar with the theory of univalent functions will readily recognize the relevancy of these topics.

The earliest paper listed in the bibliography [525] appeared in the year 1902, and the latest papers appeared in the year 1965 (with the single exception of [111] which appeared in the year 1966). Koebe's paper [688], published in 1907, is generally regarded as giving the first actual result in the theory of univalent functions.

Most of the papers listed in the bibliography have been reviewed in the Mathematical Reviews (MR) or in the Zentralblatt fur Mathematik (Zbl) or in the Jahrbuch uber die Fortschritte der Mathematik (FM), and *the volume and page number of the review is indicated at the end of each reference*. Of the approximately 150 references lacking review numbers, most of them may be classified as lecture notes, dissertations, technical and special reports, colloquiums, papers not yet published, or too recently published for inclusion in the Reviews.

A "Supplementary Bibliography" of 168 references follows the main bibliography and includes a list of some of the more recently published papers and several other references.

Approximately one thousand one hundred and twenty-one references were locatedin the Math. Reviews published during the years 1940-1964, volumes 1-28. Two hundred and forty-two references were located in the Zentralblatt published during the years 1931-1963, volumes 1-102. One hundred and eighty-one references were located in the Jahrbuch published during the years 1900-1937, volumes 31-63.

A volume-by-volume search was made of twenty-two mathematical journals which are readily available to the American public. A list of these journals and the papers on schlicht functions which they contain is given. Following this is a list of expository papers which are devoted wholly or in part to a general survey of the development of the theory of univalent and p-valent functions.

The various results in the theory of schlicht functions have been classified into sixty-eight topics. Each topic is followed by a list of those references inthe bibliography that contain information pertaining to that topic. Moreover, at the end of each reference in the bibliography the numbers enclosed in brackets are topic references indicating the topics which are discussed in the reference. The following two examples illustrate the principal use of the bibliography.

Example 1. Reference [363] lists "A proof of the Bieberbach conjecture for the fourth coefficient," by P. R. Garabedian and M. Schiffer. The notation MR 17-24 indicates that a review of this paper may be found in the Mathematical Reviews, volume 17, page 24. The numbers in brackets [9, 17, . . . , 54] indicate that the paper contains information regarding topic T9 (Meromorphic Univalent (p-valent) Functions), topic T17 (Coefficient Bounds), . . . , topic T54 (Coefficients Bounds for the Class (S)).

Example 2. The reader who is interested in close-to-convex functions will find this topic, T5. The references [129, 282, 656, . . . , A143] indicate that information regarding this topic may be found, respectively, in the papers by A. Bielecki and Z. Lewandowski, T. G. Erzohi, W. Kaplan, . . . , T. J. Suffridge. Numbers in brackets preceded by the letter "A" indicate reference to the Supplementary Bibliography.

The classification of the various papers into topics was based on a reading of approximately five hundred reprints (of the full-length papers), and in most cases on a reading of the (brief) abstracts or reviews. Therefore, in fairness to the authors, we emphasize that the classifications do not indicate the complete scope of the papers.

It is difficult in a work of this nature to cover every paper published on the subject of schlicht functions. However, it is felt that this bibliography does include a major portion of the publications in this field.

I gratefully acknowledge the aid given me by my graduate students Gerald Bierman and Victor Stanionis who assisted me with some of the numerous tasks of filing, tracking down certian ambiguous references, and the translation of some foreign papers. The replies to my requests for reprints from numerous authors were indeed gratifying and I take this opportunity to extend to them my heartfelt thanks. I owe many thanks also to the most cooperative personnel of the mathematics library at the Courant Institute of Mathematical Sciences. To New York University I am indebted for the financial aid given me by the Arts and Science Research Fund that helped pay the cost of typing the manuscript. Finally, I wish to express my sincere appreciation to the chairman of my department, Professor F. A. Ficken, who cooperated most graciously in arranging my teaching schedule so as to give me sufficient and convenient intervals of time to enable me to complete this bibliography.

S. D. Bernardi
May 1966

BIBLIOGRAPHY OF SCHLICHT FUNCTIONS (PART I)

1. Abe, H. *A note on subordination.* J. Gakugei Tokushima Univ. Nat. Sci. Math 7(1956), 47–51. MR 19-401. [1]
2. Abe, H. *On conformal mapping of a ring-shaped domain.* Sugaku 8(1956/57), 25–27. (Japanese) MR 20-664 [7, 10, 18, 48]
3. Abe, H. *On p-valent functions.* J. Gakugei Tokushima Univ. 8(1957), 33–40. MR 20-405. [9, 14, 16, 17, 19]
4. Abe, H. *On some analytic functions in an annulus.* Kodai Math Sem. Rep. 10(1958), 38–45. MR 20-544. [2, 6, 10, 13, 15, 48, 66]
5. Abe, H. *On univalent functions in an annulus.* Math. Japon. 5(1958/59), 25–28. MR 21-783. [9, 13, 15, 17, 25, 27, 48, 58]
6. Abramesco, N. *Sur le cercle d'univalence d'une fonction holomorphe $f(x)$ et sur la plus petite distance entre deux zeros d'une equation $f(x) = A$.* C.R. Acad. Sci. Paris 194(1932), 834–836. Zbl 4-10. [43]
7. Abramesco, N. *Sur le cercle d'univalence d'une fonction holomorphe $f(x)$ et sur la plus petite distance entre deux zeros d'une equation $f(x) = A$.* Rend. Circ. Mat. Palermo 58(1934), 49–54. Zbl 9-76. [43]

8. Ahlfors, L. *Untersuchungen zur Theorie der konformen Abbildung und der ganzen Funktionen.* Acta Soc. Sci. Fenn. A. 1(1930), No. 9 FM 56-984. [59]
9. Ahlfors, L. *Sur une generalisation du theoreme de Picard.* C.R. Acad. Sci. Paris 194(1932), 245-246. Zbl 3-407. [59]
10. Ahlfors, L. *An extension of Schwarz's lemma.* Trans. Amer. Math. Soc. 43(1938), 359-364. Zbl 18-410. [19, 37]
11. Ahlfors, L. *Bounded Analytic Functions.* Duke Math. J. 14(1947), 1-11. MR 9-24. [37, 60]
12. Ahlfors, L. *Open Riemann surfaces and extremal problems on compact subregions.* Comment. Math. Helvet. 24(1950), 100-134. MR12-90. [59]
13. Ahlfors, L. *Conformal mapping.* Lecture Notes, Oklahoma A.and M. College, 1951. [59]
14. Ahlfors, L. *Development of the theory of conformal mapping and Riemann surfaces through a century.* Contributions to the theory of Riemann surfaces. Ann. of Math. Stud., No. 30 (1953), 3-13. MR 14-1050. [59]
15. Ahlfors, L. *Complex Analysis.* New York, 1953, 247. MR 14-857. [59]
16. Ahlfors, L. *Extremal Probleme in der funktionen Theorie.* Ann. Acad. Sci. Fenn. Ser. A.I. No. 249 1(1958), 9. MR 19-845. [24]
17. Ahlfors, L.; Beurling, A. *Invariants conformes et problemes extremaux.* Dixieme Congres des Math. Scandinaves, (1946), 341-351. MR 9-23. [59]
18. Ahlfors, L.; Beurling, A. *Conformal invariants and function-theoretic nul-sets.* Acta Math. 83(1950), 101-129. MR 12-171. [59]
19. Ahlfors, L.; Beurling, A. *Conformal Invariants.* Construction and Applications of Conformal Maps, Proceedings of a Symposium ; 243-245. Nat. Bureau of Standards, Appl. Math. Ser. No. 18, Wash., 1952. MR 14-861. [59]
20. Ahlfors, L.; Grunsky, H. *Uber die Blochsche Konstante.* Math. Z. 42(1937), 671-673. FM 63-300. [19]
21. Aikawa, S. *On extension of Schwarz's theorem.* Kodai Math Sem. Rep. 1952(1952), 104-106. MR 15-210. [37]
22. Aksentev, L. A. *Sufficient conditions for univalence of regular functions.* Izv. Vyss. Ucebn. Zaved. Matematika (1958), No. 3(4), 3-7. (Russian) MR 26-278. [4]
23. Aksent'ev, L. A. *Elementary criteria for univalence in terms of boundary characteristics.* Izv. Vyss. Ucebn. Zaved. Matematika (1959), No. 6(13), 3-8. (Russian) Zbl 96-55. [4, 6, 10, 62]
24. Aksent'ev, L. A. *Integral representations of univalent functions.*

Izv. Vyss. Ucebn. Saved. Matematika (1959), No. 4(11), 3-8. (Russian) MR 24A-37. [4, 16, 23]

25. Aksent'ev, L. A. *On the univalence of partial sums of power series.* Izv. Vyss. Ucebn. Aved. Matematika (1960), No. 5(18), 12-15. (Russian) MR 24A-153. [4, 20, 23, 43]

26. Aksent'ev, L. A. *Univalent variation of the profile of a dam.* Issledovanija po Souremennym Problemam Teorii Funkcii Kompleksnogo Permennogo, 335-340. Gosvdarstv. Izdat. Fiz-Mat. Lit., Moscow, 1960. (Russian) MR23A-52. [4]

27. Aksent'ev, L. A. *On the univalence of the solution of the inverse problem of hydromechanics.* Izv. Vyss. Zaved. Matematika (1961), No. 4(23), 3-7. (Russian) MR 25-611. [59]

28. Aleksandrov, I. A. *On bounds for convexity and star-likeness for functions univalent and regular in a circle.* Dokl. Akad. Nauk SSSR (N.S.) 116(1957), 903-905. (Russian) MR 20-161. [6, 10, 11, 12]

29. Aleksandrov, I. A. *Conditions for convexity of the image region under mapping by functions regular and univalent in the unit circle.* Izv. Vyss. Ucebn. Zaved. Matematika (1958), No. 6(7), 3-6. (Russian) MR 23A-731. [10, 12, 35]

30. Aleksandrov, I. A. *On the star-shaped character of the mappings of a domain by functions regular and univalent in the circle.* Izv. Vyss. Ucebn. Zaved. Matematika (1959), No. 4(11), 9-15. (Russian) MR 26-500. [6]

31. Aleksandrov, I. A. *Domains of definition of some functionals on the class of functions univalent and regular in a circle.* Issledovaniya po Souremennym Problemam Teorii Funkcii Kompleksnogo Peremennogo. (Russian) MR 22A(1)-965. [6, 10, 12, 29]

32. Aleksandrov, I. A. *Variational problems for functions univalent and star-shaped in the circle.* Izv. Akad. Nauk Armjan. SSRSer. Fiz. Mat. Nauk 14(1961), No. 4, 7-19. (Russian, Armenian summary) MR25-426. [6, 17, 29, 36]

33. Aleksandrov, I. A. *Boundary values of the functional* $J = J(f, \bar{f}, f', \bar{f}')$ *on the class of holomorphic functions univalent in a circle.* Sibirsk. Mat. Z. 4(1963), 17-31. (Russian) MR 26-744. [29, 49]

34. Aleksandrov, I. A.; Černikov, V. V. *Extremal properties of star-like mappings.* Sibirsk. Mat. Z. 4(1963), 241-267. (Russian) MR 25-1210. [2, 6, 10, 24, 29, 39, 49]

35. Alenicyn, Y. E. *On Univalent Majorants.* Mat. Sbornik N.S. 26(68) (1950), 57-74. MR 11-589. [1]

36. Alenicyn, Y. E. *On functions p-valent in the mean.* Mat. Sbornik N.S. 27(69) (1950) 285-296. MR 12-491. [9, 14, 27, 33]

37. Alenicyn, Y. E. *On the estimation of the coefficients of univalent functions.* Mat. Sbornik N.S. 28(70) (1951), 401–406. MR 13–640. [17, 45, 52]
38. Alenicyn, Y. E. *On univalent functions in multiply connected domains.* Dokl. Akad. Nauk SSSR (N.S.) 102 (1955), 861–863. MR 17–25. [22, 31]
39. Alenicyn, Y. E. *On univalent functions in multiply connected domains.* Mat. Sb. N.S. 39(81) (1956), 315–336. (Russian) MR 18–292. [15, 19, 22, 60]
40. Alenicyn, Y. E. *A contribution to the theory of univalent and Bieberbach-Eilenberg functions.* Dokl. Akad. Nauk SSSR (N.S.) 109(1956), 247–249. (Russian) MR 18–293. [15, 26, 34, 49]
41. Alenicyn, Y. E. *On functions without common values and the outer boundary of the domain of values of a function.* Dokl. Akad. Nauk SSSR (N.S.) 115(1957), 1055–1057. MR 20–664. [15, 26, 34]
42. Alenicyn, Y. E. *On functions without common values and the outer boundary of the domain of values of a function.* Mat. Sb. N.S. 46(88) (1958), 373–388. (Russian) MR 20–1074. [26]
43. Alenicyn, Y. E. *An extension of the principle of subordination to multiply connected regions.* Dokl. Akad. Nauk SSSR 126(1959), 231–234. (Russian) MR 22A(1)–126. [1, 37]
44. Alenicyn, Y. E. *Functions without common values.* Issledovaniya po Souremennym Problemam Teorii Funckcii Kompleksnogo Peremennogo, 34–38. Gosodarstv. Izdat. Fix. –Mat. Lit., Moscow 1960. MR 22A(1)–1899. [26, 34]
45. Alenicyn, Y. E. *Some extremal properties of functions multivalent in multiply connected domains.* Dokl. Akad. Nauk SSSR 146(1962), 267–269. (Russian) MR 26–743. [14, 17, 27, 33, 60]
46. Alenicyn, Y. E. *On the ranges of systems of coefficients of functions representable as a sum of Stieltjes integrals.* Vestnik Leningrad Univ. 17(1962), No. 7, 25–41. (Russian. English summary) MR 25–426. [2, 13, 23, 29]
47. Alenicyn, Y. E.; Havinson, S. Y. *The radius of p-valence for bounded analytic functions in multiply connected domains.* Mat. Sb. (N.S.) 52(94) (1960), 653–657. (Russian) MR 22A(1)–1628. [43, 60]
48. Alenitzyn, G. *On the coefficients of p-valent functions.* Rec. Math. (Mat. Sbornik) N.S. 10(52), 1942, 51–58. (Russian. English summary) MR 4–138. [14, 15, 17, 50, 65]
49. Alenitzyn, G. *On the coefficients of "schlicht" functions.* Rec. Math. (Mat. Sbornik) N.S. 15(57), (1944), 131–138. (Russian. English summary) MR 7–54.) [15, 17, 39, 45, 52, 67]

50. Alenitzyn, G. *On the locally univalent functions.* Rec. Math. (Mat. Sbornik) N.S. 18(60), (1946), 115–123. (Russian. English summary) MR 7-515. [57]
51. Alenitzyn, G. *On mean p-valent functions.* Rec. Math. (Mat. Sbornik) N.S. 20(62), (1947), 113–124. MR 9-23. [27, 33]
52. Alexander, II, J. W. *Functions which map the interior of the unit circle upon simple regions.* Annals of Math. 17(1915), 12–22. FM 45-672. [4, 6, 10, 32]
53. Anderson, B. J. *A note on the constant of Koebe.* Ark. Mat. 2(1953) 415–416. MR 15-114. [19]
54. Antonjuk, G. K. *About the extension of some extreme properties of Schlicht functions in a ring to non-Schlicht functions.* Vestnik Leningrad Univ. 14(1959), No. 13, 71–82. (Russian. English summary) MR 23A-55. [48]
55. Artemiadis, N. K. *Quelques théorèmes sur les transformees de Fourier et sur les coefficients des fonctions typiquement reeles.* C. R. Acad. Sci. Paris 244(1957), 544–547. MR 18-575. [6, 13, 17, 51, 36]
56. Artemiadis, N. K. *Sur les coefficients de Taylor de certaines classes de fonctions.* C. R. Acad. Sci. Paris 244(1957), 713–715. MR 19-22. [8, 17, 49]
57. Artemiadis, N. K. *Généralisation d'un theoreme de M. S. Mandelbrojt.* C. R. Acad. Sci. Paris 244(1957), 834–836. MR 19-130. [13]
58. Artemiadis, N. K. *Sur les transformées de Fourier et leur applications aux series et sur les fonctions typiquement reeles d'ordre p.* Ann. Sci. Ecole Norm. Sup. III, Ser. 74 (1957), 269–318. Zbl 89-53. [13]
59. Arwin, A. *The Poisson integral and an analytic function in its circle of convergence.* Ann. Math. 23, (1921-22) 141–143. FM 48-395. [59]
60. Ašnevic, I. Ya; Ulina, G. V. *On regions of values of analytic functions represented by a Stieltjes integral.* Vestnik Leningrad Univ. 10(1955), No. 11, 31–42. (Russian) Amer. Math. Soc. Transl. (2) 22, 81–94. MR 17-599. [2, 6, 13, 23, 29]
61. Atkins, H. P. *On fractional derivatives of univalent functions.* Bull. Amer. Math. Soc. 52(1946) 1060–1064. MR 8-326. [15, 42]
62. Aumann, Georg. *Die Mittelpunktsverzerrung bei konvexen konformen Abbildungen.* Math. Z. 46(1940) 80–82. MR 2-83. [59]
63. Aumann, Georg. *Uber die Streckenverzerrung bei konvexen konformen Abbildung.* S. -B. Math. Nat. Ki. Bayer. Akad. Wiss. 1948 (1949), 303–308. MR 11-507. [10, 29]
64. Ballieu, R. *Sur les fonctions localement univalentes dans le cercle unite* C. R. Acad. Sci. Paris 206(1938), 413–415. Zbl 18-143. [15, 68]

65. Ballieu, R. *Sur les familles de fonctions localement univalentes dans le cercle unite.* Ann. Soc. Sci. Bruxelles, Ser. I. 58(1938), 103-114. Zbl 19-69. [15, 68]
66. Banach, S. *Sur les lignes rectificables et les surfaces dout l'aire est finie.* Fund. Math. 7(1925), 225-236. FM 51-199. [59]
67. Basilewitsch, J., (Basilevitch) *Zum Koeffizientenproblem der schlichten Funktionen.* Mat. Sb. (Rec. Math.) (N. S.) 1(1936), 211-236. FM 62-372. [17, 49]
68. Basilewitsch, J., (Basilevitch) *Sur les théorèmes de Koebe-Bieberbach* Rec. Math. Moscou. N. S. 1(1936), 283-291. Zbl 15-71. [15, 68]
69. Basilewitsch, J., (Basilevitch) *Sur la théorie des fonctions univalentes.* Rec. Math. Moscou N. S. 3(1938), 359-366. (Russian) Zbl 19-271. [17, 24]
70. Basilewitsch *Sur un théorème de Littlewood et Paley.* Rec. Math. (Mat. Sbornik) N. S. 6(48), 337-344 (1939). (Russian, French summary) MR 1-308. [15, 17, 45, 52, 67]
71. Basilewitsch, J., (Basilevitch) *D'une propriété extremale de la fonction $f(z) = z/(l-z)^2$.* Rec. Math. (Mat. Sbornik) N. S. 12(54) (1943), 315-319. (Russian. French summary) MR 5-259. [15, 24, 39, 68]
72. Bazilevič, I. E. *Improvement of estimates for the coefficients of univalent functions I.* Mat. Sbornik N. S. 22(64) (1948) 381-390. MR 9-186. [15, 17, 42, 54]
73. Bazilevič, I. E. *On distortion theorems and the coefficients of univalent functions.* Doklady Akad. Nauk SSSR (N. S.) 65(1949), 253-255. MR 10-602. [9, 15, 19, 42, 58]
74. Bazilevič, I.E. *On distortion theorems for univalent functions.* Uspehi Matem. Nauk (N. S.) 4 No. 3(31) (1949), 128-130. MR 11-508. [9, 15, 19, 42, 58]
75. Bazilevič, I. E. *On distortion theorems and coefficients of univalent functions.* Mat. Sbornik N. S. 28(70) (1951), 147-164. MR 12-600. [9, 15, 17, 18, 19, 24, 42, 45, 54, 58, 67, 68]
76. Bazilevič, I. E. *On distortion theorems in the theory of univalent functions.* Mat. Sbornik N. S. 28(70), (1951), 283-292. (Russian) MR 13-640. [9, 15, 24, 58, 68]
77. Bazilevič, I. E. *On a case of integrability in quadratures of the Loewner-Kufarev equation.* Mat. Sb. N. S. 37(79) (1955), 471-476. (Russian) MR 17-356. [4, 6, 24, 42]
78. Bazilevič, I. E. *Regions of the initial coefficients of bounded univalent functions with p-fold symmetry.* Mat. Sb. N. S. 43(85) (1957), 409-428. (Russian) MR 20-664. [22, 24, 29, 49]
79. Bazilevič, I. E. *On an estimate of the mean modulus in a class of*

bounded univalent functions. Mat. Sb. (N. S.) 48(90) (1959), 93-104. (Russian) MR 22A(1)-295. [3, 19, 22]
80. Bazilevič, I. E. *On some properties of univalent conformal mappings.* Mat. Sbornik. N. S. 32(74) (1954) 209-218. MR 14-632. [6, 7, 10]
81. Bazilevič, I. E.; Korickii, G. V. *Certain properties of level curves under univalent conformal mappings.* Dokl. Akad. Nauk. SSSR 149(1961) 279-280. (Russian); Soviet Math. Doklady 2(1961), No. 5, 1180-1181. MR 24A-370. [6, 7, 9, 10]
82. Bazilevič, I. E.; Korickii, G. V. *Some properties of level curves on univalent conformal mappings.* Mat. Sb. (N. S.) 58(100) (1962), 249-280. (Russian) MR 26-500. [6, 7, 9, 10]
83. Beatty, S. A note on Schlicht functions. Trans. Roy. Soc. Canada III Math. Sci. III S. 25(1931), 83-85. Zbl 3-262. [27, 42]
84. Beckenbach, E. F. *A relative of the lemma of Schwarz.* Bull. Amer. Math. Soc. 44(1938), 698-707. Zbl 19-350 [59]
85. Beckenbach, E. F. *On a theorem of Fejer and Riesz.* J. London Math. Soc. 13(1938), 82-86. Zbl 19-67. [59]
86. Beckenbach, E. F. *A property of mean values of an analytic function.* Rend. Circ. Mat. Palermo (2) 1(1952), 157-163. MR 14-629. [3, 7]
87. Beckenbach, E. F.; Graham, E. W. *On subordination in complex variable theory.* Proc. of a symposium, 247-254. Nat. Bureau of Standards Appl. Math. Ser. (1952) No. 18 U.S. Govt. Print. Office. MR 14-632. [1, 6, 10, 37]
88. Beckenbach, E. F.; Gustin, W.; Shniad, H. *On the mean modulus of an analytic function.* Bull. Amer. Math. Soc. 55(1949), 184-190. MR 10-441. [3, 7]
89. Beesack, P. R. *Non oscillation and disconjugacy in the complex domain.* Trans. Amer. Math. Soc. 81(1956), 211-242. MR 18-483. [14, 31]
90. Beesack, P.R. *Linear differential equations and convex mappings.* Duke Math. J. 27(1960), 483-495. MR 22A(1)-1629. [6, 10, 14, 31]
91. Beesack, P. R.; Schwarz, B. *On the zeros of solutions of second order linear differential equations.* Canadian J. of Math. 8(1956), 504-515. MR 18-211. [31]
92. Bender, J. *Some extremal theorems for multivalently star-like functions.* Duke Math. J. 29(1962), 101-106. MR 24A-370. [6, 14, 15, 17, 23, 36, 50, 63, 65]
93. Berg, P. W. *On univalent mappings by solutions of linear elliptic partial differential equations.* Trans. Amer. Math. Soc. 84(1957), 310-318. MR 18-741. [59]
94. Bergmann, S. *Uber die Entwicklung der harmonischen Funktionen*

der Ebene und des Raumes nach Orthogonalfunktionen. Math. Ann. 86(1922), 237–271. FM 48–1236. [59]
95. Bergmann, S. *The kernel function and conformal mapping.* Amer. Math. Soc. Surveys, No. 5, New York, 1950, 161 pp. MR 12–402. [59]
96. Bergmann, S.; Schiffer, M. M. *Kernel functions and conformal mapping.* Composito Math. 8(1951), 205–249. MR 12–602. [59]
97. Bermant, A. *Sur quelques properiétés des fonctions régulières.* C. R. (Dokl.) Acad. Sci. U. R. S. S., N. S. 18(1938), 137–140. Zbl 18–143. [6, 18, 19]
98. Bermant, A. *Remarque sur le lemme de Schwarz.* C. R. Acad. Sci. (Paris) 207(1938), 31–33. Zbl 19–124. [6, 19, 37]
99. Bermant, A. *Sur la variation de la dilation d'une fonction reguliere.* C. R. (Doklady) Acad. Sci. URSS (N. S.) 45(1944), 271–273. MR 7–150. [6, 19]
100. Bermant, A. *Dilation d'une fonction modulaire et problemes de recouvrement.* Rec. Math. (Mat. Sbornik) N. S. 15(57) (1944), 285–324. (Russian. French summary) MR 7–150. [1, 19]
101. Bermant, A. *On a generalization of Koebe's theorem.* C. R. (Doklady) Acad. Sci. URSS (N. S.) 52(1946), 379–381. MR 8–202. [19]
102. Bermant, A. *On certain generalizations of E. Lindelof's principle and their applications.* Rec. Math. (Mat. Sbornik) N. S. 20(62) (1947), 55–112. MR 9–138. [1, 37]
103. Bernardi, S. D. *Two theorems on Schlicht functions.* Duke Math. J. 19(1952), 5–21. MR 13–733. [17, 39]
104. Bernardi, S. D. *A survey of the development of the theory of Schlicht functions.* Duke Math. J. 19(1952), 263–287. MR 14–35. [44]
105. Bernardi, S. D. *Note on an inequality of Prawitz.* Duke Math. J. 23(1956), 385–391. MR 17–1193. [27, 39]
106. Bernardi, S. D. *The centroid of analytic mappings.* Amer. Math. Mon. 64(1957), 259–261. MR 19–258. [49]
107. Bernardi, S. D. *A determinant inequality for univalent functions.* Amer. Math. Monthly 64(1957), 495–497. MR 19–539. [17, 27, 54]
108. Bernardi, S. D. *Functions with positive real part and Schlicht functions.* Duke Math. J. 26(1959), 541–548. MR 26–63. [2, 4, 6, 10]
109. Bernardi, S. D. *Convex, starlike, and level curves.* Duke Math J. 28(1961), 57–72. MR 23A–177. [6, 10, 13, 35]
110. Bernardi, S. D. *Circular regions covered by Schlicht functions.* Duke Math. J. 32(1965), 23–36. MR 30–416. [19]
111. Bernardi, S. D. *Special classes of subordinate functions.* Duke

Math. J. 33(1966), 55-67. [1, 6, 10, 13, 20]. MR 32-992
112. Besicovitch, A. S. *Almost Periodic Functions.* Cambridge, 1932. Zbl 4-253. [59]
113. Besicovitch, A. S. *On two problems of Lowner.* J. London Math. Soc. 27(1952), 141-144. MR 13-831. [59]
114. Beurling, A. *Études sur un probleme de majoration.* Thesis, Upsala, 1933. FM 59-1042. [59]
115. Bieberbach, L. *Zur Theorie und Praxis der konformen Abbildung.* Rc. Circolo Mat. Palermo 38(1914), 98-112. FM 45-670. [59]
116. Bieberbach, L. *Über einige Extremalprobleme im Gebiete der konformen Abbildung.* Math. Ann. 77(1916), 153-172. FM 46-549. [9, 19, 22]
117. Bieberbach, L. *Über die Koeffizientem derjenigen Potenzreihen, welche eine schlichte Abbildung des Einheitskreises vermitteln.* Preuss. Akad. Wiss. Sitzungsb. (1916), 940-955. FM 46-552. [9, 17, 27, 29, 34, 42, 54]
118. Bieberbach, L. *Augstellung und Beweis des Drehungssatzes fur schlichte konforme Abbildungen.* Math. Zeit. 4(1919), 295-305. FM 47-327. [10, 15, 64, 68]
119. Bieberbach, L. *Neu Forschungen im Gebiete der konformen Abbildung.* Glasnik hrv. prirod. drustva, 33(1921). FM 48-403. [6, 42]
120. Bieberbach, L. *Über die Kreisabbildung von nahen kreisformigen Bereichen.* Preuss. Akad. Wiss. Sitzungsber. (1924), 181-188. FM 50-640. [19]
121. Bieberbach, L. *Eine hinreichende Bedingung fur schlichte Abbildungen des Einheitskreises.* J. F. M. 157(1927), 189-192. FM 53-322. [4, 15]
122. Bieberbach, L. *Zur Theorie der schlichten Abbildungen.* Bull. Calcutta Math. Soc. 20(1928), 17-20. FM 56-297. [19, 42, 43]
123. Bieberbach, L. *Über einige Extremalprobleme in Gebiete der konformen Abbildung.* Math. Ann. 25(1935), 267-272. [17, 49]
124. Bieberbach, L. *Über schlichte Abbildungen des Einheitskreises durch meromorphe Funktionen.* II. S. -B. Preuss. Akad. Wiss. (1937), 359-365. Zbl 18-29. [19]
125. Bieberbach, L. *Lehrbuch der Funktionentheorie.* V. 2, 2nd Edition. Chelsea Publ. Co., New York, 1945, 370. MR 6-261. [44]
126. Bieberbach, L. *Conformal Mapping.* Chelsea Co. (Translated by F. Steinhardt), 1953, 234. Zbl 50-84. [44]
127. Bieberbach, L. *Analytische Fortsetzung.* Ergebnisse der Mathematik und ihrer Grenzgebiete. N. F. Heft 3. Springer-Verlag, Berlin-Gottingen-Heidelberg, 1955. MR 16-913. [59]
128. Bielecki, A.; Lewandowski, Z. *Sur certaines familles de fonctions*

a-étoilées. Ann. Univ. Mariae Curie-Sklodowska Sect. A15(1961), 45-55. (Polish and Russian summaries) MR 26-63. [6]
129. Bielecki, A.; Lewandowski, Z. *Sur un théorème concernant les fonctions univalentes linéairement accessibles de M. Biernacki.* Ann. Polon. Math. 12(1962), 61-63. MR 26-980. [5]
130. Bielecki, A.; Lewandowski, Z. *Sur une généralisation de quelques théorèmes de M. Biernacki sur les fonctions analytiques.* Ann. Polon. Math. 12(1962), 65-70. MR 26-743. [1]
131. Bielecki, A.; Krzyz, J.; Lewandowski, Z. *On typically-real functions with a preassigned second coefficient.* Bull. Acad. Polon. Sci. Ser. Sci. Math. Astronom. Phys. 19(1962), 205-208. MR 25-427. [13, 15]
132. Biernacki, M. *Sur quelques majorantes de la théorie des fonctions univalentes.* C. R. Acad. Sci. Paris 201(1935), 256-258. Zbl 12-171. [1, 6, 10]
133. Biernacki, M. *Sur la représentation conforme des domaines linéairement accessibles.* Prace Mat. -Fix. 44(1936), 293-314. Zbl 14-24. [29, 49]
134. Biernacki, M. *Sur les fonctions univalentes.* Mathematica 12(1936), 49-64. FM 62-375. [1, 6, 10]
135. Biernacki, M. *Sur les fonctions multivalentes d'ordre p.* C. R. Acad. Sci. Paris 203(1936), 449-451. Zbl 14-319. [3, 14, 17, 50]
136. Biernacki, M. *Les Fonctions Multivalentes.* Hermann et Cie., Editors, 1938. Zbl 22-153. [14]
137. Biernacki, M. *Sur la représentation conforme des domaines étoilés.* Mathematica, Cluj. 16(1940), 44-49. MR 2-84. [6, 18, 19]
138. Biernacki, M. *Sur les fonctions univalentes et k-symetriques.* Bull. Sci. Math. (2) 69(1945), 204-214. MR 8-145. [16, 17, 33, 49]
139. Biernacki, M. *Sur le moyennes de module des fonctions holomorphes.* Ann. Univ. Mariae Curie-Sklodowska. Sect. A. 1, 1-8(1946). (French. Polish summary). MR 9-576. [1, 3, 19, 27]
140. Biernacki, M. *Sur les domaines couverts par des fonctions multivalentes.* Bull. Sci. Math. (2) 70(1946), 45-51. MR 8-326. [1, 14, 19]
141. Biernacki, M. *Sur les fonctions en moyenne multivalentes.* Bull. Sci. Math. (2) 70(1946), 51-76. MR 8-326. [9, 14, 15, 19, 27, 33, 53]
142. Biernacki, M. *Sur une inégalité entre les moyennes des dérivées logarithmiques.* Mathematica, Timisoara 23(1948), 54-59. MR 10-186. [6, 10, 15, 63, 64, 68]
143. Biernacki, M. *Sur quelques applications de la formule de Parseval.* Ann. Univ. Mariae Curie-Sklodowska. Sect. A. 4, (1950), 23-40. (French. Polish summary) MR 13-123 [47]

144. Biernacki, M. *Über schwach p-wertige Funktionen*. Dodatek Rocznika Polsk. Towarz. Mat. 22, 39(1951). (Polish) Zbl 45-187. [14]
145. Biernacki, M. *Sur quelques applications de la formule de Parseval*. II. Ann. Univ. Mariae Curie-Sklodowska. Sect. A. 7(1953), 5-14(1954). (Polish. Russian summaries) MR 16-808. [3, 7, 33, 47]
146. Biernacki, M. *Sur les coefficients de Taylor des fonctions univalentes*. II. Ann. Univ. Mariae Curie-Sklodowska. Sect. 9. (1955), 127-133(1957. (Polish. Russian summaries) MR 19-540. [8, 14, 16, 33]
147. Biernacki, M. *Sur les fonctions en aire multivalentes*. Ann. Univ. Mariae Curie-Sklodowska. Sect. A. 8(1954), 71-79(1956). (Polish, Russian summary) MR 18-387. [3, 17, 33]
148. Biernacki, M. *Sur les moyennes et les extrema des modules des fonctions analytiques*. Ann. Univ. Mariae Curie-Sklodowska. Sect. A. 19(1956), 127-136(1958). (Polish. Russian summaries) MR 20-291. [3, 15]
149. Biernacki, M. *Sur les coefficients Tayloriens des fonctions univalentes*. Bull. Acad. Polon. Sci. Cl. IV 4(1956), 5-8. MR 17-957. [8]
150. Biernacki, M. *Sur les coefficients de Taylor des fonctions univalentes*. II. Ann. Univ. Mariae Curie-Sklodowska. Sect. A 9(1957), 127-133. Zbl 89-50. [8, 17]
151. Biernacki, M. *Sur l'integrale des fonctions univalentes*. Bull. Acad. Poon. Sci. Ser. Sci. Math. Astr. Phys. 8(1960), 29-34. (Russian summary) MR 22A(1)-1376. [4, 16, 24]
152. Biernacki, M. Krzyz, J. *On the monotonity of certain functionals in the theory of analytic functions*. Ann. Univ. Mariae Curie-Sklodowska. Sect. A. 9(1955), 134-147(1957). (Polish. Russian summaries) MR 19-736. [3, 7, 16, 18]
153. Bilimovitch, A. *Sur la mesure de deflexion d'une fonction non analytique par rapport à une fonction analytique*. C. R. Acad. Sci. Paris 237(1953), 694-695. MR 15-521. [59]
154. Bilimovitch, A. *Sur les transformations des fonctions non analytiques*. C. R. Acad. Sci. Paris 247(1958), 1954-1956. MR 23A-50. [59]
155. Birnbaum, Z. W. *Beiträge zur Theorie schlichten Funktionen*. Studia Math. 1(1929), 159-190. FM 55-210. [9, 15, 27, 58]
156. Birnbaum, Z. W. *Uber schlichte Funktionen*. Arch. Towarz. Nauk Lwow. Wydz. -Mat. Przyr. 5(1931), 233-237. Zbl 1-343. [6, 15, 68]
157. Blakeley, G. R. *Classes of p-valent starlike functions*. Proc. Amer. Math. Soc. 13(1962), 152-157. MR 24A-253. [6, 14, 15, 63, 65]

158. Bloch, A. *Les théorèmes de M. Valiron sur les fonctions entières et la theorie de l'uniformisation.* C. R. Acad. Sci. Paris 178(1924), 2051-2052. FM 50-217. [19]
159. Bloch, A. *Les fonctions holomorphes et méromorphes dans le cercle-unite.* Paris, Gauthier-Villars (1926), FM 52-324. [59]
160. Blumenthal, L. M. *Some relationships involving subordination.* Proc. Amer. Math. Soc. 10(1959), 502-510. MR 21-830. [59]
161. Boas, R. P. *Univalent derivatives of entire functions.* Duke Math. J. 6(1940), 719-721. MR 2-82. [16, 28]
162. Bochner, S. *Über orthogonale Systeme analytischer Funktionen.* Math. Z. 14(1922), 180-207. FM 48-376. [59]
163. Bohr, H. *Über streckentreue und konforme Abbildung.* Math. Z. 1(1918), 403-420. FM 46-558. [59]
164. Bohr, H. *Über einen Statz von Edmund Landau.* Scr. Bibl. Univ. Hierosolym 1(1923), No. 2, FM 49-711. [59]
165. Bonnesen, T.; Fenchel, W. *Theorie der konvexen Korper.* Berlin, 1934. FM 60-673. [59]
166. Breusch, R. *On the distribution of the values of $|f(z)|$ in the unit circle.* Bull. Amer. Math. Soc. 54(1948), 1109-1114. MR 10-363. [59]
167. Brown, R. K. *On the partial sums of a typically-real function.* Master of Science Thesis (1950), Rutgers Univ. Library. [13]
168. Brown, R. K. *Some mapping properties of the mean sums of power series and a problem in typically-real functions of order p.* Ph.D. Thesis (1952), Rutgers Univ. [13]
169. Brown, R. K. *Typically-real functions.* Canad. J. Math. 11(1959), 122-130. MR 21-662. [13]
170. Brown, R. K. *A class of typically-real meromorphic functions.* USASRDL Tech. Report 2121(1960). [13, 15, 17, 23, 29, 51, 66]
171. Brown, R. K. *Univalence of Bessel functions.* Proc. Amer. Math. Soc. 11(1960), 278-283. MR 22A(1)-455. [4, 28, 31]
172. Brown, R. K. *Univalent solutions of $w'' + pw = 0$.* Canad. J. Math. 14(1962), 69-78. MR 23A-159. [4, 16, 28, 31]
173. Cacridis-Theodorakopulos, P. *Über die untere Grenze der Rundungschranken der beschrankten Funktionen $f(z)$, deren $f'(0)$ vorgegeben ist.* Math. Ann. 113(1936), 656-664. FM 62-370. [11, 12, 22]
174. Cacridis-Theodorakopulos, P. *Über die Krummung der Niveaukurven der beschrankten Funktionen.* Math. Ann. 114(1937), 275-283. FM 63-287. [11, 12, 22]
175. Čakalov (Tchakaloff), L. *Sur l'univalence de certaines classes de*

fonctions. Ann. Univ. Sofia Fac. Phys. -Math. 33(1937), 243–252. Zbl 17–270. [4, 32, 43]
176. Čakalov (Tchakaloff), L. *Sur une classe de fonctions analytiques univalentes.* C. R. Acad. Sci. Paris 242(1956), 437–439. MR 17–724. [4, 9, 43]
177. Čakalov (Tchakaloff), L. *Maximal regions of univalence of certain classes of analytic functions.* Ukrain. Mat. Z. 11(1959), 408–412. (Russian. English summary) MR 23A–476. [32]
178. Čakalov (Tchakaloff), L. *Sur les domaines d'univalence de certaines classes de fonctions analytiques.* Bulgar. Akad. Nauk Izv. Mat. Inst. 4(1960), No. 2, 43–55. MR 23A–177. [29, 62]
179. Čakalov (Tchakaloff), L. *Domains of univalence of certain classes of analytic functions.* Soviet Math. —Doklady 1(1960), No. 2, 781–783. MR 24A–50. [4, 32, 43]
180. Čakalov (Tchakaloff), L. *Sur les domaines d'univalence des polynomes alge-briques.* Ann. Univ. Budapest Eotvos Sect. Math. 3–4(1960/61), 357–361. MR 24A–494. [4, 32, 43]
181. Callahan, Jr., F. P. *An extremal problem for polynomials.* Proc. Amer. Math. Soc. 10(1959), 754–755. MR 22A(1)–959. [3, 32]
182. Călugăreano, G. *Sur les conditions nécessaires et suffisantes pour l'univalence d'une fonction holomorphe dans un cercle.* Mathematica 6(1932), 75–79. Zbl 5–18. [4, 43]
183. Călugăreano, G. *Sur les fonctions univalentes.* II. Acad. R. P. Romane. Fil. Cluj. Stud. Cerc. Sti. 5(1954), 15–26. MR 17–472. [4, 9, 27]
184. Capelli, P. *Some observations on univalent functions and particular class of them.* Fac. Ci. Mat. Univ. Nac. Litoral. Publ. Inst. Mat. 8(1948), 195–223. MR 11–339. [3, 7, 16, 17, 18, 42, 45, 52, 54]
185. Carathéodory, C. *Über den Variabilitatsbereich der Koeffizienten von Potenzreihen, die gegebene werte nicht annehmen.* Math. Ann. 64(1907), 95–115. FM 38–448. [59]
186. Carathéodory, C. *Uber den Variabilitatsbereich der Fourier'schen Konstanten von positiven harmonischen Funktionen.* Rend. Cir. Mat. Palermo 32(1911), 193–217. FM 42–429. [2]
187. Carathéodory, C. *Über die Begrenzung einfach zusammenhangender Geibiete.* Math. Ann. 73(1912), 323–370. FM 44–757. [59]
188. Carathéodory, C. *Elementarer Beweis fur den Fundamentalsatz der donformen Abbildung.* Schwarz-Festschrift, Springer, Berlin, (1914), 19–41. FM 45–667. [59]
189. Carathéodory, C. *Über die Studysche Rundungsschranke.* Math.

Ann. 79(1919), 402. FM 47-327. [10]
190. Carathéodory, C. *Conformal Representation.* Cambridge, 1932. Zbl 5-168. [59]
191. Carathéodory, C. *A generalization of Schwarz's lemma.* Bull. Amer. Math. Soc. 43(1937), 231-241. Zbl 16-216. [15, 22, 37]
192. Carathéodory, C. *Conformal Representation.* 2nd Ed. Cambridge Tracts in Math. -Phys., No. 28. Cambridge, at the University Press, 1952. 115 pp. MR 13-734. [59]
193. Carathéodory, C. *Theory of Functions.* I, II. Chelsea Publ. Co., New York, 1954. MR 15-612. [19, 22, 31, 46]
194. Carlson, F. *Sur les coefficients d'une fonction bornée dans le cercle unité.* Ark. Mat. Astr. Fys. 27A, No. 1, 8 pp, (1940). MR 2-185. [17, 22, 30]
195. Cartwright, M. L. *On analytic functions regular in the unit circle.* I. Quart. J. Math., Oxford Ser. 4(1933), 246-257. FM 59-325. [59]
196. Cartwright, M. L. *On analytic functions regular in the unit circle.* II. Quart. J. Math., Oxford Ser. 6(1935), 94-105. Zbl 11-358. [2, 3]
197. Cartwright, M. L. *Some inequalities in the theory of functions.* Math. Ann. 111(1935), 98-118. FM 61-351. [14, 15, 65]
198. Charzyński, Z. *Sur les fonctions extremales dans les familles de fonctions univalentes.* Colloq. Math. 1(1948), 168-170. [22, 24]
199. Charzyński, Z. *Sur les fonctions univalentes bornees.* Rozprawy Mat. 2(1953), 58pp. MR 15-23. [24]
200. Charzyński, Z. *Sur les fonctions univalentes algebriques bornées.* Rozprawy Mat. 10(1955), 41 pp. MR 17-25. [28]
201. Charzyński, Z. *Methodes variationelles dans la theorie des fonctions univalentes.* Bull. Math. Soc. Sci. Math. Phys. R. P. Roumaine. (N. S.) 1(49) (1957), 259-264. MR 20-544. [22, 28]
202. Charzyński, Z. *Fonctions univalentes inverses. Polynomes univalentes.* Soc. Sci. Lettres Lodz. Bull. Cl. III 9(1958), No. 7, 21 pp. MR 22A(1)-807. [24, 32]
203. Charzyński, Z.; Janowski, W. *Sur l'èquation générale des fonctions extrèmales dans la famille des fonctions univalentes bornees.* Ann. Univ. Mariae Curie-Sklodowska Sect. A 4(1950), 41-56. (French. Polish summary) MR 13-122. [22]
204. Charzyński, Z.; Janowski, W. *Domaine de variation des coeffcients A_2 et A_3 des fonctions univalentes et bornees.* Bull. Soc. Sci. Lettres Lódź. 19(1959), No. 4, 29 pp. MR 23A-732. [22, 24, 29]
205. Charzyński, Z.; Schiffer, M. *A new proof of the Bieberbach conjecture for the fourth coefficient.* Arch. Rational Mech. Anal. 5(1960), 187-193. MR 22A(1)-965. [4, 17, 24, 27, 42, 53, 54]
206. Charzyński, Z.; Schiffer, M. *A geometric proof of the Bieberbach*

conjecture for the fourth coefficient. Scripta Math. 25(1960), 173–181. MR 22A(1)–1629. [17, 24, 54]
207. Chen, K. *On the theory of schlicht functions.* Proc. Imp. Acad. Jap. 9(1933), 465–467. Zbl 8–168. [17, 42, 45, 52]
208. Chen, K. *On the theory of schlicht functions.* Tohoku Math J. 40(1935), 160–174. FM 61–1153. [14, 17, 39, 45, 52]
209. Chen, K. *Contributions to the theory of schlicht functions.* Tohoku Math. J. 41(1935), 125–147. Zbl 12–214. [9, 15, 58]
210. Churchill, R. V. *Introduction to Complex Variables and Applications.* New York, McGraw-Hill, 1948. MR 10–439. [59]
211. Čin, Yan' Šin *On the arguments of the coefficients in the expansion of a univalent function.* Acta Math. Sinica 4(1954), 81–86. MR 17–142. [4, 39]
212. Cioránescu, N. *Sur les zéros des fonctions méromorphes et certaines classes de polynômes.* Bull. Math. Soc. Romaine Sci. 37(1936), 9–11. Zbl 14–220. [1]
213. Clunie, J. *Univalent regions of integral functions.* J. Math., Oxford Ser(2) 5(1954), 291–296. MR 16–809. [28]
214. Clunie, J. *On functions meromorphic in the unit circle.* J. London Math. Soc. 32(1957), 65–67. MR 18–884. [59]
215. Clunie, J. *On schlicht functions.* Ann. Math. (2) 69(1959), 511–519. MR 21–1194. [9, 17, 25]
216. Clunie, J. *On meromorphic schlicht functions.* J. London Math. Soc. 34(1959), 215–216. MR 21–1064. [6, 9, 17, 25]
217. Clunie, J.; Keogh, F. R. *On starlike and convex schlicht functions.* J. London Math. Soc. 35(1960), 229–233. MR 22A(1)–295. [4, 6, 10, 17, 18, 36, 38]
218. Combes, J. *Sur le théorème de Landau-Carathéodory.* C. R. Acad. Sci. Paris 228(1949), 41–42. MR 9–363. [19]
219. Connell, E. N. *On the properties of analytic functions.* Duke Math. J. 28(1961), 73–81. MR 23A–50. [59]
220. Copson, E. T. *An Introduction to the Theory of Functions of a Complex Variable.* Oxford, 1935. Zbl 12–169. [59]
221. Cotlar, M. *Functions which are univalent on a subset of the boundary of a domain of regularity.* Publ. Inst. Mat. Univ. Nac. Litoral 4(1942), 47–96. (Spanish) MR 4–155. [49, 62]
222. Courant, R. *Über eine Eigenschaft der Abbildungsfunktionen bei konformer Abbildung.* Gott. Nachr. (1914), 101–109. FM 45–668. [59]
223. Courant, R. *Über direkte Methoden in der Variationsrechnung und uber verwandte Fragen.* Math. Ann. 97(1927), 711–736. FM 53–488. [59]

224. Courant, R. *Plateau's problem and Direchlet's principle.* Ann. Math. 38(1937), 679–724. Zbl 17-268. [59]
225. Courant, R. *Dirichlet's Principle, Conformal Mapping and Minimal Surfaces.* Interscience Pub., N.Y., 1950, 330 pp. MR 12-90. [59]
226. Cowling, V. F. *On analytic functions having a positive real part in the unit circle.* Amer. Math. Mon. 63(1956), 329–330. MR 17-1070. [2, 17, 21]
227. Cowling, V. F.; Royster, W. C. *Some applications of the Weierstrass mean value theorem.* J. Math. Soc. Japan 13(1961), 104–108. MR 23A-732. [4]
228. Crum, M. M. *A property of schlicht functions.* J. London Math. Soc. 31(1956), 493–494. MR 18-121. [34]
229. Cyu, F. *Some properties of functions analytic in a circle.* Acta Math. Sinica 9(1959), 382–388. (Chinese. Russian summary) MR 23A-335. [47]
230. Daiovitch, V. *Sur l'existence de valeurs limites de la resultante des fonctions appartenant a la classe H_δ, $\delta > 1$.* Bull. Soc. Math. Phys. Serbie 8(1956), 23–28. MR 20-874. [3, 47]
231. Danilevskii, A. M. *On single-valued univalent functions on an annulus.* Har'kov Gos. Univ. Uc. Zap. 34 -Zap. Mat. Otd. Fiz. -Mat. Fak i Har'kov Mat. Obsc. (4) 22(1950), 51–53(1951). Russian) MR 17-1069. [1, 15, 18, 48]
232. Darwin, C. *On some transformations involving elliptic functions.* Phil. Magazine (7) 41(1950), 1–11. MR 11-341. [59]
233. Davis, P. J. *Packing inequalities for circles.* Mich. Math. J. 10(1963), 25–31. MR 26-981. [2]
234. Davis, P. J.; Pollak, H. *On the zeros of total sets of polynomials.* Trans. Amer. Math. Soc. 72(1952), 82–103. MR 13-552. [9, 15, 19, 49, 53, 68]
235. de Bruijn, N. G. *Ein Satz über schlichte Funktionen.* Nederl. Akad. Wetensch., Proc. 44(1941), 47–49. MR 2-274. [6, 9, 17, 25, 36, 38, 41]
236. Denjoy, A. *Sur la représentation conforme.* C. R. Acad. Sci. Paris 219(1944), 11–14. MR 7-287. [18, 62]
237. Denjoy, A. *Sur la représentation conforme des aires planes.* Mathematica Tim. 20(1944), 73–89. MR 6-262. [18, 62]
238. Detwiler, B.; Royster, W. C. *A variational formula for functions convex in the direction of the imaginary axis.* Abstract 568-2, Notices Amer. Math. Soc. 7(1960), 242. [24, 41]
239. Dienes, P. *The Taylor Series.* Clarendon Press, Oxford, 1931. Zbl 3-155. [44]

240. Dieudonné, J. *Sur le rayon d'univalence des polynomes.* C. R. Acad. Sci. Paris 192(1931), 79-81. Zbl 1-18. [32, 43]
241. Dieudonné, J. *Sur les fonctions univalentes.* C. R. Acad. Sci. Paris 192(1931), 1148-1150. Zbl 1-344. [4, 8, 17, 39, 45, 52, 54]
242. Dieudonné, J. *Recherches dur quelques problemes relatifs aux polynomes et aux fonctions bornees d'une variable complexe.* Ann. École Norm. (3) 48(1931), 247-358. Zbl 3-119. [6, 11, 14, 15, 17, 22, 32, 39, 42, 43, 45, 52, 54, 61]
243. Dieudonné, J. *Sur les rayons d'étoilement et de convexité de certaines fonctions.* C. R. Acad. Sci. Paris 196(1933), 37-39. Zbl 6-64. [6, 10, 11, 12, 28]
244. Dieudonné, M. J. *La Theorie Analytique des Polynomes d'une Variable.* Gauthier-Villars, Paris, 1938. [59]
245. Dinghas, A. *A simple proof of a formula in the theory of functions.* Math. Student. 22(1954), 101-102. MR 16-231. [59]
246. Dinghas, A. *Verzerrungsschranken bei schlichten Abbildungen des Einheitskreises.* Arch. Math. 8(1957), 413-416. MR 21-517. [14, 15, 42, 65, 68]
247. Dinghas, A. *Über einige Monotoniesätze in der Theorie der schlichten Funktionen.* Avhdl. Norske Vid. Akad. Oslo I(1959), No. 1, 18 pp. Zbl 85-295. [6, 10, 15, 35, 42, 64, 68]
248. Doob, J. L. *A minimum problem in the theory of analytic functions.* Duke Math. J. 8(1941), 413-424. MR 3-76. [59]
249. Douglas, J. *Solution of the problem of Plateau.* Trans. Amer. Math. Soc. 33(1931), 263-321. Zbl 3-328. [59]
250. Dufresnoy, J. *Sur l'aire sphérique décrite par les valeurs d'une fonction méromorphe.* Bull. Sci. Math. (2), 65(1941), 214-219. MR 7-56. [18, 46]
251. Dufresnoy, J. *Sur une nouvelle démonstration d'un théorème d'Ahlfors.* C. R. Acad. Sci. Paris 212(1941), 662-665. MR 3-81. [46]
252. Dufresnoy, J. *Sur les cercles de remplissage des fonctions méromorphes.* C. R. Acad. Sci. Paris 214(1942), 467-469. MR 4-138. [46]
253. Dufresnoy, J. *Sur les fonctions méromorphes et univalentes dans le cercle unite.* Bull. Sci. Math. (2) 69(1945), 21-36. MR 7-56. [62]
254. Dufresnoy, J. *Le problème des coefficients pour certaines fonctions méromorphes dans le cercle unité.* Ann. Acad. Sci. Fenn. Ser. A. I., No. 250/9 (1958), 7 pp. MR 20-968. [22]
255. Dundučenko, L. E. *Certain extremal properties of analytic functions given in a circle and in a circular ring.* Ukrain. Mat. Z. 8(1956), 377-395. (Russian) MR 19-25. [6, 10, 14, 15, 17, 22, 30, 35, 36, 38, 48, 49, 50, 61, 63, 64, 65]
256. Dundučenko, L. E. *Über einige Klasse Funktionen, welche im Kreis*

$|z| \leq 1/\sqrt{2}$ *schlicht sind*. Bul Inst. Politehn. Iasi (N. S.) 3(1957), 37-38. (Russian. German. Roumanian summaries) MR 20-771. [4, 15, 17, 23, 43]

257. Dundučenko, L. E. *Some properties of schlicht functions with axial symmetry*. Izv. Vyss. Ucebn. Zaved. Matematika (1958), No. 3(4), 84-95. (Russian) MR 24A-254. [15, 17, 23, 41]

258. Dundučenko, L. E. *Deformation theorems for starlike functions*. Acad. R. P. Romine. An. Romino-Soviet. Ser. Mat. -Fiz. (3) 13(1959), No. 3(30), 63-75. (Roumanian) MR 22A(1)-646. [6, 14, 15, 23, 63, 65]

259. Dundučenko, L. E. *A generalization of classes of analytic functions considered by L. Čakalov*. Bulgar. Akad. Nauk Izv. Mat. Inst. 5(1961), 35-41. (Russian. Bulgarian. English summaries) MR 23A-731. [4, 15, 17, 43]

260. Dundučenko, L. E.; Kas'janjuk, S. A. *On spiral functions univalent in an annulus*. Usephi, Mat. Nauk 15(1960), No. 5(95), 165-170. (Russian) MR 23A-475. [6, 15, 49, 63]

261. Dundučenko, L. O. *On some properties of analytic functions belonging to special classes*. Dopovidi Akad. Nauk Ukrain. RSR (1956), 119-123. MR 17-1069. [6, 10, 14, 15, 17, 22, 30, 35, 36, 38, 49, 50, 63, 64, 65]

262. Dundučenko, L. O. *On univalent functions parabolically convex in a circle*. Dopovidi Akad. Nauk Ukrain. R. S. R. (1958), 128-130. (Ukrainian. Russian. English summaries) MR 20-968. [10, 15, 23]

263. Dundućenko, L. O. *On a class of functions univalent in the circle* $|z| < (\sqrt{2})^{-1}$. Dopovidi Akad. Nauk Ukrain. RSR (1958), 595-597. (Ukrainian. Russian. English summaries) MR 20-968. [15, 17, 43]

264. Duren, P. L. *Coefficient estimates for univalent functions*. Proc. Amer. Math. Soc. 13(1962), 168-169. MR 28-258. [9, 17, 25, 53]

265. Duren, P. L. *Distortion in certain conformal mappings of an annulus*. Mich. Math. J. 10(1963), 431-441. MR 28-43. [48]

266. Duren, P. L.; Schiffer, M. *A variational method for functions schlicht in an annulus*. Arch. Rational Mech. Anal. 9(1962), 260-272. MR 25-39. [48]

267. Duren, P. L.; Schiffer, M. *The theory of the second variation in extremum problems for univalent functions*. Jour. d'Analyse Math. 10(1963), 193-252. MR 27-60. [24, 42, 53]

268. Dvorák, O. *Sur les fonctions univalentes*. Cas. Mat. Fys. 63(1933), 9-16. Zbl 8-74. [2, 42, 45]

269. Dvoretzky, Aryeh. *Sur une classe de fonctions univalentes*. C. R. Acad. Sci. Paris 221, (1945), 605-607. MR 8-22. [17, 42, 54]

270. Dvoretzky, Aryeh. *Les coefficients d'une fonction univalente et le domaine etale.* C. R. Acad. Sci. Paris 225(1947), 447-449. MR 9-23. [19]
271. Dvoretzky, Aryeh. *Bounds for the coefficients of univalent functions.* Proc. Amer. Math. Soc. 1(1950), 629-635. MR 12-327. [17, 19, 54]
272. Dziubiński, I. *Equation generale de Lowner.* Bull. Soc. Sci. Lettres Łódź 11(1960), No. 4, 8 pp. MR 23A-732. [2, 22, 24]
273. Effertz, F. H.; Meuffels, W. *Uber das Koeffizientproblem der rationalen Funktionen mit positivem Realteil.* Arch. Math. 12(1961), 51-60. MR 23A-335. [59]
274. Egerváry, E. *Abbildungseigenschaften der arithmetischen Mittle der geometrischen Reihen.* Math. Z. 42(1937), 221-230. Zbl 15-306. [6, 10, 20]
275. Eilenberg, S. *Sur quelques proprietés topologiques de la surface de sphère.* Fund. Math. 25(1935), 267-272. Zbl 12-228. [34]
276. Elyash, E. S.; Levine, N. *A note on the function $W = AZ + B$.* Amer. Math. Mon. 66(1959), 803. MR 21-1190. [59]
277. Epstein, B. *Some inequalities relating to conformal mapping upon canonical slit-domains.* Bull. Amer. Math. Soc. 53(1947), 813-819. MR 9-180. [9, 17, 18, 25, 27, 60]
278. Epstein, B.; Schoenberg, I. J. *On a conjecture concerning schlicht functions.* Bull. Amer. Math. Soc. 65(1959), 273-275. MR 21-1350. [16, 24, 47]
279. Erdös, P.; Herzog, F.; Piranian, G. *Schlicht Taylor series whose convergence on the unit circle is uniform but not absolute.* Pacific J. Math. 1(1951), 75-82. MR 13-335. [62]
280. Erwe, F. *Uber die Schlichtheitsschranken gewisser Funktionenfamilien.* Math. Z. 56(1952), 57-64. MR 14-260. [9, 27, 39]
281. Erysov, S. P. *On the relation between the modulus of the derivative and the modulus of the function for univalent conformal mappings.* Summary of dissertation, Saratov Univ. 1954. (Russian) [15, 68]
282. Èrzohi, T. G. *On some classes of p-valent functions.* Dopovidi Akad. Nauk Ukrain. RSR (1962), 1560-1564. (Ukrainian. Russian. English summaries). [5, 6, 14, 15, 63, 64, 65]
283. Evgrafov, M. A. *Power series with integer coefficients.* II. Mat. Sbornik. N. S. 29(71) (1951), 121-132. (Russian) MR 13-335. [39]
284. Evgrafov, M. A. *Power series with integral coefficients.* I. Mat. Sb. N. S. 29(70) (1951), 715-722. (Russian) MR 13-335. [39]
285. Faber, G. *NeuerBeweis eine Koebe-Bieberbachschen Satzes über konforme Abbildung.* Sitzgsber. Bayer. Akad. Wiss. Munchen (1916), 39-42. FM 46-550. [9, 18]

286. Faber, G. *Über Potentialtheorie und konforme Abbildung.* Sitzgsber. Math. -phys. Kl Bayer. Akad. Wiss. Munchen (1920), 49-64. FM 47-324. [59]
287. Faber, G. *Über den Hauptsatz auf der Theorie der konformen Abbildung.* Sitzgsber. Math. -phys. Kl. Bayer. Akad. Wiss. Munchen (1922), 91-100. FM 48-1231. [59]
288. Farrell, O. J.; Bolster, M. R. *On approximation by functions of lesser valence.* Proc. Amer. Math. Soc. 13(1962), 158-162. MR 25-45. [14]
289. Federov, V. S. *On the derivative of a complex function.* Dokl. Akad. Nauk SSSR (N. S.) 63(1948), 357-358. MR 10-288. [4]
290. Fejér, L. *Über die Konvergenz der Potenzreihe an der Konvergenzgrenze in Fallen der konformen Abbildung auf die schlichte Ebene.* Schwarz-Festschr. (1914), 42-53. FM 45-670. [22]
291. Fejér, L. *Fourierreihe und Potenzreihe.* Montsh. Math. Phys. 28(1917), 64-76. FM 46-452. [59]
292. Fejér, L. *Über die Positivitat von Summen, die nach trigonometrischen oder Legendreschen Funktionen fortschreiten.* Acta Litt. Ac Scient. Szeged (1925), 75-86. FM 51-219. [59]
293. Fejér, L. *Über die Koeffizientensumme einer beschrankten und schlichten Potenzreihe.* Acta Math. 49(1926), 183-190. FM 52-311. [20, 22]
294. Fejér, L. *Über gewisse Minimumprobleme der Funktionen-theorie.* Math. Ann. 97(1926), 104-123. FM 52-310. [3, 7, 17]
295. Fejér, L. *Neue Eigenschaften der Mittelwerte bei den Fourierreihen.* J. London Math. Soc. 8(1933), 53-62. FM 59-298. [4, 10, 20, 39, 45]
296. Fejér, L. *Gestaltliches über die Partialsummen und ihre Mittelwerte bei der Fourierreihe und Potenzreihe.* Z. Angew. Math. Mech. 13(1933), 80-88. FM 59-298. [4, 10, 20, 39, 45]
297. Fejér, L. *Untersuchungen über Potenzreihen mit mehrfach monotoner Koeffizientenfolge.* Acta Litterarum ac Scientiarum, vol. 8 (1936), 89-115. Zbl 16-108. [4, 6, 39]
298. Fejér, L. *Hatvanysorok tobbszorosen monoton egyutthatosorozattal.* Matematikai es Termeszettudomanyi Ertesito, 55(1936), 1-29. [59]
299. Fejér, L. *Trigonometrische Reihen und Potenzreihen mit mehrfach monotoner Koeffizientenfolge.* Trans. Amer. Math. Soc. 39(1936), 18-59. [59]
300. Fejér, L.; Riesz, F. *Über einige funktionentheoretische Ungleichungen.* Math. Z. 11(1921), 305-314. FM 48-327. [3, 7]

301. Fejér, L.; Szegö, G. *Special conformal mappings.* Duke Math. J. 18(1951), 535-548. [6, 10, 16, 20, 41, 49]
302. Fekete, M. *Über die Verteilung der Wirzein bei gewissen algebraischen Gleichungen mit ganzzahligen Koeffizienten.* Math. Z. 17(1923), 228-249. FM 49-47. [59]
303. Fekete, M. *Über den transfiniten Durchmesser ebener Punktmengen, II.* Math. Z. 32(1930), 215-221. FM 56-112. [59]
304. Fekete, M.; Szegö, G. *Eine Bemerkung über ungerade schlichte Funktionen.* J. of London Math. Soc. 8(1933), 85-89. Zbl 6-353. [16, 17, 45, 52]
305. Fel'dman, Y. S. *Some estimates for p-valent functions.* Dokl. Akad. Nauk SSSR (N. S.) 92(1953), 239-242. MR 15-413. [3, 14, 17, 49, 50]
306. Fel'dman, Y. S. *On certain extremal domains connected with univalent functions.* Vestnik Leningrad. Univ. Ser. Mat. Meh. Astronom. 18(1963), No. 2, 67-85. MR 25-1211. [15, 49, 68]
307. Fenchel, W. *Bemerkungen über die im Einheitskreis meromorphen schlichten Funktionen.* Sitzsber. Preuss. Akad. Wiss. Phys. -Math. Kl. H 22/23(1931), 431-436. Zbl 2-269. [9, 19]
308. Ferrand, J. (Lelong) *Étude de la représentation conforme au voisinage de la frontière.* Ann. Sci. École Norm. Sup. (3) 59(1942), 43-106. MR 6-207. [46, 62]
309. Ferrand, J. (Lelong) *Sur un théorème de M. Golusin.* C. R. Acad. Sci. Paris 215(1942), 254-255. MR 5-94. [19]
310. Ferrand, J. (Lelong) *Sur les fonctions holomorphes dans une couronne.* Bull. Sci. Math. (2) 67(1943), 42-49. MR 6-262. [19]
311. Ferrand, J. (Lelong) *Sur l'inégalité d'Ahlfors et son application au problème de la dérivée angulaire.* Bull. Soc. Math. France 72(1944), 178-192. MR 7-55. [46]
312. Ferrand, J. (Lelong) *Sur la déformation analytique d'un domain.* C. R. Acad. Sci. Paris 221(1945), 132-134. MR 7-201. [46]
313. Ferrand, J. (Lelong) *Extension d'une inégalité de M. Ahlfors.* C. R. Acad. Sci. Paris 220(1945), 873-874. MR 7-201. [46]
314. Ferrand, J. (Lelong) *Représentation conforme et transformations a integrale de dirichlet bornee.* (1955) (Paris), 259 pp. MR 16-1096. [59]
315. Fischer, E. *Über das Carathéodorysche Problem, Potenzreihen mit positivem reellen Teil betreffend.* Rend. Circ. Mat. Palermo 32(1911), 240-256. FM 42-277. [2]
316. Flett, T. M. *Note on a function—theoretic identity.* J. London Math. Soc. 29(1954), 115-118. MR 15-303. [59]

317. Flett, T. M. *Some remarks on schlicht functions and harmonic functions of uniformly bounded variation.* Quart. J. Math. Oxford Ser. (2) 6(1955), 59-72. MR 16-916. [23, 62]
318. Flett, T. M. *On the radial order of a univalent function.* J. Math. Soc. Japan 11(1959), 1-3. MR 20-662. [62]
319. Ford, L. R. *Automorphic Functions.* New York, McGraw-Hill, (1929). FM 55-810. [59]
320. Ford, L. R. *On properties of regions which persist in subregions bounded by level curves of the Green's function.* Duke Math. J. 1(1935), 103-104. Zbl 11-261. [1, 37, 60]
321. Fourès, L. *Sur les domaines d'univalence de certaines fonctions entières.* C. R. Acad. Sci. Paris 226-1948), 1157-1159. MR 9-507. [28]
322. Fourès, L. *Fonctions analytiques admettant une fonction d'automorphie donnee.* C. R. Acad. Sci. Paris 232(1951), 1894-1895. MR 13-222. [49]
323. Frame, J. S. *Power series expansions for inverse functions.* Amer. Math. Monthly 64(1957), 236-240. MR 19-22. [59]
324. Frank, P.; Löwner, K. *Eine Anwendung des Koebeschen Verzerrungssatzes auf ein Problem der Hydrodynamik.* Math. Z. 3(1919), 78-86. FM 47-326. [59]
325. Frazer, H. *Further inequalities in the theory of functions.* J. London Math. Soc. 10(1935), 143-150. FM 61-351. [14, 15, 65]
326. Frazer, H. *On functions regular in a convex region.* J. London Math. Soc. 20(1945), 199-204. MR 8-19. [59]
327. Freud, G. *Uber einen Reihentheoretischen Satz von Fejer.* Acta Math. Acad. Sci Hungar 3(1952), 173-176. (Russian summary) MR 14-737. [4, 22]
328. Fridman, G. A. *On the problem of the coefficients of functions of the class H_δ.* Dokl. Akad. Nauk SSSR (N. S.) 65(1949), 805-808. (Russian) MR 10-602. [3, 17]
329. Friedman, Bernard *Two theorems on schlicht functions.* Duke Math. J. 13(1946), 171-177. MR 8-22. [17, 39, 54]
330. Fritsch, W. *Uber konvexe und schlichte Abbildungen von Kreisbereichen.* Deutsche Math. 2(1937), 421-445. Zbl 17-122. [14, 22, 32, 43]
331. Frostman, O. *Sur les produits de Blaschke.* Kungl. Fysiografiska Sallskapets i Lund Forhandlingar (Proc. Roy. Physiog. Soc. Lundj) 12, No. 15, (1942), 169-182. MR 6-262. [59]
332. Fuchs, B. A.; Levin, V. I. *Functions of a complex variable and some of their applications.* (English trans.) Pergamon Press, 1961, 286 pp. Zbl 98-276. [59]

333. Fuchs, H. J.; et al. *Research Problems.* Colloquium held at Cornell Univ. August 17–August 21, 1961. Amer. Math. Soc. Notices 8, No. 6(1961), pp. 483–485; Bull. Amer. Math. Soc. 68, No. 1, Jan. 1962, pp. 21–24. [16]
334. Fuchs, W. H. J. *A uniqueness theorem for mean values of analytic functions.* Proc. London Math. Soc. (2) 48(1943), 35–47. MR 5–36. [59]
335. Gabriel, R. F. *The Schwarzian derivative and convex functions.* Proc. Amer. Math. Soc. 6(1955), 58–66. MR 16–807. [4, 9, 10, 13, 31]
336. Gabriel, R. F. *A generalized Schwarzian derivative and convex functions.* Duke Math. J. 24(1957), 617–626. MR 19–1045. [4, 6, 10, 14, 31]
337. Gabriel, R. M. *Some results concerning the integrals of moduli of regular functions along certain curves.* J. London Math. Soc. 2(1927), 112–117. FM 53–306. [3]
338. Gabriel, R. M. *Some results concerning the integrals of moduli of regular functions along curves of certain types.* Proc. London Math. Soc. (2), 28(1928), 121–127. FM 54–331. [3, 10]
339. Gabriel, R. M. *Concerning integrals of moduli of regular functions along convex curves.* Proc. London Math. Soc. (2), 39(1935), 216–231. Zbl 11–358. [3, 10, 16]
340. Gabriel, R. M. *A note upon functions positive and subharmonic inside and on a closed convex curve.* J. London Math. Soc. 21(1946), 87–90. MR 8–461. [59]
341. Gaier, D. *Schlichte Potenzreihen, die aug $|z| = 1$ gleichmässig, aber nicht absolut konvergieren.* Mat. Z. 57(1952), 349–350. MR 14–737. [7, 62]
342. Gaier, D. *Schlichte Potenzreihen an der Konvergenzgrenze.* Math. Z. 58(1953), 456–458. MR 15–113. [62]
343. Gaier, D. *Über ein Iterationsverfähren von Komatu zur konformen Abbildung von Ringgebieten.* J. Math. Mech. 6(1957), 865–885. MR 20–18. [59]
344. Gaier, D. *Untersuchungen zur Durchführung der konformen Abbildung mehrfach zusammenhangender Gebiete.* Arch. Rat. Mech. Anal. 3(1959), 149–178. MR 21–728. [59]
345. Gaier, D. *On conformal mapping of nearly circular regions.* Pacific J. Math. 12(1962), 149–162. MR 25–1208. [59]
346. Gaier, D.; Huckemann, F. *Extremal problems for functions schlicht in an annulus.* Arch. Rational Mech. Anal. 9(1962), 415–421. MR 25–40. [7, 16, 48, 62]

347. Galbraith, A. S.; Green, J. W. *A note on the mean value of the Poisson kernel.* Bull. Amer. Math. Soc. 53(1947), 314–320. MR 8-511. [59]
348. Galbraith, A. S.; Seidel, W.; Walsh, J. L. *On the growth of derivatives of functions omitting two values.* Trans. Amer. Math. Soc. 67(1949), 320–326. MR 11-344. [14, 19, 43, 62]
349. Gal'perin, I. M. *On a class of univalent functions.* Kiev-Avtomobil-Doroz Inst. Trudy 2(1955), 189–191. MR 17-957. [10, 17, 23, 28, 38]
350. Gal'perin, I. M. *The theory of univalent functions with bounded rotation.* Izv. Vyss. Ucebn. Zaved. Matematika (1958), No. 3(4), 50–61. (Russian) MR 24A-38. [2, 4, 7, 18, 20]
351. Gal'perin, I. M. *Finite sections of the Taylor series of two special classes of univalent functions.* Uspehi Mat. Nauk 13(1958), No. 5(83), 171–178. (Russian) MR 21-25). [20, 43]
352. Gal'perin, I. M. *On the arg f' (z) of a p-leaved star-like function.* Ukrain. Mat. Z. 11(1959), 207–210. (Russian) MR 22A(1)-128. [6, 24, 49]
353. Gal'perin, I. M. *On the theory of p-valent spiral functions.* Dopovidi Akad. Nauk Ukrain. RSR (1959), 1051–1053. (Ukrainian. Russian. English summaries) MR 22A(2)-1628. [6, 14, 15, 23, 63, 65]
354. Gal'perin, I. M. *On the theory of univalent functions with bounded rotation.* Dopovidi Akad. Nauk Ukrain. RSR (1960), 1465–1468. (Ukrainian. Russian. English summaries) MR 23A-330. [4, 20, 43]
355. Gal'perin, I. M. *On the theory of special classes of functions schlicht in the unit circle with k-tuple rotational symmetry.* Ukrain. at. Z. 13(1961), No. 4, 88–92. (Russian) MR 26-277. [10, 15, 17, 23, 38, 64]
356. Gal'perin, I. M. *On the theory of p-valent functions.* Dopovidi Akad. Nauk Ukrain. RSR (1962), 1555–1560. (Ukrainian. Russian. English summaries) MR 26-744. [5, 6, 14, 15, 63, 64, 65]
357. Garabedian, P. R. *Schwarz's lemma and the Szego kernel function.* Trans. Amer. Math. Soc. 67(1949), 1–35. MR 11-340. [59]
358. Garabedian, P. R. *Distortion of length in conformal mapping.* Duke Math. J. 16(1949), 439–459. MR 11-21. [59]
359. Garabedian, P. R.; Royden, H. L. *A remark on cavitational flow.* Proc. Nat. Acad. Sci. U.S.A. 39(1952), 57–61. MR 14-102. [59]
360. Garabedian, P. R.; Royden, H. L. *The one-quarter theorem for mean univalent functions.* Ann. of Math. (2) 59(1954), 316–324. MR 15-613. [16, 19, 24, 33]

361. Garabedian, P. R.; Schiffer, M. *Identities in the theory of conformal mapping.* Trans. Amer. Math. Soc. 65(1949), 187–238. MR 10-522. [59]
362. Garabedian, P. R.; Schiffer, M. *Convexity of domain functionals.* J. d'Analyse Math. 2(1952–53), 281–368. MR 15-627. [59]
363. Garabedian, P. R.; Schiffer, M. *A proof of the Bieberbach conjecture for the fourth coefficient.* J. Rational Mech. Anal. 4(1955), 427–465. MR 17-24. [9, 17, 24, 25, 27, 42, 45, 52, 53, 54]
364. Garabedian, P. R.; Schiffer, M. *A coefficient inequality for schlicht functions.* Ann. of Math. (2) 61(1955), 116–136. MR 16-579. [9, 16, 17, 24, 25, 39, 54]
365. Garcia, F. *On the coefficients of functions univalent in the unit circle.* Actas Acad. Ci. Lima 4(1941), 76–85. (Spanish) MR 3-78. [17, 45, 52, 54]
366. Gehring, F.W.; Hayman, W.K. *An inequality in the theory of conformal mapping.* J. Math. Pures Appl. (9) 41(1962), 353–361. MR 26-1209. [7, 15, 24]
367. Gel'fer, S. (Guelfer) *Zur Theorie der multivalenten Funktionen.* Rec. Math. (Mat. Sbornik) N.S. 8(50) (1940), 239–250. (Russian. German summary) MR 2-185. [1, 14, 15, 49, 65]
368. Gel'fer, S. (Guelfer) *Sur les bornes de l'étoilement et de la convexité des fonctions p-valentes.* Rec. Math. (Mat. Sbornik) N.S. 16(58) (1945), 81–86. (Russian. French summary) MR 7-55. [9, 11, 12, 14]
369. Gel'fer, S. (Guelfer) *On a property of bounded functions.* Redc. Math. (Mat. Sbornik) N. S. 16(58) (1945), 291–294. (Russian. English summary) MR 7-288. [11, 15, 22, 37, 61]
370. Gel'fer, S. (Guelfer) *On the class of regular functions which do not take on any pair of values w and $-w$.* Rec. Math. (Mat. Sb.) N. S. 19(1946), 33–46. MR 8-573. [17, 34]
371. Gel'fer, S. (Guelfer) *The variation of multivalent functions.* Dokl. Akad. Nauk SSSR (N. S.) 98 pp. (1954), 885–888. Amer. Math. Soc. Trans. (2)26, (1963), 1–4. MR 16-459. [24]
372. Gel'fer, S. (Guelfer) *On coefficients of typically real functions.* Dokl. Akad. Nauk SSSR (N. S.) 94, (1954), 373–376. (Russian) MR 15-786. [9, 13, 17, 23, 25, 51]
373. Gel'fer, S. (Guelfer) *On the coefficient problem for p-valent functions.* Dokl. Akad. Nauk SSSR (N. S.) 106(1956), 955–958. Amer. Math. Soc. Transl. (2) 26(1963), 5–10. MR 17-957. [14, 24, 42]
374. Gel'fer, S. (Guelfer) *On the coefficients of typically real functions.* Dokl. Akad Nauk SSSR (N. S.) 115(1957), 211–213. (Russian) MR 20-161. [13, 17, 45, 51, 52]

375. Gel'fer, S. (Guelfer) *On typically real functions of order p.* Amer. Math. Soc. Transl. (2) 18(1961), 15-36. MR 23A-330. [13, 15, 17, 39, 51, 66]
376. Gel'fer, S. (Guelfer) *The method of variations in the theory of p-valent functions.* Amer. Math. Soc. Transl. (2) 18(1961), 37-43. MR 23A-330. [14, 15, 17, 24, 50, 65]
377. Gel'fer, S. (Guelfer) *An extension of the Goluzin-Schiffer variational method to multiply-connected regions.* Dokl. Akad. Nauk SSSR 142(1962), 503-506. (Russian) MR 24A-612. [24, 60]
378. Geronimus, J. *Sur le probleme des coefficients pour les fonctions bornees.* C. R. Acad. Sci. URSS N. S. 14(1937), 97-98. Zbl 16-125. [3, 22]
379. Geronimus, Y. L. *Some estimates for the coefficients of bounded functions.* Izv. Akad. Nauk SSSR Ser. Mat. 24(1960), 203-212. (Russian) MR 22A-646. [17, 22, 30]
380. Goldberg, J. L. *Functions with positive real part in a half-plane.* Duke Math. J. 29(1962), 333-339. Zbl 101-297. MR 29-264. [2]
381. Goluzin, G. *On some estimates for functions which map the circle conformally and univalently.* Rec. Math. Moscow 36(1929), 152-172. FM 55-789. [6, 9, 10, 11, 15, 39, 42]
382. Goluzin, G. *Zum Majorationsprinzip der Funktionentheorie.* Re. Math. Moscou 42(1935), 647-649. Zbl 14-69. [1]
383. Goluzin, G. *Zur Theorie der schlichten konformen Abbildungen.* Mat. Sb. (N. S.) 42(1935), 169-190. Zbl 12-409. [9, 10, 15, 39, 49, 64, 68]
384. Goluzin, G. *Sur la representation conforme.* Mat. Sb. (N. S.) 43(1936), 273-281. Zbl 16-216. [19]
385. Goluzin, G. *Sur les theoremes de rotation dans la theorie des fonctions univalentes.* Mat. Sb. 43(1936), 293-296. Zbl 15-71. [6, 15, 68]
386. Goluzin, G. *On distortion theorems of schlicht conformal mappings.* Rec. Math. Moscou (2), 1(1936), 127-35. (Russian. German summary) Zbl 14-221. [15, 68]
387. Goluzin, G. *On distortion under univalent conformal mappings of multiply connected domains.* Mat. Sb. (N. S.) 2(1937), 37-63. (Russian. German summary) Zbl 17-121. [9, 12, 15, 29, 58, 60]
388. Goluzin, G. *Some bounds for the coefficients of univalent functions.* Mat. Sb. N. S. 3(1938), 321-330. (Russian. German summary) Zbl 19-171. [14, 15, 17, 50, 65]
389. Goluzin, G. *Some estimates of the coefficients of schlicht functions.* Mat. Sb. 3(1938), 321-330. (Russian) Zbl 19-171. [14, 15, 17, 50, 65]
390. Goluzin, G. *Ergänzung zur Arbeig "Über die Verzerrungssätze der*

schlichten konformen Abbildungen." Rec. Math. Moscou (2) 2(1937), 685–688. FM 63-292. [9, 15, 24, 58, 68]
391. Goluzin, G. *Some covering theorems for functions regular in a circle.* Mat. Sb. (N. S.) 2(1937), 617–618. (Russian. German summary) FM 63(1)-300. [19]
392. Goluzin, G. *Zur theorie der schlichten Funktionen.* Rec. Math. (Mat. Sbornik) N. S. 6(48) (1939), 383–388. (Russian. German summary) MR 1-308. [1, 15, 16, 24]
393. Goluzin, G. *Über p-valente Funktionen.* Rec. Math. (Mat. Sbornik) N. S. 8(50) (1940), 277–284. (Russian. German summary) MR 2-185. [9, 14, 17, 19, 33, 53]
394. Goluzin, G. *Über Koeffizienten der schlichten Funktionen.* Rec. Math. (Mat. Sbornik) N. S. 12(54) (1943), 40–47. (Russian. German summary) MR 5-93. [15, 17, 54, 68]
395. Goluzin, G. *Zur Theorie der schlichten Funktionen.* Rec. Math. (Mat. Sbornik) N. S. 12(54) (1943), 48–55. (Russian. German summary) MR 5-93. [9, 15, 58]
396. Goluzin, G. *Some estimates of derivates of bounded functions.* Rec. Math. (Mat. Sbornik) N. S. 16(58) (1945), 295–306. (Russian. English summary) MR 7-202. [11, 14, 15, 17, 22, 50, 61]
397. Goluzin, G. *On the theory of univalent functions.* Rec. Math. (Mat. Sbornik) N. S. 18(60) (1946), 167–179. (Russian. English summary) MR 7-515. [9, 15, 58, 68]
398. Goluzin, G. *On the problems of Carethéodory-Fejér and similar problems.* Rec. Math. Mat. Sbornik) N. S. 18(60) (1946), 213–226. (Russian. English summary) MR 8-22. [59]
399. Goluzin, G. *On the distortion theorems for schlicht conform representation.* Rec. Math. (Mat. Sbornik) N. S. 18(60), (1946), 379–390. (Russian. English summary) MR 8-574. [15, 29, 45, 67, 68]
400. Goluzin, G. *On distortion theorems and coefficients of univalent functions.* Rec. Math. (Mat. Sbornik) N. S. 19(61) (1946), 183–202. (Russian. English summary) MR 8-325. [6, 8, 9, 15, 17, 24, 36, 54, 58, 68]
401. Goluzin, G. *Method of variations in the theory of conform representation.* Rec. Math. (Mat. Sbornik) N. S. 19(61) (1946), 203–236. (Russian. English summary) MR 8-325. [24]
402. Goluzin, G. *Estimates for analytic functions with bounded mean of the modulus.* Trav. Inst. Math. Stekloff 18(1946), 87 pp. (Russian. English summary) MR 8-573. [59]
403. Goluzin, G. *Interior Problems of the Theory of Schlicht Functions.* Transl. by T. C. Doyle, A. C. Schaeffer, and D. C. Spencer. Office

of Naval Research, Navy Dept., Wash. D.C., 1947. 138. MR 8-575. [44]
404. Goluzin, G. *Method of variations in the theory of conform representation*. II. Rec. Math. (Mat. Sbornik) N. S. 21(63) (1947), 83-117. MR 9-421. [9, 15, 19, 24, 29, 58]
405. Goluzin, G. *Method of variations in conformal mapping*. III. Mat. Sb., N. S. 21(1947), 119-130. (Russian. English summary) MR 9-421. [24]
406. Goluzin, G. *Some covering theorems in the theory of analytic functions*. Mat. Sbornik N. S. 22(64) (1948), 353-372. MR 10-241. [14, 19]
407. Goluzin, G. *On the coefficients of univalent functions*. Mat. Sbornik N. S. 22(64) (1948), 373-380. MR 10-186. [3, 9, 15, 17, 42, 54, 58]
408. Goluzin, G. *On distortion theorems and the coefficients of univalent functions*. Mat. Sbornik N. S. 23(65) (1948), 353-360. MR 10-602. [9, 15, 17, 24, 25, 39, 54, 58, 68]
409. Goluzin, G. *Some inequalities for analytic functions*. Izv. Akad. Nauk Kazah. SSR 60, Ser. Mat. Meh. 3(1949), 101-105. (Russian. Kazak summary) MR 13-639. [3, 17]
410. Goluzin, G. *On mean values*. Mat. Sbornik N. S. 25(67) (1949), 307-314. MR 11-339. [3, 14]
411. Goluzin, G. *Some questions of the theory of univalent functions*. Trudy Mat. Inst. Steklov. 27(1949), 111 pp. (Russian) MR 13-123 [44]
412. Goluzin, G. *Some estimates for bounded functions*. Mat. Sbornik N. S. 26(68) (1950), 7-18. MR 11-426. [2, 10, 15, 17, 20, 21, 22, 30, 61]
413. Goluzin, G. *On typically real functions*. Mat. Sbornik N. S. 27(69) (1950), 201-218. MR 12-490. [3, 13, 15, 17, 45, 51, 52, 66]
414. Goluzin, G. *On subordinate univalent functions*. Trudy Mat. Inst. Steklov 38(1951), 68-71. Izdat. Akad. Nauk SSSR, Moscou. MR 13-733. [1]
415. Goluzin, G. *On the theory of univalent functions*. Mat. Sbornik N. S. 28(70) (1951), 351-358. MR 13-639. [15, 24, 68]
416. Goluzin, G. *On the theory of univalent functions*. Mat. Sbornik N. S. 29(71) (1951). (Russian) MR 13-223. [9, 13, 15, 24, 39, 53, 58, 66]
417. Goluzin, G. *On majorants of subordinate analytic functions*. I. Mat. Sbornik N. S. 29(71) (1951), 209-224. (Russian) MR 13-223. [1, 3, 6, 18]
418. Goluzin, G. *Variational method in conformal mapping*. IV. Mat. Sbornik N. S. 29(71) (1951), 455-468. MR 13-454. [9, 15, 24, 26, 58]

419. Goluzin, G. *On the parametric representation of functions univalent in a ring.* Mat. Sbornik N. S. 29(71) (1951), 469-476. MR 13-930. [23, 24]
420. Goluzin, G. *On majoration of subordinate analytic functions. II.* Mat. Sbornik N. S. 29(71) (1951), 593-602. MR 13-454. [1]
421. Goluzin, G. *On the problem of coefficients of univalent functions.* Doklady Akad. Nauk SSSR (N. S.) 81(1951), 721-723. MR 13-546. [4, 17, 54]
422. Goluzin, G. *A variational method in the theory of analytic functions.* Leningrad Gos. Univ. Uc. Zap. 144, Ser. Mat. Nauk, 23(1952), 85-101. MR 17-1070. [2, 6, 9, 13, 15, 23, 24, 58]
423. Goluzin, G. *On mean values.* Amer. Math. Soc. Translation No. 61 (1952), 10 pp. MR 13-639. [3]
424. Goluzin, G. *Geometric theory of functions of a complex variable.* GITTL, Moscow, (1952), 540 pp. (Russian) MR 15-112. [4]
425. Goluzin, G. *Geometrische Funktiontheorie.* Berlin, 1957, 438 pp. MR 19-735. [59]
426. Goluzin, G. *On a variational method in the theory of analytic functions.* Amer. Math. Soc. Transl. (2) 18(1961), 1-14. MR 23A-332. [2, 6, 9, 13, 15, 23, 24, 58]. A translation of 422
427. Goluzina, E. G. *On typically real functions with fixed second coefficient.* Vestnik Leningrad. Univ. 17(1962), No. 7, 62-70. (Russian. English summary) MR 25-426. [13, 15, 66]
428. Goodman, A. W. *On some determinants related to p-valent functions.* Trans. Amer. Math. Soc. 63(1948), 175-192. MR 9-421. [4, 14, 16, 17, 42, 50]
429. Goodman, A. W. *Note on regions omitted by univalent functions.* Bull. Amer. Math. Soc. 55(1949), 363-369. MR 10-601. [18, 19]
430. Goodman, A. W. *Sur les coefficients des fonctions p-valentes.* C. R. Acad. Sci. Paris 228(1949), 1917-1918. MR 11-92. [6, 10, 14, 17, 36, 38, 39, 50]
431. Goodman, A. W. *On the Schwarz-Christoffel transformation and p-valent functions.* Trans. Amer. Math. Soc. 68(1950), 204-223. MR 11-508. [6, 10, 14, 15, 16, 17, 36, 38, 39, 42, 50, 63, 64, 65]
432. Goodman, A. W. *Typically-real functions with assigned zeros.* Proc. Amer. Math. Soc. 2(1951), 349-357. MR 13-22. [13, 16, 17, 51]
433. Goodman, A. W. *Inaccessible boundary points.* Proc. Amer. Math. Soc. 3(1952), 742-750. MR 14-367. [6, 10]
434. Goodman, A. W. *The rotation theorem for starlike univalent functions.* Proc. Amer. Math. Soc. 4(1953), 278-286. MR 14-739. [6, 9, 15, 49, 58, 63]

435. Goodman, A. W. *Almost bounded functions.* Trans. Amer. Math. Soc. 78(1955), 82–97. MR 16–685. [15, 17, 18, 27, 34]
436. Goodman, A. W. *Functions typically real and meromorphic in the unit circle.* Trans. Amer. Math. Soc. 81(1956), 92–105. MR 17–724. [9, 13, 15, 17, 23, 51, 66]
437. Goodman, A. W. *Univalent functions and nonanalytic curves.* Proc. Amer. Math. Soc. 8(1957), 598–601. MR 19–260. [4, 6, 10]
438. Goodman, A. W. *Variation formulas for multivalent functions.* Trans. Amer. Math. Soc. 89(1958), 129–148. MR 20–544. [14, 16, 24]
439. Goodman, A. W. *Variation of the branch points for an analytic function.* Trans. Amer. Math. Soc. 89(1958), 277–284. MR 20–969. [24]
440. Goodman, A. W. *On variation formulas for univalent functions.* Trans. Amer. Mth. Soc. 89(1958), 285–294. MR 20–969. [24, 42]
441. Goodman, A. W. *On the critical points of a multivalent functions.* Trans. Amer. Math. Soc. 89(1958), 295–309. MR 20–969. [14, 16, 24]
442. Goodman, A. W. *Conformal mapping onto certain curvilinear polygons.* Univ. Nac. Tucuman Rev. Ser. A 13(1960), 20–26. MR 26–743. [6, 24]
443. Goodman, A. W. *Analytic functions that take values in a convex region.* Proc. Amer. Math. Soc. 14(1963), 60–64. MR 26–63. [1, 10]
444. Goodman, A. W.; Reich, E. *On regions omitted by univalent functions. II.* Canad. J. Math. 7(1955), 83–88. MR 16–579. [7, 18, 19]
445. Goodman, A. W.,; Robertson, M. S. *A class of multivalent functions.* Trans. Amer. Math. Soc. 70(1951), 127–136. MR 12–691. [6, 10, 13, 14, 16, 17, 39, 42, 50, 51]
446. Goodman, R. E. *On the Bloch-Landau constant for schlicht functions.* Bull. Amer. Math. Soc. 51(1945), 234–239. MR 6–262. [1, 19]
447. Goursat, E. *Cours d'analyse mathématique, 1924.* FM 50–150. [59]
448. Grad, A. *The region of values of the derivative of a schlicht function.* Proc. Nat. Acad. Sci. U.S.A. 36(1950), 198–202. MR 11–508. [29]
449. Greenstein, D. S. *On the analytic continuation of functions which map the upper half plane into itself.* J. Math. Anal. Appl. 1(1960), 335–362. MR 23A–618. [59]
450. Grenander, U.; Szegö, G. *Toeplitz Forms and their Applications.* Univ. of Calif. Press, Berkeley, 1958, 245 pp. MR 20–223. [59]
451. Gronwall, T. *On a theorem of Fejér's.* Trans. Amer. Math. Soc. (1912), 445–468. FM 43–321. [59]

452. Gronwall, T. *Some remarks on conformal representation*. Ann. of Math. (2) 16(1914-15), 72-76. FM 45-672. [9, 18, 19, 22]
453. Gronwall, T. *Sur la deformation dans la representation conforme*. C. R. Acad. Sci. Paris 162(1916), 249-252. FM 46-555. [10, 15, 64, 68]
454. Gronwall, T. *Sur la déformation dans la représentation conforme sous des conditions restrictives*. C. R. Acad. Sci. Paris 162(1916), 316-318. FM 46-556. [10, 15]
455 Gronwall, T. *On distortion in conformal mapping when the second coefficient in the mapping function has an assigned value*. Nat. Acad. Proc. 6(1920), 300-302. FM 47-324. [9, 15, 58, 68]
456. Gronwall, T. *On power series with positive real part in the unit circle*. Ann. of Math. 23(1921-22), 317-322. FM 48-403. [2]
457. Gronwall, T. *Summation of series and conformal mappings*. Ann. of Math. 33(1932), 101-117. Zbl 3-55. [59]
458. Gross, W. *Über die Singularitäten analytischer Funktionen*. Mh. Math. U. Physik 29(1918), 3-47. FM 46-512. [49]
459. Gross, W. *Zum Verhalten analytischer Funktionen in der Umgebung singularer Stellen*. Math. Z. 2(1918), 242-294. [59]
460. Grötzsch, H. *Über einige Extremalprobleme der konformen Abbildung*. Leipzig Berichte 80(1928), 367-376. FM 54-378. [15]
461. Grötzsch, H. *Über einige Extremalprobleme der konformen Abbildung*. II. Leipzig Berichte 80(1928), 497-502, FM 54-378. [15]
462. Grötzsch, H. *Über die Verzerrung bei schlichten nicht konformen Abbildungen und uber eine damit zusammenhangende Erweiterung des Picardschen Satzes*. Leipzig Berichte 80(1928), 503-507. FM 54-378. [59]
463. Grötzsch, H. *Über die Verzerrung bei schlichter konformer Abbildung mehrfach zusammenhängender schlichter Bereiche*. Leipzig Verichte 81(1929), 38-50. FM 55-793. [24]
464. Grötzsch, H. *Über konforme Abbildung unendlich vielfach zusammenhangender schlichter Bereiche mit endlich vielen Haufungsrandkomponenten*. Leipzig Berichte 81(1929), 51-86. FM 55-792. [59]
465. Grötzsch, H. *Über die Verzerrung bei schlichter konformer Abbildung mehrfach zusammenhängender schlichter Bereiche*. II. Leipzig Berichte 81(1929), 217-221. FM 55-794. [24]
466. Grötzsch, H. *Über ein Variationsprobleme der konformen Abbildung*. Leipzig Berichte 82(1930), 251-263. FM 56-298. [59]
467. Grötzsch, H. *Zum Parallelschlitztheorem der konformen Abbildung schlichter unendlich-vielfach zusammenhängender Bereiche*.

Leipzig Berichte 83(1931), 185–200. Zbl 3-14. [16, 60]
468. Grötzsch, H. *Über die Verschiebung bei schlichter konformer Abbildung schlichter Bereiche.* Leipzig Berichte. (1931), pp 254–279. Zbl 3-261. [19, 60]
469. Grötzsch, H. *Über die Verzerrung bei schlichter konformer Abbildung merhfach zusammenhangender schlichter Bereiche.* III. Leipzig Berichte 83(1931), 283–297. [60]
470. Grötzsch, H. *Über Extremalprobleme bei schlichter konformer Abbildung schlichter Bereiche.* Leipzig Berichte 84(1932), 3–14. Zbl 5-68. [19, 60]
471. Grötzsch, H. *Über das Parallelschlitztheorem der konformer Abbildung schlichter Bereiche.* Leipzig Berichte 84(1932), 15–36. Zbl 5-68. [19, 57, 60]
472. Grötzsch, H. *Über die Verschiebung bei schlichter konformer Abbildung schlichter Bereiche.* II. Leipzig Berichte 84(1932), 269–278. Zbl 6-171. [29, 60]
473. Grötzsch, H. *Über die Geometrie der schlichten konformen Abbildung.* Sitzgsber. Preuss. Akad. Wiss., Phys.-Math. Kl. (1933), 654–671. Zbl 7-312. [24]
474. Grötzsch, H. *Über die Geometrie der schlichten konformen Abbildung.* II. Sitzgsber. Presuss. Akad. Wiss., Phys.-Math. Kl. (1933), 893–908. Zbl 8-319. [15, 49]
475. Grötzsch, H. *Über zwei Verschiebungsprobleme der konformen Abbildung.* Sitzgsber. Preuss. Akad. Wiss., Phys.-Math. Kl. (1933), 87–100. FM 59-349. [24]
476. Grötzsch, H. *Die werte des Doppelverhaltnisses bei schlichter konformer Abbildung.* Preuss. Akad. Wiss. Sitzgsber. (1933), 87–100. Zbl 7-214. [59]
477. Grötzsch, H. *Verallgemeinerung eines Bieberbachschen Satzes.* Jber. Deutsch. Math. Ver. 43(1933), 143–145. FM 59-351. [60]
478. Grötzsch, H. *Über Flachensatze der konformen Abbildung.* Jber. Deutsch. Math. -Ver. 44(1934), 266–269. FM 60-288. [27, 60]
479. Grötzsch, H. *Einige Bemerkungen zur schlichten konformen Abbildung.* Jber. Deutsch. Math. -Ver. 44(1934), 270–275. Zbl 10-308. [19, 49]
480. Grötzsch, H. *Über die Geometrie der schlichten konformen Abbildung.* III. Sitzgsber. Preuss. Akad. Wiss., Phys. -math. Kl. (1934), 434–444. Zbl 11-313. [49]
481. Grötzsch, H. *Zur Theorie der Verschiebung bei schlichter konformer Abbildung.* Comment. Math. Helv. 8(1936), 382–390. Zbl 14-267. [6, 9, 15, 58]

482. Grunsky, H. *Neue Abschatzungen zur konformen Abbildung ein- und mehrfach zusammenhangender Bereiche.* Schr. Inst. Angew. Math. Univ. Berl. 11(1932), 95-140. Zbl 5-362. [9, 15, 22, 60, 61]
483. Grunsky, H. *Einige Analoga zum Schwarzschen Lemma.* Math. Ann. 108(1933), 190-196. Zbl 6-262. [37]
484. Grunsky, H. *Zwei Bemerkungen zur konformen Abbildung.* Jber. Deutsch. Math. Ver. 43(1933), 1. Abt. 140-3. Zbl 8-119. [11, 15, 68]
485. Grunsky, H. *Koeffizientenbedingungen für schlicht abbildende meromorphe Funktionen.* Math. Z. 45(1939), 29-61. Zbl 22-151. [4, 9, 53]
486. Grunsky, H. *Über konforme Abbildungen die gewisse Gebietsfunktionen in elementare Funktionen transformieren.* I. Math. Z. 67(1957), 129-132. MR 19-538. [59]
487. Grunsky, H. *Eine Grundaufgabe der Uniformisierungstheorie als Extremalproblem.* Math. Ann. 139(1959/60), 204-216. MR 22A(1)-805. [24, 49]
488. Grunwald, G.; Turán, P. *Über den Blochschen Satz.* Acta Litt. Sci. Univ., Szeged, Sect. Sci. Math. 8(1937), 236-240. FM 63-300. [19]
489. Gwilliam, A. E. *Cesàro means of power series.* J. London Math. Soc. 10(1935), 248-253. Zbl 13-68. [3]
490. Gwilliam, A. E. *Cesàro means of power series.* II. Proc. London Math. Soc. IIS. 40(1935), 345-352. Zbl 13-68. [3]
491. Hadamard, J. *Leçons sur le Calcul des Variations.* Paris, 1910, FM 41-432. [59]
492. Haimo, D. T. *A note on convex mappings.* Proc. Amer. Math. Soc. 7(1956), 423-428. MR 18-471. [4, 9, 10, 31]
493. Hamdi, O. *Uber starkschlichte Abbildung des Einheitkreises.* Universite d'Istanbul. Faculte des Sciences. Recueil de Memoires Commemorant la Pose de la Premiere Pierre des Nouveaux Instituts de la Faculte des Sciences Istanbul, (1948), 39-44. MR 10-524. [4, 17, 19]
494. Hardy, G. H. *A theorem concerning Taylor's series.* Quart. J. Math. 44(1913), 147-160. FM 44-476. [59]
495. Hardy, G. H. *The mean value of the modulus of an analytic function.* Proc. London Math. Soc. (2) 14(1915), 269-277. FM 45-1331. [3]
496. Hardy, G. H.; Littlewood, J. E. *A theorem concerning series with applications to the theory of functions.* Meddl. Kobenhavn 7(1925), No. 4, 16 pp. FM 51-272. [14, 17, 18, 42, 50]
497. Hardy, G. H.; Littlewood, J. E. *Theorem concerning Cesaro means of power series.* Proc. London Math. Soc. (2) 36(1934),

516-531. FM 60-257. [59]
498. Hardy, G. H.; Littlewood, J. E. *Theorems concerning mean values of analytic or harmonic functions.* Quart. J. Math., Oxford Ser. 12(1941), 221-256. MR 4-8. [3, 47]
499. Hardy, G. H.; Littlewood, J. E. *Polya, G. Inequalities.* Cambridge Univ. Press 1934, 314 pp. FM 60-169. [59]
500. Haruki, H. *Two theorms on the univalent function.* Proc Phys.-Math. Soc. Japan (3) 25(1943), 622-623. MR 7-287, [4, 6, 11, 15, 43, 68]
501. Hausdorff, F. *Mengenlehre, 3rd Ed.* Walter de Gruyter, Berlin, 1935. Zbl 12-203. [59]
502. Havinson, S. J. *The radii of univalence, starlikeness, and convexity of a class of analytic functions in multiply connected domains.* Izv. Vyss. Ucebn. Zaved. Mat. (1958), No. 3(4), 233-240. (Russian) MR 25-40. [11, 12, 43, 60]
503. Hayman, W. K. *Some inequalities in the theory of functions.* Proc. Cambridge Phil. Soc. 44(1948), 159-178. MR 10-186. [1]
504. Hayman, W. K. *Inequalities in the theory of functions.* Proc. London Math. Soc. (2) 51(1949), 45-473. MR 11-22 [59]
505. Hayman, W. K. *Symmetrization in the theory of functions.* Tech. Rep. No. 11, Navy Contract N6-ORI-106 Task Order 5, Stanford Univ., Calif. (1950), 38 pp. MR 12-401. [14, 15, 17, 19, 22, 24, 33, 50, 61, 65]
506. Hayman, W. K. *A characterization of the maximum modulus of functions regular at the origin.* J. Analyse Math. 1(1951), 135-154. MR 13-545. [59]
507. Hayman, W. K. *Some applications of the transfinite diameter to the theory of functions.* J. Analyse Math. 1(1951), 155-179. MR 13-545. [9, 14, 15, 19, 33, 65, 68]
508. Hayman, W. K. *The maximum modulus and valency of functions meromorphic in the unit circle.* I. II. Acta Math. 86(1951), 89-191, 193-257. MR 13-546. [9, 14]
509. Hayman, W. K. *Functions with values in a given domain.* Proc. Amer. Math. Soc. 3(1952), 428-432. MR 14-156. [17, 19, 27]
510. Hayman, W. K. *On Nevanlinna's second theorem and extensions.* Rend. Circ. Mat. Palermo (2) 2(1953), 346-392. MR 16-122. [59]
511. Hayman, W. K. *La regularite des fonctions univalentes.* C. R. Acad. Sci. Paris 237(1953), 1624-1625. MR 15-516. [3, 17, 42, 54, 62]
512. Hayman, W. K. *The asymptotic behavior of p-valent functions.* Proc. London Math. Soc. (3) 5(1955), 257-284. MR 17-142. [14, 16, 17, 33]

513. Hayman, W. K. *Uniformly normal familites.* Lectures on Functions of a Complex Variable, 199–212. Univ. Mich. Press, Ann Arbor, 1955. MR 17-25. [59]
514. Hayman, W. K. *The coefficients of schlicht and allied functions.* Proc. of Int. Congress of Math. 1954, Amsterdam, vol. III, 102–108. Erven P. Noordhoff N. V. Groningen, H. Holland Publ. Co., Amsterdam, 1956. MR 19-404. [44]
515. Hayman, W. K. *Multivalent Functions.* Cambridge Tracts in Math and Math. -Phys., No. 48. Cambridge University Press, Cambridge, (1958), 151 pp. MR 21-1349. [44]
516. Hayman, W. K. *Some solved and unsolved coefficient problems for schlicht functions.* Sem. Analytic Functions 1(1958), 264–277. Zbl 94-275. [16]
517. Hayman, W. K. *Bounds for the large coefficients of univalent functions.* Ann. Acad. Sci. Fenn. Ser. A. I, No. 250/13 (1958), 13 pp. MR 20-544. [42]
518. Hayman, W. K. *On the coefficients of univalent functions.* Proc. Cambridge Phil. Soc. 55(1959), 373–374. MR 22A(2)-964. [16, 47]
519. Hayman, W. K. *On functions with positive real part.* J. London Math. Soc. 36(1961), 35–48. Zbl 97-61. MR 27-65. [2, 3, 6, 7, 15, 16, 17, 21, 22, 23, 45, 52, 67]
520. Hayman, W. K. *On the limits of moduli of analytic functions.* Ann. Polon. Math. 12(1962), 143–150. MR 26-65. [59]
521. Hayman, W. K. *On successive coefficients of univalent functions.* J. London Math. Soc. 38(1963), 228–243. MR 26-1209. [8, 17, 33, 42, 54]
522. Hayman, W. K.; Kennedy, P. B. *On the growth of multivalent functions.* J. London Math. Soc. 33(1958), 333–341. MR 20-969. [33, 62]
523. Heinhold, J. *Zur konforme Abbildung schlichter Gebiete.* Math. Z. 67(1957), 133–138. MR 19-401. [16, 29]
524. Heins, M. *Selected Topics in the Classical Theory of Functions of a Complex Variable.* Holt, Rinehart and Winston, New York, 1962, 160 pp. [19]
525. Hensel, K.; Landsberg, G. *Theorie der algebraischen funktionen einer Variablen und ihre Anwendung auf algebraische Kurven und abelsche Integrale.* Leipzig, 1902. FM 33-427. [59]
526. Herglotz, G. *Über Potenzreihen mit positivem, reellen Teil im Einheitskreis.* Ber. Verh. Sachs. Akad. Wiss. Leipzig (1911), 501–511. FM 42-438. [2]
527. Hersch, J. *Longueurs extrèmales et théorie des fonctions.* Comment. Math. Helvet. 29(1955), 301–337. MR 17-835. [59]

528. Herzig, A. *Die Winkelderivierte und das Poisson-Stieltjes-Integral.* Math. Z. 46(1940), 129-156. MR 1-213. [2, 37, 46]
529. Herzog, F.; Piranian, G. *Sets of convergence of Taylor series.* I. Duke Math. J. 16(1949), 529-534. MR 11-91. [59]
530. Herzog, F.; Piranian, G. *On the univalence of functions whose derivative has a positive real part.* Proc. Amer. Math. Soc. 2(1951), 625-633. MR 13-223. [2, 4]
531. Herzog, F.; Piranian, G. *Sets of convergence of Taylor series.* II. Duke Math. J. 20(1953), 41-54. MR 14-738. [59]
532. Heuser, P. *Über die Teilsummen beschränkter Potenzreihen.* Math. Z. 39(1935), 660-662. FM 61-346. [15, 20, 22, 61]
533. Hibbert, L. *Univalence et automorphie pour les polynomes et les fonctions entieres.* Bull. Soc. Math. France 66(1938), 81-154. Zbl 20-140. [28, 32, 35]
534. Hilbert, D. *Zur Theorie der konformen Abbildung. Nachr. Kgl. Ges. Wiss. Gottingen,* Math. -Phys. Kl. (1909), 314-323. FM 40-732. [59]
535. Hille, E. *A note on regular singular points.* Arkiv. Form. Mat., Astr., och Fysik, 19A(1925), 1-21. FM 51-335. [59]
536. Hille, E. *Remarks on a paper by Zeev Nehari.* Bull. Amer. Math. Soc. 55(1949), 552-553. MR 10-697. [28, 31]
537. Hille, E. *Analytic Function Theory.* I. II. Ginn and Co., Boston, vol. I(1959), 308 pp.; vol. II(1962), 496 pp. MR 21-1196. [44]
538. Hoh, C. *The Szego problem in the theory of schlicht functions.* Sci. Record (N.S.) 2(1958), 86-91. MR 20-877. [19, 22]
539. Hopf, H.; Rinow, W. *Uber den Begriff der vollstandigen differentialgeometrischen Flace.* Comment. Math. Helvet. 3(1931), 209-225. Zbl 2-350. [59]
540. Hornich, H. *Der Schlichtheitsradius bei ganzen Funktionen.* Akad. WIss. Wien. S. -B. IIa. 154(1945), 59-65. MR 9-420. [23, 28, 43]
541. Hornich, H. *Zur Frage der isolierten schlichten Funktionen.* Math. Ann. 135(1958), 189-191. MR 20-969. [28]
542. Hornich, H. *Zur Struktur der schlichten Funkitonen.* I. II. Abh. Math. Sem. Univ. Hamburg 22(1958), 38-49, 176-179. MR 20-161, 664. [4, 28]
543. Hsu, T. -F. *On the third coefficient of bounded schlicht functions.* Advancement in Math. 2(1956), 279-289. (Chinese) MR 20-968. [22]
544. Hu, Ke. *On the distortion of schlicht functions.* Math. Sinica 4(1954), 259-262. MR 17-142. [9, 15, 58]
545. Huber, A. *On an inequality of Fejér and Riesz.* Ann. Math. (2) 63(1956), 572-587. MR 18-296. [7]

546. Huber, H. *Ein Mittelwertsatz für Funktionen einer komplexen Veranderlichen.* Comment. Math. Helv. 21(1948), 58–66. MR 9-506. [10, 49]
547. Huber, H. *Über analytische Abbildungen von Ringgebieten in Ringgebiete.* Compos. Math. 9(1951), 161–168. MR 13-337. [59]
548. Huber, H. *Über analytische Abbildungen Riemannscher Flachen in sich.* Comment Math. Helv. 27(1953), 1–72. MR 14-862. [59]
549. Huckemann, F. *Über einige Extremalprobleme bei konformer Abbildung eines Kreisringes.* Math. Z. 80(1962), 200–208. MR 26-1211. [48]
550. Hue, T. *The segment covered by the image of a function with convex image.* Advancement in Math. 3(1957), 594–596. (Chinese) MR 23A-329. [10, 19]
551. Hummel, J.A. *The coefficient regions of starlike functions.* Pacific J. Math. 7(1957), 1381–1389. MR 20-292. [6, 24, 29]
552. Hummel, J.A. *A variational method for starlike functions.* Proc. Amer. Math. Soc. 9(1958), 82–87. MR 20-292. [6, 10, 17, 24, 29, 36, 38]
553. Hummel, J.A. *Extremal problems in the class of starlike functions.* Proc. Amer. Math. Soc. II(1960), 741–749. MR 22A(2)–1899. [6, 24]
554. Hummel, J.A. *The Grunsky coefficients of schlicht functions.* Univ. of Maryland, NSF-G 11592. Proc. Amer. Math. Soc. 15, no. 1, Feb. 1964, 142–150. MR 28-258. [4, 17, 53]
555. Hurwitz, A. *Über die Anwendung der elli tischen Modulfunktionen auf einen Satz der allgemeinen Funktiontheorie.* Vjschr. Naturforsch. Ges. Zurich 49(1904), 242–253. FM 35-401. [59]
556. Hurwitz, A.; Courant, R. *Funktiontheorie.* Springer, 1929. [59]
557. Ilieff (Iliev), L. *Zur Theorie der schlichten Funktionen.* Annuaire (Godisnik) Univ. Sofia Fac. Sci. Livre 1. 45(1949), 115–135. MR 12-816. [15, 20, 43, 45, 67, 68]
558. Ilieff (Iliev), L. *Anwendung eines Satzes von G.M. Golusin über die schlichten Funktionen.* C.R. Acad. Bulgare Sci. Math. Nat. 2(1949), No. 1, 21–24. MR 11-92. [15, 20, 43, 68]
559. Ilieff (Iliev), L. *Application of a theorem of G.M. Golusin on equivalent functions.* Dokl. Akad. Nauk SSSR (N.S.) 69(1949), 491–494. MR 11-92. [15, 20, 43, 68]
560. Ilieff (Iliev), L. *On finite sums of univalent functions.* Dokl. Akad. Nauk SSSR (N.S.) 70(1950), 9–11. MR 11-508. [15, 20, 43]
561. Ilieff (Iliev), L. *Sätze über die Abschnitte der schlichten Funktionen.* Annuaire (Godisnik) Univ. Sofia Fac. Sci. Livre 1, 46(1950), 147–151. MR 13-832. [20, 45]

562. Ilieff (Iliev), L. *Über die 3-symmetrischen schlichten Funktionen.* Anuaire (Godisnik) Univ. Sofia Fac. Sci. Livre 1, 46(1950), 161–165. MR 13–832. [17, 39, 43, 49]
563. Ilieff (Iliev), L. *Über die Abschnitte der 3-symmetrischen schlichten Funktionen.* C.R. Acad. Bulgare Sci. 3(1950), No. 1, 9–12(1951). (German. Russian summary) MR 13-336. [39, 43]
564. Ilief (Iliev), L. *Über die Abschnitte der schlichten Funktionen, die den kreis $\|z\| < 1$ konvex abbilden.* Annuaire (Godisnik) Univ. Sofia Fac. Sci. Livre 1, 46(1950), 153–159. MR 13–832. [10, 20, 43]
565. Ilief (Iliev), L. *Über die Abschnitte der schlichten Funktionen.* Acta Math. Acad. Sci. Hungar 2(1951), 109–112. MR 13–640. [20, 45]
566. Ilieff (Iliev), L. *On three fold symmetric univalent functions.* Dokl. Akad. Nauk SSSR (N.S.) 79(1951), 9–11. MR 13–123. [17, 39, 43, 49]
567. Ilief (Iliev), L. *Theorems on triply symmetric univalent functions.* Dokl. Akad. Nauk SSSR (N.S.) 84(1952), 9–12. MR 13–832. [17, 20, 43, 54]
568. Ilief (Iliev), L. *On triply symmetric univalent functions.* Bulgar. Akad. Nauk Izv. Math. Inst. 1(1953), 27–34. MR 15–114. [17, 20, 43, 54]
569. Ilief (Iliev), L. *Analytically noncontinuable series of Faber polynomials.* Bulgar Akad. Nauk Izv. Mat. Inst. 1(1953), 35–56. MR 15–301. [53]
570. Ilief (Iliev), L. *Schlichte Funktionen, die den Einheitskreis konvex abbilden.* C.R. Acad. Bulgare Sci. 5(1952), No. 2–3, 1–4(1953). MR 15–948. [10, 15, 20, 43, 49, 64]
571. Ilief (Iliev), L. *Series of Faber polynomials whose coefficients assume a finite number of values.* Dokl. Akad. Nauk SSSR (N.S.) 90(1953), 499–502. (Russian) MR 15–23. [53]
572. Ilief (Iliev), L. *Theorem on the univalence of finite sums of triply symmetric univalent functions.* Dokl. Akad. Nauk SSSR (N.S.) 100(1953), 621–622. MR 16–809. [20, 43, 49]
573. Ilief (Iliev), L. *On the difference quotient for bounded univalent functions.* Dokl. Akad. Nauk SSSR (N.S.) 100(1955), 861–862. MR 16–809. [15, 22, 49, 61]
574. Imura, H. *A note on bounded functions.* Mem. COll. Sci. Univ. Kyoto Ser. A. Math. 27(1952), 245–248. MR 15–208. [6, 14, 22]
575. Itahara, T. *On the multivalency of power series.* Japan J. Math. 10(1933, 71–78. FM 59–326. [14, 20]
576. Itahara, T. *Note on a certain multivalent function.* Proc. Imp. Acad. Japan 10(1934), 544–545. Zbl 10–308. [14, 17, 27, 50]

577. Ito, J. *One the function whose imaginary part on the unit circle changes its sign 2p times.* Sci. Rep. Tokyo Bunrika Diag. Sect. A 4(1944), 107–114. MR 14-34. [2, 15, 17, 23, 49]
578. Ito, J. *On the function whose real part on the unit circle changes its sign finite times.* Bull. Nagoya Inst. Tech. 3(1951), 293–305. Zbl 87-286. [49]
579. Ito, J. *The variation of the sign of the real part of a meromorphic function on the unit circle.* Trans. Amer. Math. Soc. 89/1958), 60–78. MR 20-405. [9, 49]
580. Ito, J. *The coefficient problem of regular functions.* Proc. Amer. Math. Soc. 12(1961), 53–60. MR 22A(2)-2094. [13, 17, 49, 51]
581. Jacobsthal, E. *Über das Schwarzsche Lemma.* J. Reine Angewandte Math. 165(1931), 59–63 [37]
582. Jacobsthal, E. *Über die Kreise, die durch eine gegebene lineare Funktion auf einen konzentrischen Kreis abgebildet werden.* Norske Vid. Selsk. Skr., Trondheim (1953), No. 3, 22 pp. (1954). MR 16-581. [59]
583. Jakubowski, Z. J. *Sur le maximum de la fonctionnelle $|A_3 - aA_2^2|$ ($0 \leq a \leq 1$) dans la famille de fonctions F_M.* Bull. Soc. Sci. Lettres Łódź 13(1962), No. 1, 19 pp. MR 25-796. [17, 22, 30]
584. Janikowski, J. *Methode algebrique et equation de Lowner.* Bull. Soc. Sci. Lettres Łódź 12(1961), No. 16, 9 pp. MR 25-611. [24]
585. Janowski, W. *Le maximum d'argument des fonctions univalentes bornees.* Ann. Univ. Mariae Curie-Sklodowska Sect. A. 4(1950), 57–72. (French. Polish summary) MR 13-122. [15, 22, 61]
586. Janowski, W. *Le maximum des coefficients A_2 and A_3 des fonctions univalentes bornées.* Ann. Polon. Math. 2(1955), 145–160. MR 17-598. [17, 22, 30]
587. Janowski, W. *Le maximum des coefficients B_2 et B_3 des fonctions univalentes k-symmétriques bornées.* Ann. Polon. Math. 2(1955), 161–169 (1956). MR 17-598. [17, 22, 30, 49]
588. Janowski, W. *Le maximum de la patrie imaginaire des fonctions univalentes bornées.* Ann. Polon. Math. 2(1955), 182–200 (1956). MR 17-599. [22, 49]
589. Janowski, W. *Sur les fonctions univalentes k-symmétriques.* Ann. Polon. Math. 2(1955), 201–208 (1956). MR 17-599. [22, 49]
590. Janowski, W. *Sur les valeurs extrémales du module de la derivee des fonctions univalentes bornees.* Bull. Acad. Polon. Sci. Ser. Sci. Math. Astronom. Phys. 6(1958), 255–259. (Russian summary) MR 25-427. [15, 22, 29, 61]
591. Janowski, W. *Sur les valeurs extrémales du module de la derivee de*

fonctions univalentes bornées. Lódzkie Towarzystwo Naukowe, Prace Wydziata III, No. 53 Panstwowe Wydawnictwo Naukowe, Lódź, 1958. 50 pp. MR 26-62. [15, 22, 61]

592. Jenkins, J.A. *Some problems in conformal mapping.* Trans. Amer. Math. Soc. 67(1949), 327-350. MR 13-341. [24]
593. Jenkins, J.A. *Positive quadratic differentials in triply-connected domains.* Ann. of Math. 53(1951), 1-3. MR 12-400. [59]
594. Jenkins, J.A. *On a theorem of Spencer.* J. London Math. Soc. 26(1951), 313-316. MR 13-338. [59]
595. Jenkins, J.A. *On an inequality of Golusin.* Amer. J. Math. 73(1951), 181-185. MR 12-816. [9, 15, 16, 20, 43, 45, 58]
596. Jenkins, J.A. *Remarks on "Some problems in conformal mapping."* Proc. Amer. Math. Soc. 3(1952), 147-151. MR 13-642. [60]
597. Jenkins, J.A. *Another remark on "Some problems in conformal mapping."* Proc. Amer. Math. Soc. 4(1953), 978-981. MR 15-414. [59]
598. Jenkins, J.A. *On values omitted by univalent functions.* Amer. J. Math. 75(1953), 406-408. MR 14-967. [1, 16, 19, 24]
599. Jenkins, J.A. *Symmetrization results for some conformal invariants.* Amer. J. Math. 75(1953), 510-522. MR 15-115. [15, 17, 24, 54, 68]
600. Jenkins, J.A. *Some results related to extremal length. Contrib. to theory of Riemann surfaces.* 87-94. Ann. of Math. Studies No. 30. Princeton Univ. Press, Princeton, N.J. (1953). MR 15 115. [24]
601. Jenkins, J.A. *Various remarks on univalent functions.* Proc. Amer. Math. Soc. (1953), 595-599. MR 15-114. [9, 17, 24, 25, 27, 29]
602. Jenkins, J.A. *On the local structure of the trajectories of a quadratic differential.* Proc. Amer. Math. Soc. 5(1954), 357-362. MR 15-947. [24]
603. Jenkins, J.A. *On a problem of Gronwall.* Ann. of Math. (2) 59(1954), 490-504. MR 15-786. [24]
604. Jenkins, J.A. *A recent note of Kolbina.* Duke Math. J. 21(1954), 155-162. MR 15-694. [26]
605. Jenkins, J.A. *On Bieberbach-Eilenberg functions.* Trans. Amer. Math. Soc. 76(1954), 389-396. MR 16-24. [1, 15, 17, 34]
606. Jenkins, J.A. *A general coefficient theorem.* Trans. Amer. Math. Soc. 77(1954), 262-280. MR 16-232. [24]
607. Jenkins, J.A. *On circularly symmetric functions.* Proc. Amer. Math. Soc. 6(1955), 620-624. MR 17-249. [13, 39, 49]

608. Jenkins, J.A. *On a lemma of R. Huron.* J. London Math. Soc. 30(1955), 382–384. MR 17–251. [59]
609. Jenkins, J.A. *Some uniqueness results in the theory of symmetrization.* Ann. Math. (2) 61(1955), 106–115. MR 16–460. [59]
610. Jenkins, J.A. *On circumerentially mean p-valent functions.* Trans. Amer. Math. Soc. 79(1955), 423–428. MR 17–143. [7, 17, 33]
611. Jenkins, J.A. *On Bieberbach-Eilenberg functions II.* Trans. Amer. Math. Soc. 78(1955), 510–515. MR 16–684. [1, 34]
612. Jenkins, J.A. *On a result of Keogh.* J. London Math. Soc. 31(1956), 391–399. MR 18–121. [7, 22]
613. Jenkins, J.A. *Some theorems on boundary distortion.* Trans. Amer. Math. Soc. 81(1956), 477–500. MR 17–956. [7, 22, 62]
614. Jenkins, J.A. *On a conjecture of Spencer.* Ann. of Math. 65(1957), 405–410. MR 19–25. [14, 16, 19, 24, 33]
615. Jenkins, J.A. *Some new canonical mappings for multiply-connected domains.* Ann. Math. 65(1957), 179–195. MR 18–568. [60]
616. Jenkins, J.A. *On the existence of certain general extremal metrics.* Ann. of Math. 66(1957), 440–453. MR 19–845. [59]
617. Jenkins, J.A. *On a canonical conformal mapping of J.L. Walsh.* Trans. Amer. Math. Soc. 88(1958), 207–213. MR 19–538. [24, 60]
618. Jenkins, J.A. *Univalent Functions and Conformal Mappings.* Springer-Verlag-Berlin (1958). MR 20–543. [44]
619. Jenkins, J.A. *On certain coefficients of univalent functions.* Analytic Functions, 159–194. Princeton Univ. Press, Princeton, N.J., 1960. MR 22A(2)–1377. [9, 17, 24, 25, 54]
620. Jenkins, J.A. *On weighted distortion in conformal mapping.* Illinois J. Math. 4(1960), 28–37. MR 22A(1)–807. [24, 29, 46]
621. Jenkins, J.A. *On univalent functions with real coefficients.* Ann. of Math. (2) 71(1960), 1–15. MR 22A(2)–964. [15, 19, 24, 39]
622. Jenkins, J.A. *On the global structure of the trajectories of a positive quadratic differential.* Illinois J. of Math. 4(1960), 405–412. MR 23A–330. [59]
623. Jenkins, J.A. *The general coefficient theorem and its applications.* Contributions to function theory (Internat. Colloq. Function Theory, Bombay, 1960), pp. 211–218. Tata Institute of Fundamental Research, Bombay, 1960. MR 26–1002. [9, 17, 24, 25]
624. Jenkins, J.A. *An extension of the general coefficient theorem.* Trans. Amer. Math. Soc. 95(1960), 387–407. MR 22A(2)–1377. [9, 17, 24, 25]
625. Jenkins, J.A. *On certain coefficients of univalent functions.* II.

Trans. Amer. Math. Soc. 96(1960), 534-545. MR 23A-52. [9, 17, 24, 25, 45, 49]
626. Jenkins, J.A. *Some problems for typically-real functions.* Canad. J. Math. 13(1961), 299-304. MR 22A(2)-2093. [13, 15, 23, 66]
627. Jenkins, J.A. *On the schlicht Bloch constant.* J. Math. Mech. 10(1961), 729-734. MR 23A-332. [15, 19, 22, 34]
628. Jenkins, J.A. *The general coefficient theorem and certain applications.* Bull. Amer. Math. Soc. 68(1962), 1-9. MR 24A-371. [24]
629. Jenkins, J.A. *On a conjecture of Goodman concerning meromorphic univalent functions.* Michigan Math. J. 9(1962), 25-27. MR 24A-370. [9, 13, 16, 17, 25, 42]
630. Jenkins, J.A. *An addendum to the general coefficient theorem.* Trans. Amer. Math. Soc. 107(1963), 125-128. MR 26-980. [24]
631. Jenkins, J.A.; Morse, M. *Topological methods on Riemann surfaces.* Pseudoharmonic functions. Contributions to the theory of Riemann surfaces. Ann. of Math. Stud. No. 30(1953), 111-139. MR 15-210. [59]
632. Jenkins, J.A.; Morse, M. *Curve families F* locally the level curves of a pseudoharmonic function.* Acta Math. 91(1954), 1-42. MR 15-956. [59]
633. Jenkins, J.A.; Spencer, D.F. *Hyperelliptic trajectories.* Ann. of Math. 53(1951), 4-35. MR 12-400. [59]
634. Jensen, J.L.W.V. *Investigation of a class of fundamental inequalities in the theory of analytic functions.* Ann. Math. 21(1919-20), 1-29. FM 47-271. [59]. A translation from Danish
635. Jerrard, Richard *Curvature and length.* Bull. Amer. Math. Soc. 67(1961), 113-114. MR 23A(1)-471. [59]
636. Joh, K. *Einige Bemerkung über schlichte Funktionen.* Proc. Imp. Acad. Jap. 9(1933), 561-564. Zbl 8-215. [10]
637. Joh, K. *Über die Abschnitte der Ungeraden schlichten Potenzreihen.* Proc. Imp. Acad. Jap. 11(1935), 407-409. Zbl 15-71. [20, 43, 45]
638. Joh, K. *Theorems on the 'schlicht' functions.* Proc. Physico-Math. Soc. Japan 19(1937), 1-12. Zbl 17-408. [17, 20, 39, 43, 45]
639. Joh, K. *Theorems on 'schlicht' functions.* II. Proc. Phys.-Math. Soc. Jap. III 20(1938), 591-610. Zbl 22-151. [16, 17, 22, 24, 42]
640. Joh, K. *On a theorem of Lowner on univalent functions.* Proc. Imp. Acad. Jap. 14(1938), 41-44. Zbl 19-271. [16, 17, 24]
641. Joh, K. *Theorems on 'schlicht' functions.* III. Proc. Physico-Math. Soc. Japan 21(1939), 191-208. Zbl 21-143. [15, 20, 22, 24, 43, 45, 61]

642. Joh, K. *Theorems on 'schlicht' functions.* IV. Proc. Phys.-Math. Soc. Japan (3) 22(1940), 329-343. MR 1-308. [6, 10, 16, 19]
643. Joh, K. *Theorems on 'schlicht' functions.* V. On the coefficient problem. Proc. Phys.-Math. Soc. Japan (3) 23(1941), 409-423. MR 3-201. [16, 24, 42]
644. Joh, K.; Hukusima, Y. *On the "Verzerrungssatz" of p-valent functions.* Proc. Phys.-Math. Soc. Japan (3) 25(1943), 377-383. MR 7-288. [14, 15, 65]
645. Joh, K.; Takahashi, S. *Ein Beweis für Szegösche Vermutung über schlichte Potenzreihen.* Proc. Imperial Acad. Japan, 10(1934), 137-139. Zbl 9-75. [16, 17, 39]
646. Jorgensen, V. *On conformal mapping on a surface of a sphere.* Mat. Tidsskr. B. (1950), 131-137. (Danish) MR 12-401. [27]
647. Jorgensen, V. *A remark on Bloch's theorem.* Tidsskr. B. (1952), 100-103. MR 15-21. [19, 46]
648. Julia, G. *Extension nouvells d'un lemme de Schwarz.* Acta Math. 42(1920), 349-355. FM 47-272. [46]
649. Julia, G. *Sur une équation aux dérivées fonctionelles liée a là resprésentation conforme.* Ann. Scient. École Norm. SUp. 39 (3rd series) (1922), 1-28. FM 48-404. [59]
650. Julia, G. *Principes Geometriques d'Analyse.* Paris, 1930. FM 56-294. [46]
651. Jung, H.P. *Beiträge zur Theorie der schlichten Funktionen.* Mitt. Math. Sem. Giessen No. 52(1955), 29 pp. MR 16-1010. [17, 23, 24, 32, 42, 54]
652. Kakehashi, T. *On schlicht functions.* I. Proc. Japan Acad. 35(1959), 134-136. MR 21-660. [4, 28]
653. Kakeya, S. *On the star-shaped representation of an analytic function.* Sci. Rep. Tokyo Bunrika Daig. A. (1932), 237-240. Zbl 6-120. [6, 20, 43]
654. Kakeya, Soichi. *On the function whose imaginary part on the unit circle changes its sign only twice.* Proc. Imp. Acad. Tokyo 18(1942), 435-439. MR 8-22. [17, 49, 54]
655. Kaplan, W. *On Gross' star theorem, schlicht functions, logarithmic potentials and Fourier series.* Ann. Acad. Sci. Fennicae. Ser. A.I. Math.-Phys No. 86, (1951), 23 pp. MR 13-337. [4, 62]
656. Kaplan, W. *Close to convex schlicht functions.* Mich. Math. J. 1(1952), 169-185. MR 14-966. [4, 5]
657. Kaplan, W. *Extensions of the Gross star theorem.* Mich. Math. J. 2(1954), 105-108. MR 16-232. [59]
658. Kaplan, W.; et al. *Lectures on Functions of a Complex Variable.*

The Univ. of Mich. Press, 1955. MR 16-1097. [59]
659. Kasjanjuk, S.A. *On a generalization of the locally E-starlike and locally E-convex analytic functions of J.S. Maksimov.* Izv. Vyss. Ucebn. Zaved. Mathematika (1958), No. 1(2), 103-113. (Russian) MR 23A-732. [6, 10, 15, 23]
660. Kasjanjuk, S.A. *Functions regular in an annulus and convex in the direction of the imaginary axis.* Izv. Vyss. Ucebn. Zaved. Matematika (1958), No. 6(7), 105-110. (Russian) MR 23A-731. [10, 17, 23, 38, 40, 41, 49]
661. Kasjanjuk, S.A. *Some subclasses of convex and star-shaped conformal mappings of an annulus.* Izv. Vyss. Ucebn. Zaved. Matematika (1960), No. 6(19), 126-139. (Russian) MR 25-796. [6, 10, 15, 17, 35, 63, 64]
662. Kennedy, P.B. *Conformal mapping of bounded domains.* J. London Math. Soc. 31(1956), 332-336. MR 17-1191. [7, 22]
663. Keogh, F.R. *Some inequalities for convex and star-shaped domains.* J. London Math. Soc. 29(1954), 121-123. MR 16-302. [6, 7, 10, 35]
664. Keogh, F.R. *A property of bounded schlicht functions.* J. London Math. Soc. 29(1954), 379-382. MR 15-862. [7, 22]
665. Keogh, F.R. *Some theorems on conformal mapping of bounded star-shaped domains.* Proc. London Math. Soc. (3) 9(1959), 481-491. MR 22A(1)-295. [6, 7, 22, 23]
666. Keogh, F.R.; Peterson, G.M. *A strengthened form of a theorem of Wiener.* Math. Z. 71(1959), 31-35. MR 23A-371. [59]
667. Kerekjarto, B. *Von Vorlesungen über Topologie.* Berlin, 1923. FM 49-396. [59]
668. Kimura, K. *On the univalency and multivalency of an analytic function.* Proc. Phys.-Math. Soc. Japan (3) 19(1937), 241-245. FM 63-293. [4, 14]
669. Knopp, K. *Mehrfach monotome Zahlenfolgen.* Math. Z. 22(1925), 75-85. FM 51-178. [59]
670. Knopp, K. *Elements of the theory of functions.* Trans. by F. Bagemihl. New York: Dover Co. 1952. 140 pp. Zbl 48-308. [59]
671. Kober, H. *Dictionary of Conformal Representations.* Dover Publ., Inc., New York, N.Y., 1952, 208 pp. MR 14-156. [59]
672. Kobori, A. *Über die Schlichtheit der Abschnitte gewisser Potenzreihen.* Mem. Coll. Sci. Kyoto A. 14(1931), 251-262. Zbl 5-250. [4, 20, 43]
673. Kobori, A. *Über sternige und konvexe Abbildung.* Mem. COll. Sci. Kyoto A 15(1932), 267-278. Zbl 5-362. [6, 10, 11, 12, 16, 20]

674. Kobori, A. *Über die notwendige und hinreichende Bedingung dafur, dass Potenzreihe den Kreisbereich auf den schlichten Sternigen bzw. konvexen Bereich abbildet.* Mem. Coll. Sci. Kyoto Imp. Univ. A 15(1932), 279-291. Zbl 5-362. [4, 6, 10, 16]
675. Kobori, A. *Über sternige und konvexe Abbildung.* II Mem. Coll. Sci. Kyoto A 16(1933), 127-135. Zbl 7-120. [6, 10, 11, 20, 39, 45]
676. Kobori, A. *Zwei Sätze über die Abschnitte der schlichten Potenzreihen.* Mem. Coll. Sci. Kyoto A 17(1934), 171-186. Zbl 9-361. [6, 10, 11, 12, 20]
677. Kobori, A. *Eine Klasse von Potenzreihen.* Jap. J. Math. 13(1936), 49-60. Zbl 14-318. [1, 6, 10, 11, 12, 15, 17, 20, 36, 63]
678. Kobori, A. *Sur la multivalence d'une famille des fonctions analytiques.* Proc. Imp. Acad. Jap. 14(1938), 157159. Zbl 20-142. [6, 14, 32]
679. Kobori, A. *Sur les fonctions multivalentes.* Proc. Phys.-Math. Soc. Japan (3) 23(1941), 423-431. MR 3-79. [14, 15, 17, 50, 65]
680. Kobori, A. *Sur les fonctions multivalentes.* Proc. Imp. Acad. Tokyo 20(1944), 216-217. MR 7-288. [14, 15, 65]
681. Kobori, A. *An evaluation in the theory of multivalent functions.* Proc. Japan Acad. 22 No. 1-4 (1946), 75-77. MR 12-172. [14]
682. Kobori, A. *Zur Theorie der mehrwertigen Funktionen.* Japan J. Math. 19(1947), 301-319. MR 10-362. [9, 11, 14, 15, 16, 17, 25, 27, 43, 50, 65]
683. Kobori, A. *An evaluation in the theory of multivalent functions.* Japan J. Math. 19(1948), 275-285. MR 11-340. [14, 15, 65]
684. Kobori, A. *Une remarque sur les fonctions multivalentes.* Mem. Coll. Sci. Univ. Kyoto. Ser. A. Math. 27(1952), 1-5. MR 14-156. [14, 15, 16, 17, 50, 65]
685. Kobori, A.; Abe, H. *Une remarque sur un théorème de M. Hayman.* Japan J. Math. 29(1959), 32-34. MR 23A-731. [1, 14, 17, 19, 24]
686. Kocur, M.F. *On a class of univalent functions of V.A. Zmorovic.* Dopovidi Akad. Nauk Ukrain. RSR (1959), 1060-1063. (Ukrainian. Russian. English summaries) MR 22A(2)-1629. [17, 23, 28]
687. Kocur, M.F. *On a class of univalent functions in the circle.* Uspehi Mat. Nauk 17(1962), No. 4(106), 153-156. (Russian) MR 26-62. [4, 5, 15, 17, 18, 38, 64]
688. Koebe, P. *Über die Uniformisierung beliebiger analytischer Kurven.* Nachr. Ges. Wiss. Gottingen, (1907), 191-210. FM 38-455. [59]
689. Koebe, P. *Über die Uniformisierung der algebraischen Kurven durch automorphe Funktionen mit imaginarer Substitutions-*

gruppe. Nachr. Kgl. Ges. Wiss. Gottingen, Math. -Phys. Kl. (1909), 68-76. FM 40-468. [59]

690. Koebe, P. *Über die Uniformisierung der algebraischen Kurven.* II. Math. Ann. 69(1910), 1-81. FM 41-480. [59]

691. Koebe, P. *Zur konformen Abbildung unendlich-vielfach zusammenhängender schlichter Bereiche auf Schlitzbereiche.* Nachr. Kgl. Ges. Wiss. Gottingen. Math. -Phys. Kl. (1918), 60-71. FM 46-546. [60]

692. Koebe, P. *Abhandlungen zur Theorie der konformen Abbildung.* IV. Abbildung mehrfach zusammenhangender schlicter Bereiche auf Schlitzbereiche. Acta Math. 41(1918), 305-344. FM 46-545. [60]

693. Koebe, P. *Abhandlungen zur Theorie der konformen Abbildung.* V. Abbildung mehrfach zusammenhangender schlicter Bereiche auf Schlitzbereiche (Fortsetzung). Math. Z. 2(1918), 198-236. FM 46-546. [60]

694. Koebe, P. *Über das Schwarzsche Lemma und einige damit zusammenhangende ungleichheitsbeziehungen der Potentialtheorie und Funktionentheorie.* Math. Z. 6(1920), 52-84. FM 47-271. [37]

695. Kolbina, L. I. *Conformal mappings of the unit circle on nonoverlapping domains.* Vestnik Leningradskogo Univ. No. 5, (1955), 37-43. (Russian) MR 17-26. [59]

696. Kolbina, L. I. *Some extremal problems in conformal mapping.* Doklady Akad. Nauk SSSR (N.S.) 84(1952), 865-868. MR 14-35. [24, 26]

697. Kolbina, L. I. *On the theory of univalent functions.* Doklady Akad. Nauk SSSR (N.S.) 84(1952), 1127-1130. MR 14-35. [9, 15, 58]

698. Kolbina, L. I. *On distortion theorems for certain classes of p-valent functions.* Vestnik Leningrad Univ. II. No. 7, (1956), 71-76. MR 17-1070. [14, 15, 65]

699. Komatu, Y. *Über einen Satz von Herrn Löwner.* Proc. Imp. Acad. Tokyo 16(1940), 512-514. MR 2-276. [59]

700. Komatu, Y. *Über eine Verschärfung des Löwnerschen Hilfssatzes.* Proc. Imp. Acad. Tokyo 18(1942), 354-359. MR 7-286. [7, 22, 24, 62]

701. Komatu, Y. *Untersuchungen über konforme Abbildung von zweifach zusammenhangenden Gebieten.* Proc. Phys. -Math. Soc. Japan (3) 125 (1943), 1-42. MR. 7-514. [59]

702. Komatu, Y. *Untersuchungen über konforme Abbildung von zweifach zusammenhangenden Gebieten.* Proc. Phys. -Math Soc. Japan 25 (3rd series), (1943), 1-42. MR 7-514. [59]

703. Komatu, Y. *Einige Darstellungen analytischer Funktionen und ihre Anwendungen auf konforme Abbildung.* Proc. Imp. Acad. Tokyo 20(1944), 536-541. MR 7-287. [2, 10, 23]
704. Komatu, Y. *Ein alternierendes Approximationsverfahren fur conforme Abbildung von einem Ringgebiete auf einen Kreisring.* Proc. Japan Acad. 21(1945), 146-155. MR 11-341. [59]
705. Komatu, Y. *Note on the theory of conformal representation by meromorphic functions.* I, II. Proc. Imp. Acad. Tokyo 21(1945), 269-277, (1949), 278-284. MR 11-170. [9, 17, 19]
706. Komatu, Y. *Darstellungen der in einem Kreisringe analytischen Funktionen nebst den Anwendungen auf konforme Abbildung uber Pologonalringgebiete.* Japan J. Math. 19(1945), 203-215. MR 7-287. [59]
707. Komatu, Y. *On conformal slit-mapping of a circular ring.* Math. Japonicae 1(1948), 24-27. MR 10-186. [24, 60]
708. Komatu, Y. *Ein alternierendes Approximationsverfahren für konforme Abbildung von einem Ringgebiete auf einen Kreisring.* Proc. Japan Acad. 21(1943), 146-155 (1949). MR 11-341. [60]
709. Komatu, Y. *Note on the theory of conformal representation by meromorphic functions.* I, II. Proc. Japan Acad. 21(1945), 269-277(1949), 278-284(1949). MR 11-170. [9, 15, 17, 19, 25, 58]
710. Komatu, Y. *Zur konformen Abbildung zweifach zusammenhangender Gebiete.* I, II, III, IV, V. Proc. Japan Acad. 21(1945), 285-295, 296-307, 337-339, 372-377, 401-406. (1949). MR 11-169. [60]
711. Komatu, Y. *Fundamental differential equations in the theory of conformal mapping.* Proc. Japan Acad. 25(1949), No. 1, 1-10. MR 12-490. [15, 24, 68]
712. Komatu, Y. *On Robin's constant and a distortion theorem.* Kodai Math. Sem. Rep. (1950), 37-39 (1950), MR 12-250. [9, 15, 58]
713. Komatu, Y. *On conformal slip mapping of multiply connected domains.* Proc. Japan Acad. 26(1950), No. 7. 26-31. MR 13-734. [24]
714. Komatu, Y. *A coefficient problem for functions univalent in an annulus.* Kodai Math. Sem. Rep. 8(1956), 49-70. MR 18-292. [16, 17, 48]
715. Komatu, Y. *On the coefficients of typically-real Laurent series.* Kodai Math. Sem. Rep. 9(1957), 42-48. MR 19-404. [9, 13, 16, 17, 51]
716. Komatu, Y. *On conformal mapping of a domain with convex or star-like boundary.* Kodai Math. Sem. Rep. 9(1957), 105-139. MR 19-949. [6, 10, 15, 23, 63, 64]
717. Komatu, Y. *On analytic functions with positive real part in a circle.*

Kodai Math Sem. Rep. 10(1958), 64-83. MR 20-665. [2, 7, 18]
718. Komatu, Y. *On analytic functions with positive real part in an annulus.* Kodai Math. Sem. Rep. 10(1958), 84-100. MR 20-665. [2, 7, 18, 48]
719. Komatu, Y. *On convolution of power series.* Kodai Math. Sem. Rep. 19(1958), 141-144. MR 21-388. [2, 4, 6, 10, 47]
720. Komatu, Y. *On the range of analytic functions with positive real part.* Kodai Math. Sem. Reports 10, (1958), 145-160. Zbl 87-77. [2, 7, 18]
721. Komatu, Y. *On convolution of Laurent series.* Proc. Japan Acad. 34(1958), 649-652. MR 21-1064. [2, 9, 47]
722. Komatu, Y. *On coefficient problems for some particular classes of analytic functions.* Kodai Math. Sem. Rep. II (1959), 124-130. MR 22A(1)-18. [6, 9, 10, 24]
723. Komatu, Y. *On starlike and convex mappings of a circle.* Kodai Math. Sem. Rep. 13(1961), 123-126. MR 24A-253. [6, 10, 15, 23, 29, 64]
724. Komatu, Y.; Nagura, S. *Theory of univalent functions.* Sugaku 1 (1949), 286-302. MR 14-1075. [15, 24, 68]
725. Komatu, Y.; Nishimiya, H. *On distortion in schlicht mappings.* Kodai Math. Sem. Rep. (1950), 47-50. MR 12-250. [10, 15, 46]
726. Komatu, Y.; Ozawa, M. *Conformal mapping of multiply connected domains.* Kodai Math. Sem. Rep. 3(1951), 81-95. MR 13-734. [59]
727. König, K. *Die ersten Koeffizienten schlichter Funktionen.* Mitt. Math. Ges. Hambg. 7(1931), 9-12. Zbl 1-214. [17, 54]
728. König, K. *Einige Koeffizientenproblem aus der Funktionentheorie.* Tohoku Math. J. 39(1933), 374-379. Zbl 7-351. [17, 27, 31, 39, 54]
729. König, K. *Eine Bemerkung über ungerade schlichte Funktionen.* Mitt. Math. Ges. Hamburg. 7(1934), 238-239. Zbl 9-25. [45]
730. Koosis, P. *Proof of a theorem of the brothers Riesz.* Studia Math. 17(1958), 295-298. MR 21-131. [59]
731. Korickii, G. V. *On curvature of level lines and of their orthogonal trajectories in conformal mappings.* Matt. Sb. N.S. 37(79) (1955), 103-116. (Russian) MR 17-26. [9, 10, 35]
732. Korickii, G. V. *On the curvature of level curves of univalent conformal maps.* Uspehi Mat. Nauk 15(1960), No. 5(95), 179-182. (Russian) MR 23A-330. [9, 35, 49]
733. Koritzky, G. *Der Satz von Herrn Szegö fur einige spezielle Klassen der schlichten Funktionen.* Rec. Math. Moscou 36(1929), 91-98. FM 55-791. [6, 15, 20, 43, 45, 67]

734. Koritzky, G. U. *Curvature of level curves in univalent conformal mappings.* Dokl. Akad. Nauk SSSR (N.S.) 115(1957), 653-654. (Russian) MR 19-845. [6, 9, 35, 45, 49]
735. Koseki, K. *Über die Koeffizienten der schlichten Funktionen.* Math. J. Okayama Univ. 9(1959/60), 173-197. MR 22A(2)-1168. [24, 42]
736. Koseki, K. *Über die Koeffizienten der schlichten Funktionen.* II. Math. J. Okayama Univ. 10(1960/61), 125-142. MR 23A-330. [16, 24, 42]
737. Koseki, K. *Über die p-wertigen Funktionen.* Math. J. Okayama Univ. 10(1960/61), 87-99. MR 23A-619. [14, 15, 17, 34, 65]
738. Koseki, K. *Beiträge zur Theorie der schlichten Funktionen.* Math. J. Okayama Univ. 10(1960/61), 1-9. MR 24A-495. [28]
739. Koseki, K. *Über die Koeffizienten der schlichten Funktionen.* III. Math. J. Okayama Univ. 11(1962), 27-42. MR 23A-427. [24, 42]
740. Koseki, K. *Über die Ableitungen der schlichten Funktionen.* Math. J. Okayama Univ. 11(1962), 43-50. MR 25-254. [17, 54]
741. Koseki, K. *Über die Koeffizienten der schlichten und meromorphen Funktionen.* I. Math. J. Okayama Univ. 11(1962), 51-58. MR 25-254. [9, 17, 25]
742. Kössler, M. *Ein Beitrag zur Theorie der schlichten Potenzreihen.* Mem. Soc. Roy. Sci. Boheme 1932 No. 5 (1933), 1-7. Zbl 7-312. [15, 17, 68]
743. Kössler, M. *Über besondere Klassen von schlicht Abbildungen Potenzreihen.* I. Mem. Soc. Roy. Sci. Boheme (1934) 7 p. Zbl 10-361. [1, 4, 22]
744. Kössler, M. *Über Potenzreihen mit beschranktem Imaginarteile.* C.R. 2me Congres Math Pays slaves 150-151, Casopis Praha 64, (1935), 150-151. FM 61-347. [59]
745. Kössler, M. *Über reele Charakteristiken von Potenzreihen.* Czechoslovak Math. J. 4(79), (1954), 274-282. (Russian summary) MR 16-914. [59]
746. Krasnovidova, I.S.; Rogozin, U.S. *A sufficient condition for univalency of the solution of an inverse boundary problem.* Uspehi Matem. Nauk (N.S.) 8(1953) No. 1(53) 151-153. MR 14-740. [4]
747. Kresnjakova, L.V. *On the radii of starlikeness and convexity for functions of bounded modulus.* Izv. Akad. Nauk Armjan. SSR. Ser. Fiz.—Mat. Nauk 14(1961), No. 4, 49-55. (Russian. Armenian summary) MR 25-39. [3, 11, 12]
748. Kresnjakova, L.V. *Analytic functions with bounded mean modulus.* Izv. Vyss. Ucebn. Zaved. Matematika (1961), No. 1(20),

98-103. (Russian) MR 26-615. [3, 11, 28]
749. Kresnjakova, L.V. *Some estimates for regular functions with bounded mean modulus.* Izv. Vyss. Ucebn. Zaved. Mat. (1963), No. 1(32), 94-97. [3, 11, 15, 16, 17]
750. Kreyszig, E.; Todd, J. *The radius of univalence of the error function.* Bull. Amer. Math. Soc. 64(1958), 363-364. MR 20-969. [28]
751. Kreyszig, E.; Todd, J. *The radius of univalence of the error function.* Numer. Math. 1(1959), 78#89. MR 21#133. [10, 13, 28]
752. *Kreyszig, E.; Todd, J. On the radius of univalence of the function exp $z^2 \int_0^z$ exp$(-t^2)$ dt.* Pacific J. Math. 9, No. 1(1959), 123-127. Zbl 85-66. [28, 43]
753. Kreyszig, E; Todd, J. *The radius of univalence of Bessel functions.* Illinois J. of Math. 4(1960), 143-149. MR 22A(1)-297. [28]
754. Kronsbein, J. *Analytical expressions for some extremal schlicht functions.* J. London Math. Soc. 17(1942), 152-157. MR 4-215. [24, 60]
755. Krzywoblocki, M.Z. *A local maximum property of the fourth coefficient of schlicht functions.* Duke Math. J. 14(1947), 109-128. MR 8-508. [17, 24, 54]
756. Krzyz, B. *Sur les fonctons en moyenne (ψ) p-valentes.* Ann. Univ. Mariae Curie-Sklodowska Sect. A 12(1958), 38-44. (Polish. Russian summaries) MR 22A(2)-1900. [33]
757. Krzyz, J. *On the maximum modulus of univalent functions.* Bull. Acad. Polon. Sci. Cl. III (3) (1955), 203-206. MR 17-143. [62]
758. Krzyz, J. *On the derivative of bounded univalent functions.* Bull. Acad. Polon. Sci. Ser. Sci. Math. Astr. Phys. 6(1958), 157-159. MR 21-27. [15, 18, 22, 61]
759. Krzyz, J. *On the derivative of bounded p-valent functions.* Ann. Univ. Marie Curie-Sklodowska Sect. A 12(1958), 23-28. (Polish. Russian summaries) MR 23A-177. [14, 15, 22, 24, 61]
760. Krzyz, J. *Distortion theorems for bounded p-valent functions.* Ann. Univ. Mariae Curie-Sklodowska Sect. A 12(1958), 29-38. (Polish. Russian summaries) MR 23A-177. [14, 15, 22, 24, 61, 65]
761. Krzyz, J. *Distortion theorems for bounded convex functions.* Bull. Acad. Polon. Sci. Ser. Sci. Math. Astronom. Phys. 8(1960), 625-627. (Russian summary, unbound insert) MR 24A-153. [10, 15, 19, 22, 64]
762. Krzyz, J. *Distortion theorems for bounded convex functions. II.* Ann. Univ. Mariae Curie-Sklodowska Sect. A 14(1960), (Polish. Russian summaries) MR 25-797. [10, 15, 19, 22, 64]
763. Krzyz, J. *On univalent functions with two preassigned values.* Ann.

Univ. Mariae Curie-Sklodowska Sect. A 15(1961), 57-77. (Polish. Russian summaries) MR 25-427. [15, 24, 29, 54]
764. Krzyz, J. *On a problem of P. Montel.* Ann. Polon. Math. 12(1962), 55-60. MR 25-611. [15, 24, 68]
765. Krzyz, J. *The radius of close-to-convexity within the family of univalent functions.* Bull. Acad. Polon. Sci. Ser. Sci. Math. Astronom. Phys. 10(1962), 201-204. MR 25-1210. [5, 12]
766. Krzyz, J. *On the derivative of close-to-convex functions.* Colloq. Math. 10(1963), 143-146. MR 25-1210. [5, 15, 29, 64]
767. Krzyz, J.; Radziszewski, K. *Isoperimetrical defect and conformal mapping.* Ann. Univ. Mariae Curie-Sklodowska Sect. A. 10(1956), 49-56 (1958). (Polish and Russian summaries) MR 20-289. [7, 10, 16, 18]
768. Ku, C-H. *A note on bounded schlicht functions.* Scie. Record, Acad. Sinica 3(1950), 157-159. Zbl 40-36. [19, 22]
769. Kubo, T. *Some theorems on bounded analytic functions.* Mem. Coll. Sci. Univ. Kyoto Ser. A. Math. 27(1953), 235-243. MR 15-208. [22, 60]
770. Kubo, T. *Bergman kernel function and canonical slit-mapping.* Mem. Coll. Sci. Univ. Kyoto Ser. A. Math. 28(1953), 33-40. MR15-695. [49]
771. Kubo, T. *Kelvin principle and some inequalities in the theory of functions.* I. Mem. Coll. Sci. Univ. Kyoto Ser. A. Math. 28(1953), 299-311. MR 16-122. [59]
772. Kubo, T. *Symmetrization and univalent functions in an annulus.* J. Math. Soc. Japan 6(1954), 55-67. MR 15-948. [19, 22]
773. Kubo, T. *Theory and applications of symmetrization.* Sugaku 9(1957/58), 45-55. (Japanese) MR 20-544. [24]
774. Kubo, T. *Hyperbolic transfinite diameter and some theorems on analytic functions in an annulus.* J. Math. Soc. Japan 10(1958), 348-364. MR 21-659. [19]
775. Kudrjavcev, A.L. *On an approximate method for obtaining conformal mappings.* Izv. Akad. Nauk UZSSR Ser. Fiz.—Mat. (1959), No. 6, 78-82. (Russian) MR 22A(2)-253. [59]
776. Kufarev, P.P. *On integrals of the simplest differential equations with variable polar singularity in the right-hand side.* Uc. Zap. Tomsk. Univ. 1(1946), 35-48. (Russian) [59]
777. Kufarev, P.P. *A remark on integrals of Lowner's equation.* Dokl. Akad. Nauk SSSR 57(1947), 655-656. (Russian) MR 9-421. [24]
778. Kufarev, P.P. *On the theory of univalent functions.* Dokl. Akad. Nauk SSSR (N.S.) 57(1947), 751-754. MR 9-507. [24, 49]

779. Kufarev, P.P. *On conformal mapping of complementary regions.* Dokl. Akad. Nauk SSSR (N.S.) 73(1950), 881–884. (Russian). MR 12-401. [24, 26, 49]
780. Kufarev, P.P. *On a property of extremal regions of the problem of coefficients.* Dokl. Akad. Nauk SSSR (N.S.) 97(1954), 391–393. (Russian) MR 16-122. [42]
781. Kufarev, P.P. *Remark on the problem of coefficients.* Tomskii Gos. Univ. Uc. Zap. Mat. Meh. 25(1955), 15–18. (Russian) MR 19-404. [16, 42]
782. Kufarev, P.P. *On certain method of investigation of extremum problems in the theory of univalent functions.* Dokl. Akad. Nauk SSSR (N.S.) 107(1956), 633–635. MR 17-1069. [9, 29]
783. Kufarev, P.P.; Fales, A.E. *On an extremal problem for complementary domains.* Dokl. Akad. Nauk SSSR (N.S.) 81(1951), 995–998. MR 14-262. [59]
784. Kufarev, P.P.; Semuhina, N.V. *On an extension of Golusin's variational method to doubly connected regions.* Dokl. Akad. Nauk SSSR (N.S.) 107(1956), 505–507. MR 17-1193. [24, 29, 48]
785. Kulshrestha, P.K. *On evaluations of the measure of curvature of level curves of schlicht functions.* Ganita 7(1956), 123–137. MR 20-292. [35]
786. Kulshrestha, P.K. *On measure of curvature of level curves and orthogonal trajectories of a class of mean p-valent functions in the unit circle.* Ganita 9(1958), 1–4. MR 21-517. [31, 33, 35]
787. Kung, S. [Gun Syn]. *The function K(t) in Golusin's and Lowner's differential equation.* Acta Math. Sinica 3(1953), 225–230. MR 17-142. [24]
788. Kung, S. *Distortion theorems and coefficients of schlicht functions.* Acta Math. Sinica 3(1953), 231–250. MR 17-142. [15, 17, 42, 54, 68]
789. Kung, S. *Some theorems on symmetric schlicht functions.* Acta Math. Sinica 3(1953), 251–260. MR 17-142. [17, 49, 54]
790. Kung, S. *On coefficients of univalent functions.* Acta Math. Sinica 4(1954), 87–103. MR 17-142. [17, 54]
791. Kung, S. *The sections of schlicht functions.* Acta Math. Sinica 4(1954), 105–112. MR 17-142. [9, 16, 20, 43]
792. Kung, S. *On mean valent functions.* Acta Math. Sinica 4(1954), 245–257. MR 17-142. [3, 33]
793. Kung, S. *Contributions to the theory of schlicht functions.* I. Distortion theorems. Sci. Sinica 4(1955), 229–249. MR 19-738. [15, 17, 42, 49, 54, 68]

794. Kung, S. *On the coefficients of schlicht functions.* III. Acta Math. Sinica 6(1956), 490-499. (Chinese. English summary) MR 20-544. [42]
795. Kung, S. *On the coefficients of schlicht functions.* II. Acta Math. Sinica 6(1956), 115-125. (Chinese. English summary) MR 18-121. [17, 49]
796. Kunugui, K. *Sur une constante de la transformation conforme.* Proc. Imp. Acad. Tokyo 19(1943), 278-281. MR 7-379. [7, 49]
797. Kuramochi, Z. *A remark on the bounded analytic function.* Osaka Math. J. 4(1952), 185-190. Zbl. 48-319. [22]
798. Kuroda, I. *On properties of Friedman's functions and other functions closely related with them.* Bull. Yamagata Univ. (Nat. Sci.) 4(1957), 1-11. (Japanese. English summary) MR 20-969. [39]
799. Kuz'mina, G.V. *Determination of the least radius of univalency for a certain class of analytical functions.* Dokl. Akad. Nauk SSSR (N.S.) 117(1957), 751-754. MR 20-161. [43]
800. Kuz'mina, G.V. *Numerical determination of radii of univalence of analytical functions.* Trudy Mat. Inst. Steklov. 53(1959), 192-235. MR 22A(2)-1629. [43]
801. Kuz'mina, G.V. *Some covering theorems for univalent functions.* Dokl. Akad. Nauk SSSR (N.S.) 142(1962), 29-31. (Russian) MR 24A-254. [19]
802. Kwesselawa, D. *Zum Lindelöfschen Prinzip.* Mitt. Georg. Abt. Akad. Wiss. USSR (1940), 713-718. (Russian. German transl) MR 3-78. [1]
803. Lad, S. *Contribution a la théorie des fonctions univalentes.* Casopis Pest. Mat. Fys. 62(1936), 12-19.
804. Ladegast, K. *Beiträge zur Theorie der schlichten Funktionen.* Math. Z. 58(1953), 115-159. MR 15-24. [9, 13, 15, 17, 19, 25, 27, 31, 39, 43, 45, 54, 58, 67, 68]
805. Lai, W. *Über die Konjektur von Goodman fur die bienahe Beschrankten Funktionen.* Acta Math. Sinica 9(1959), 292-294. (Chinese. German summary) MR 22A(2)-2094. [34]
806. Lai, W. *On a conjecture of Goodman for almost bounded functions.* Sci. Sinica 11(1962), 1303-1305. MR 26-277. [34]
807. Landau, E. *Zum Koebeschen Verzerrungssatz.* R.C. CIrc. Mat. Palermo. 46(1922), 347-348. FM 48-406. [9, 19]
808. Landau, E. *Einige Bermerkungen über schlichte Abbildung.* Jber. D.M.V. 34(1926), 239-243. FM 52-349. [15, 42]
809. Landau, E. *Über schlichte Funktionen.* Math. Z. 30(1929), 635-638. FM 55-187. [42]

810. Landau, E. *Über die Blochsche Konstante und zwei verwandte Weltkonstanten.* Math. Z. 30(1929), 608-634. FM 55-770. [19]
811. Landau, E. *Darstellung und Begrundung einiger neuer Ergebnisse der funktionen Theorie.* Ed. 2, Berlin, 1929. [59]
812. Landau, E. *Über ungerade schlichte Funktionen.* Math. Z. 37(1933), 33-35. Zbl. 6-211. [16, 17, 45, 52]
813. Landau, E. *Ausgewählte Kapitel der Funktionentheorie.* Trav. Inst. Math. Tbilissi (Trudy Tbiliss. Mat. Inst.) 8(1940), 23-68. (German. Russian summary) MR 3-78. [19]
814. Landau, E.; Valiron, G. *A deduction from Schwarz's lemma.* J. London Math. Soc. 4(1929), 162-163. FM 55-769. [37]
815. Landau, H. J. *On canonical conformal maps of multiply connected domains* Trans. Amer. Math. Soc. 99(1961), 1-20. MR 22A(2)-2089. [60]
816. Landau, H. J.; Osserman, R. *Some distortion theorems for multivalent mappings.* Proc. Amer. Math. Soc. 10(1959), 87-91. MR 21-660. [14, 49]
817. Landau, H. J.; Osserman, R. *On analyticmappings of Riemann surfaces.* J. d'Analyse Math. 7(1959/1960), 249-279. MR 23A-52. [59]
818. Lavrentieff (Lavrentiev), M. A. *On the theory of conformal mapping.* Trav. Inst. Phys. -Math. Stekloff Sect. Math., Leningrad. No. 5(1934), 129-245. FM 60-1026. [59]
819. Lavrentieff (Lavrentiev), M. A. *Sur quelques propriétés des fonctions univalentes.* C. R. Acad. URSS 1(1935), 2-4. Zbl 12-214. [59]
820. Lavrentieff (Lavrentiev), M. A. *Sur quelques propriétés des fonctions univalentes.* Rec. Math. Moscou N. S. 1(1936), 815-844. Zbl 16-217. [7]
821. Lavrentieff (Lavrentiev), M. A. *Sur la continuité des fonctions univalentes.* C. R. Acad. Sci. URSS N. S. 4(1936), 215-217. Zbl 16-169. [22, 60]
822. Lavrentieff (Lavrentiev), M. A.; Chepeleff, V. M. *Sur quelques propriétés des fonctions univalentes.* Mat. Sb. 44(1937), 319-326. Zbl 17-173. [19]
823. Lavrentieff (Lavrentiev), M. A.; Kwasslava, D. *Über einen Ostrowskischen Satz.* Georg. Abt. Akad. Wiss., USSR 1(1940), 171-174. MR 2-83. [59]
824. Lebedev, N. A. *On some estimates and extremal problems in conformal mapping.* Dissertation, Leningrad State Univ., 1951. [24]
825. Lebedev, N. A. *The method of variationson conformal mapping.* Dokl. Akad. Nauk SSSR (N.S.) 76(1951), 25-27. MR 12-491. [24]

826. Lebedev, N. A. *Some estimates for functions regular and univalent in a circle.* Vestnik Leningrad Univ. 10(1955), No. 11, 3–21. Amer. Math. Soc. Transl. (2)22, 59–80. MR 17-599. [15, 22, 24, 29, 61, 68]
827. Lebedev, N. A. *On parametric representation of functions regular and univalent in a ring.* Dokl. Akad. Nauk SSSR (N.S.) 103(1955), 767–768. MR 17-356. [24]
828. Lebedev, N. A. *Majorizing region for the expression $I = ln[z^\lambda f'(z)^{1-\lambda}/f(z)^\lambda]$ in the class S.* Vestnik Leningrad Univ. 10(1955), No. 8, 29–41. Amer. Math. Soc. Transl. (2) 22, 43–57. MR 17-248. [15, 24, 29, 68]
829. Lebedev, N. A. *On the theory of conformal mappings of a circle onto non-overlapping regions.* Dokl. Akad. Nauk SSSR (N.S.) 103(1955), 553–555. (Russian) MR 17-250. [26]
830. Lebedev, N. A. *On the domains of values of a certain functional in the problem of non-overlapping domains.* Dokl. Akad. Nauk SSSR (N.S.) 115(1957), 1070–1073. (Russian) MR 19-951. [19, 24, 26, 29]
831. Lebedev, N. A. *The area principle in the problem of non-overlapping regions.* Soviet Math. -Dokl. 1(1960) No. 2, 640–644. MR 22A(2)-1896. [3, 27]
832. Lebedev, N. A. *An application of the area principle to non-overlapping domains.* Trudy Mat. Inst. Steklov. 60(1961), 211–231. (Russian) MR 24A-254. [3, 26, 27]
833. Lebedev, N. A.; Milin, I. M. *On the coefficients of certain classes of analytic functions.* Rec. Math. (Mat. Sb) N. S. 24(1949), 249–262. MR 13-640. [3, 15, 17, 34, 42, 54, 68]
834. Lebedev, N. A.; Milin, I. M. *On the coefficients of certain classes of analytic functions.* Dokl. Akad. Nauk SSSR 67(1949), 221–223. MR 11-339. [3, 15, 16, 17, 34, 42, 54, 68]
835. Lebedev, N. A.; Milin, I. M. *On the coefficients of certain classes of analytic functions.* Mat. Sb. (N.S.) 28(70) (1951), 359–400. MR 13-640. [3, 15, 16, 17, 34, 42, 68]
836. Lebedev, N. A.; Sogomonova, G. A. *A method of obtaining a certain kind of estimate for functions regular in the circle.* Vestnik Leningrad Univ. 14(1959), No. 13, 15–19. Addendum 15(1960), No. 13, 152. (Russian. English summary) MR 22A(2)-1376. [1]
837. Lehto, G. *On the distortion of conformal mapping with bounded boundary rotation.* Ann. Acad. Sci. Fenn. Ser. A 1(1952), No. 124, 14 pp. MR 14-743. [59]
838. Lehto, G. *A majorant principle in the history of functions.* Math. Scand. 1(1953), 5–17. MR 15-115. [1]
839. Lehto, G. *The spherical derivative of meromorphic functions in the*

neighborhood of an isolated singularity. Comment. Math. Helv. 33(1959), 196–205. MR 21-1063. [46]
840. Lehto, G.; Virtanen, K. I. *Boundary behavior and normal meromorphic functions.* Acta Math. 97(1957), 47–65. MR 19-403. [59]
841. Lehto, G.; Virtanen, K. I. *On the behaviour of meromorphic functions in the neighbourhood of an isloated singularity.* Ann. Acad. Sci. Fenn. Ser. A. I. (1957), No. 240, 9pp. MR 19-404. [59]
842. Levin, V. *Bemerkung zu den schlichten Abbildungen des Einheitskreises.* Jber. D. M. V. 42(1933), 68–70. [17]
843. Levin, V. *Eine Bemerkung zum Koeffizientenproblem der shclichten Funktionen.* Jber. D. M. V. 42(1933), 70–71. FM 51-367. [17]
844. Levin, V. *Ein Beitrag zum Koeffizientenproblem der schlichten Funktionen.* Math. Z. 38(1933/34), 306–311. Zbl 8-119. [17, 42, 45, 49, 52]
845. Levin, V. *Über die Koeffizientensummen einiger Klassen von Potenzreihen.* Math. Z. 38(1934), 565–590. FM 60-243. [59]
846. Levin, V. *Some remarks on the coefficients of schlicht funcns.* Proc. London Math. Soc. 39(1935), 467–480. Zbl 12-171. [3, 6, 10, 15, 17, 19, 27, 36, 42, 45, 52, 54, 68]
847. Levy, P. *Problèmes Concrets d'Analyse Fonctionelle.* Paris, 1951 (Second edition of Lecons d'Analyse Fonctionelle, Paris, 1922). MR 12-834. [59]
848. Lewandowski, Z. *Quelques remarques sur les théorèmes de Schild relatifs à une classe de fonctions univalentes.* Ann. Univ. Mariae Curie-Sklodowska, Sec. A 9(1955), 149–155 (1957). MR 19-738. [4, 12, 15, 32]
849. Lewandowski, Z. *Nouvelles remarques sur les théorèmes de Schild relatifs a une classe de fonctions univalentes (Démonstration d'une hypothèse de Schild).* Ann. Univ. Mariae Curie-Sklodowska. Sect. A 10(1956), 81–94 (1958). (Polish. Russian summaries) MR 20-292. [12, 16, 19, 32]
850. Lewandowski, Z. *Quelques remarques sur les théorèmes de Schild relatifs à une classe de fonctions univalentes.* Ann. Univ. Mariae Curie-Sklodowska Sect. A 9(1955), 149–155 (1957). (Polish. Russian summaries)MR 19-738. [4, 12, 15, 32]
851. Lewandowski, Z. *Sur l'identité de certaines classes de fonctions univalentes.* I. Ann. Univ. Mariae Curie-Sklodowska Sect. A 12(1958), 131–146. (Polish. Russian summaries) MR 24A-37. [5, 16]
852. Lewandowski, Z. *Sur certaines classes de fonctions univalentes introduites par P.* Montel et W. Rogosinski. Bull. Acad. Polon. Sci. Math. Astr. Phys. 7(1959), 261–265. (Russian summary) MR 21-1194. [15, 19, 39,, 42, 68]

853. Lewandowski, Z. *Sur certaines classes de fonctions univalentes dans de cercle-unité.* Ann. Univ. Mariae-Curie Sklodowska Sect. A 13(1959), 115-126. (Polish. Russian summaries). MR 22A(2)-1900. [6, 15, 19, 39, 42, 63]
854. Lewandowski, Z. *Sur l'identité de certaines classes de fonctions univalentes.* II. Ann. Univ. Mariae Curie-Sklodowska Sect. A 14(1960), 19-46. MR 28-44. [5, 11]
855. Lewandowski, Z. *Sur les majorantes des fonctions holomorphes dans le cercle* $|z| < 1$. Ann. Univ. Mariae Curie-Sklodowska Sect. A 15(1961), 5-11. (Polish. Russian summaries) MR 26-63. [1, 16]
856. Lewandowski, Z. *Starlike majorants and subordination.* Ann. Univ. Mariae-Sklodowska Sect. A 15(1961), 79-84. (Polish. Russian summaries) MR 25-427. [1, 6]
857. Li, E. P. *On the theory of univalent functions on a circular ring.* Dokl. Akad. Nauk SSSR (N.S.) 92(1953), 475-477. MR 15-516. [24, 60]
858. Li, E. P. *On typically real functions on a circular ring.* Dokl. Akad. Nauk SSSR (N.S.) 92(1953), 699-702. MR 15-516. [13, 15, 23, 66]
859. Libera, R. J.; Robertson, M. S. *Meromorphic close-to-convex functions.* Michigan Math. J. 8(1961), 165-175. MR 24A-370. [4, 5, 9, 10, 17, 25, 38, 39]
860. Lindelöf, E. *Memoire sur certaines ineglités dans la théorie des fonctions monogènes et sur quelques propriétés nouvelles de ces fonctions dans le voisinage d'un point singulier essentiel.* Acta Soc. Sci. Fenn. 35(1908), No. 7. [59]
861. Linis, V. *Note on univalent functions.* Amer. Math. Mon. 62(1955), 109-110. MR 16-809. [39]
862. Littlewood, J. E. *On inequalities in the theory of functions.* Proc. London Math. Soc. (2), 23(1925), 481-519. FM 51-247. [3, 42]
863. Littlewood, J. E. *On the coefficients of schlicht functions.* Quart. J. Math. 9(1938), 14-20. Zbl 18-261. [1, 16, 17, 42, 45, 52]
864. Littlewood, J. E. *Lectures on the Theory of Functions.* Oxford Univ. Press, 1944, 243 pp. MR 6-261. [44]
865. Littlewood, J. E.; Paley, R. E. *A proof that an odd schlicht function has bounded coefficients.* J. London Math. Soc. 7(1932), 167-9. Zbl 5-18. [17, 45, 52]
866. Liu, Li-Chuan. *Some inequalities derived from fundamental lemma concerning schlicht functions.* Acta Math. Sinica 7(1957), 313-326. (Chinese. English summary) MR 21-25. [9, 17, 22, 25, 30]
867. Liu, Li-Chuan. *Bounded schlicht functions in the unit circle.* Acta Math. Sinica 7(1957), 439-450. (Chinese. English summary) MR 21-26. [15, 22, 24, 61]

868. Lochs, G. *Zur Abschatzung schlichter Potenzreihen.* Monatshefte F. Math. 38(1931), 377-380. FM 57-403. [17, 19]
869. Loewner, C.; Netanyahu, E. *On some compositions of Hadamard type in classes of analytic functions.* Bull. Amer. Math. Soc. 65(1959), 284-286. MR 21-1350. [16, 23, 42, 47]. See also Lowner.
870. Lohin, I. F. *Remarks on estiamtes for regular functions.* Mat. Sbornik N. S. 24(66) (1949), 249-262. MR 11-339. [1, 10, 15, 17, 34]
871. Lohwater, A. J. *The exceptional values of meromorphic functions.* Colloq. Math. 7(1959), 89-93. MR 22A(1)-455. [59]
872. Lohwater, A. J.; Piranian, G. *Conformal mapping of a Jordan region whose boundary has positive two-dimensional measure.* Michigan Math. J. 1, (1942), 1-4. MR 14-262. [62]
873. Lohwater, A. J.; Piranian, G. *On the derivative of a univalent function.* Proc. Amer. Math. Soc. 4(1953), 591-594. MR 15-114. [62]
874. Lohwater, A. J.; Piranian, G.; Rudin, W. *The derivative of a schlicht funciton.* Math. Scand. 3(1955), 103-106. MR 17-249 [4, 62]
875. Lohwater, A. J.; Seidel, W. *An example in conformal mapping.* Duke Math. J. 15(1948), 137-143. MR 9-420. [62]
876. London, D. *On the zeros of the solution of $w'' + pw = 0$.* Pac. J. Math. 12(1962), 979-991. [4, 31]
877. Loomis, L. H. *The radius and modulus of n-valence for analytic functions whose first $n - 1$ derivatives vanish at a point.* Bull. Amer. Math. Soc. 46(1940), 496-501. MR 1-308. [4, 14, 19, 22, 37, 43]
878. Loomis, L. H. *The decomposition of meromorphic functions into rational functions of univalent functions.* Trans. Amer. Math. Soc. 50(1941), 1-14, MR 3-78. [59]
879. Loomis, L. H. *On an inequality of Seidel and Walsh.* Bull. Amer. Math. Soc. 48(1942), 908-911. MR 5-37. [14, 22]
880. Löwner, K. *Untersuchungen über die Verzerrung bei konformen Abbildungen des Einheitskreises $|z| < 1$.* Leipzig Berichte 69(1917), 89-106. FM 46-556. [15]
881. Löwner, K. *Über Extremumsätze bei der konformen Abbildung des Äusseren des Einheitkreises.* Math. Z. 3(1919), 65-77. FM 47-325. [9, 10, 15, 19, 27, 58, 64]
882. Löwner, K. *Untersuchungen über schlichte konforme Abbildungen des Einheitskreises.* I. Math. Ann. 89(1932), 103-121. FM 49-714. [15, 17, 23, 24, 42]
883. Lozovik, V. G. *Functions with bounded rotation in the unit circle.* Dopovidi Akad. Nauk Ukrain RSR (1960), 1584-1588. (Ukrainian. Russian and English summaries) MR 23A-177. [4, 17, 20, 23]
884. Lozovik, V. G. *On a class of functions which are univalent in the*

unit circle. (Russian) Izv. Vyss. Ucebn. Zaved. Matematika (1963), No. 2(33), 63-69. MR 26-980. [10, 15, 17, 23, 64]
885. Luke, Y. L. *The radius of univalence of the function* $\exp z^2 \int_0^z \exp(-f^2) \, df$. Numer. Math 3(1961), 76-78. MR 22A-1629. [28]
886. Lye, Su-cin. *Some results on univalent functions in dissertations written by students in the analysis section of the department of mathematics in the North-West University.* Advancement in Math. 3(1957), 325-334. (Chinese) MR 20-1074. [44]
887. MacGregor, T. H. *Functions whose derivative has a positive real part.* Trans. Amer. Math. Soc. 104(1962), 532-537. MR 25-797. [1, 2, 4, 5, 7, 10, 12, 15, 17, 18, 19, 20, 21, 43, 56]
888. MacGregor, T. H. *Coefficient estimates for starlike mappings.* Mich. Math. J. 10(1963), 277-281. MR 27-734. [6, 9, 10, 17, 36, 38, 39, 42, 45]
889. MacGregor, T. H. *The radius of convexity for starlike functions of order 1/2.* Proc. Amer. Math. Soc. 14(1963), 71-76. MR 27-60. [2, 6, 10, 11, 12, 23, 43]
890. MacGregor, T. H. *The radius of univalence of certain analytic functions.* Proc. Amer. Math. Soc. 14(1963), 514-520. MR 26-1210. [2, 4, 6, 10, 11, 12]
891. MacGregor, T. H. *The radius of univalence of certain analytic functions.* II. Proc. Amer. Math. Soc. 14(1963), 521-524. MR 26-1211. [2, 4, 6, 10, 11, 12]
892. MacIntyre, A. J. *Two theorems on schlicht functions.* J. London Math. Soc. 11(1036), 7-11. Zbl 13-271. [15, 17, 19, 68]
893. MacIntyre, A. J. *On Bloch's theorem.* Math. Z. 44(1938), 536-540. Zbl 19-419. [19]
894. MacIntyre, A. J.; Rogosinski, W. W. *Some elementary inequalities in funciton theory.* Edinburgh Math. Notes No. 35(1945), 1-3. MR 7-150. [3]
895. MacIntyre, A. J.; Rogosinski, W. W. *Extremum problems in the theory of analytic functions.* Acta Math. 82(1950), 275-325. MR 12-89. [59]
896. Maksimov, Y. D. *Extremal problems in certain classes of analytic functions.* Dokl. Akad. Nuak SSSR (N.S.) 100(1955), 1041-1044. (Russian) MR 16-810. [5, 6, 10, 15, 23, 63, 64]
897. Maksimov, Y. D. *On locally e-convex and locally e-starlike multivalent funcitons.* Dokl. Akad. Nauk SSSR (N.S.) 103(1955), 965-967. MR 17-357. [5, 6, 10, 11, 12, 15, 23, 63, 64]
898. Maksimov, Y. D. *Bounds for the coefficients of certain classes of analytic functions.* Dokl. Akad. Nauk SSSR 110(1956), 507-510.

(Russian) Zbl 74-56. [5, 6, 10, 17, 36, 38]
899. Maksimov, Y. D. *Extension of the structural formula for convex univalent functions to a multiply connected circular region.* Dokl. Akad. Nauk SSSR (N.S.) 136(1961), 284-287. (Russian(Transl. Soviet Math. -Dokl. 2, 55-58. MR 22A(2)-1376. [23, 60]
900. Malik, M. A. *An inequality for polynomials.* Canad. Math. Bull. 6(1963), 65-69. MR 26-742. [59]
901. Mandelbrojt, S. *Quelques remarques sur les fonctions univalentes.* Bull. Sci. Math. II. S. 58(1034), 185-200. Zbl 9-216. [3, 13, 17, 39, 43, 51]
902. Marchenko, A. R. *Sur la représentation conforme.* C. R. (Doklady) Acad. Sci. USSR 6(1935), 287-290. Zbl 11-261. [7]
903. Marchenko, A. R. *Some extremal problems in the theory of univalent functions.* Leningrad Gos. Univ. Uc. Zap. 144Ser. Mat. Nauk 23(1952), 257-269. MR 17-1069. [16, 42]
904. Marcinkiewicz, J.; Zygmund, A. *A theorem of Lusin.* Duke Math. J. 4(1938), 473-485. Zbl. 19-420. [59]
905. Marden, M. *The geometry of the zeros of a polynomial in a complex variable.* Amer. Math. Soc. Surveys No. 3 (1949). MR 11-101. [59]
906. Marty, F. *Sur les dérivées second et troisième d'une fonction holomorphe univalente dans la cercle unité.* C. R. Acad. Sci. Paris 194(1932), 1308-1310. Zbl 4-261. [15, 16, 17, 42, 54, 68]
907. Marty, F. *Sur le module des coefficients de MacLaurin d'une fonction univalente.* C. R. Acad. Sci. Paris 198(1934), 1569 1571. Zbl 9-76. [17, 24, 42, 54]
908. Marx, A. *Zwei Satze uber schlichte Funktionen.* Sitzsber. Pr. Ak. Wiss. Phys. -Math. Kl. (1929), 96-100. FM 55-210. [6, 11, 15]
909. Marx, A. *Untersuchungen uber schlichte Abbildungen.* Math. Ann. 107(1932/33), 40-67. FM 58-363. [6, 7, 10, 11, 15, 64, 68]
910. Matthies, K. *Eine Bestabschatzung des Konvergenzradius der Potenzreihe der Umkehrfunktion einer analytischen Funktion.* Arch. Math. 7(1957), 457-458. MR 19-128. [1]
911. Meisters, G. H.; Olech, C. *Locally one-to-one mappings and a classical theorem on schlicht functions.* Duke Math. J. 30(1963), 63-80. MR 26-283. [4]
912. Menchoff, D. *Sur les conditions suffisantes pour qu'une fonction univalente soit holomorphe.* Rec. Math. Moscou 40(1933), 3-21. Zbl 7-120. [4]
913. Merkes, E. P. *On typically-real functions in a cut plane.* Proc. Amer. Math. Soc. 10(1959), 863-868. MR 22A(1)-807. [2, 13, 41, 55]

914. Merkes, E. P. *Bounded J-fractions and univalence.* Mich. Math. J. 6(1959), 395-400. MR 22A(2)-1373. [11, 12, 55]
915. Merkes, E. P.; Scott, W. T. *On univalence of a continued fraction.* Pacific J. Math. 10(1960), 1361-1369. MR 22A(2)-2093. [11, 12, 55]
916. Merkes, E. P.; Scott, W. T. *Covering theorems for S-fractions.* Math. Z. 73(1960), 333-338. MR 22A(2)-964. [15, 19, 28, 43, 55]
917. Merkes, E. P.; Scott, W. T. *Starlike hypergeometric functions.* Proc. Amer. Math. Soc. 12(1961), 885-888. Zbl 102-65. [4, 6, 28]
918. Merkes, E. P.; Robertson, M. S.; Scott, W. T. *On products of starlike functions.* Proc. Amer. Math. Soc. 13(1962), 960-964. MR 26-63. [4, 6, 43]
919. Meschkowski, H. *Beziehungen zwischen den Normalabbildungsfunktionen der Theorie der konformen Abbildung.* Math. Z. 55(1951), 114-124. MR 13-734. [60]
920. Meschkowski, H. *Einige Extremalprobleme aus der Theorie der konformen Abbildung.* Ann. Acad. Sci. Fenicae Ser. A. I. Math. -Phys. No. 117(1952), 12 pp. MR 14- 367. [59]
921. Meschkowski, H. *Verzerrungssatze fur mehrfach zusammenhangende Bereiche.* Compositio Math. 11(1953), 44-59. MR 15-116. [9, 15, 48, 58]
922. Miki, Y. *A note on close-to-convex functions.* J. Math. Soc. Japon 8(1956), 256-268. MR 19-951. [5, 20]
923. Minami, U. *On the univalency and multivalency of a class of meromorphic functions.* Proc. Imp. Acad. Jap. 12(1936), 33-35. Zbl 14-355. [14]
924. Mioduszewski, J. *On certain estimations of coefficients of univalent analytic functions.* Ann. Polon. Math. 7(1960), 135-140. MR 22A(1)-295. [9, 17, 25, 39]
925. Mioduszewski, J. *On the necessary and sufficient conditions for the analytic function to be univalent or p-valent in the usual and in the generalized sense.* Ann. Polon. Math. 7(1960), 127-133. MR 22A(1)-295. [4, 9, 14, 17, 22, 25, 27, 33]
926. Mitjuk, I. P. *Univalent conformal mappings of multiply connected domains.* Dopovidi Akad. Nauk Ukrain. RSR (1961), 158-160. (Ukrainian. Russian, English summaries) MR 23A-618. [60]
927. Mitjuk, I. P. *A generalization of some theorems on univalent conformal maps of doubly connected domains.* Dopovidi Akad. Nauk Ukrain. RSR (1961), 1115-1118. (Ukrainian. Russian and English summaries) MR 24A-495. [60]
928. Mitjuk, I. P. *Quelques applications du principe de la symétrisation.* (Russian. French and Roumanian summaries). Bul. Inst. Politehn.

Iasi (N.S.) 7(11) (1961), No. 3-4, 15-18. MR 26-980. [11, 12, 24]
929. Mitjuk, I, P. *The symmetrization principle for an annulus and some of its applications.* Dopovidi Akad. Nauk Ukrain. RSR (1962), 9-11. (Ukrainian. Russian and English summaries) MR 25-426. [24]
930. Mitrovic, Dragisa. *Une généralisation du théorème de Rouché.* Hrvatsko Prirodoslovno Drustvo. Glasnik Met.-Fiz. Astr. Ser. II. 7 (1952), 19-22. (Serbo-Croatian. French summary) MR 14-32. [59]
931. Mocanu, P. T. *Une généralisation du théorème de la contraction dans la classe S des fonctions univalentes.* Acad. R. P. Romine. Fil. Cluj. Stud. Cerc. Mat. 8(1957), 303-312. MR 21-1064. [24]
932. Mocanu, P. T. *Sur une généralisation du théorème de contraction dans la classe des fonctions univalentes.* Acad. R. P. Romine. Fil. Cluj. Stud. Cerc. Mat. 9(1958), 149-159. MR 21-1064. [6, 7, 15, 49, 68]
933. Mocanu, P. T. *Sur un théorème de recouvrement dans la classe des fonctions univalentes.* Gaz. Mat. Fiz. Ser. A(N.S.) 10(63) (1958), 473-477. (Roumanian. French and Russian summaries) MR 20-877. [19]
934. Mocanu, P. T. *Sur les rayon de stellarité des fonctions univalentes.* Acad. R. P. Romine. Fil. Cluj. Stud. Cerc. Mat. 11(1960), 337-341. (Roumanian. Russian and French summaries) MR 26-744. [6, 11, 16, 42]
935. Mocanu, P. T. *Un théorème sur les fonctions univalentes.* Studia Univ. Babes-Bolyai Ser. I Math. Phys. (1960), No. 1, 91-95. (Roumanian. Russian and French summaries) MR 26-276. [24]
936. Mocanu, P. T. *Un problème extremal dans la classe de fonctions univalentes.* Acad. R. P. Romine Fil. Cluj. Stud. Cerc. Mat. II (1960), 99-106. (Roumanian. Russian and French summaries) MR 24A-38. [24]
937. Monna, A. F. *Sur les fonctions univalente.* Akad. Wetensch. Amsterdam, Proc. 45(1942), 826-832. Zbl 28-401. [19, 22]
938. Montel, P. *Leçons sur les familles normales de fonctions analytiques et leurs applications.* Paris, 1927. FM 53-303. [59]
939. Montel, P. *Leçons sur les fonctions univalentes ou multivalentes.* Gauthier-Villars, Paris, 1933. Zbl 6-351. [44]
940. Montel, P. *Sur l'univalence ou la multivalence locale.* C. R. Acad. Sci. Paris 203(1936), 579-581. Zbl 15-70. [4, 15, 68]
941. Montel, P. *Sur quelques propriétés des différences divisées.* J. Math. Pures Appl. (9) 16(1937), 219-231. Zbl 17-107. [10, 14, 49]
942. Montel, P. *Sur les fonctions localement univalentes ou*

multivalentes. Ann. Sci. École Norm. Sup. (3), 54(1937), 39-54. FM 63-290. [14, 43]
943. Montel, P. *Sur certains cas d'univalence ou de multivalence locales.* Mathematica Cluj. 14(1938), 190-195. Zbl 20-237. [15]
944. Morse, M.; Heins, M. *Topological methods in the theory of functions of a complex variable. II.* Ann. of Math. (2) 46(1945), 625-666. MR 8-21. [59]
945. Morse, M.; Heins, M. *Deformation classes of meromorphic functions and their extensions to interior transformations.* Acta Math. 79(1947), 51-103. MR 8-507. [59]
946. Mullender, P. *On some conformal mappings.* Simon Stevin 26(1949), 136-142. MR 10-697. [16, 32]
947. Mullender, P. *On some conformal mappings.* Simon Stevin 30(1954), 44-47. (Dutch) MR 15-787. [32]
948. Muller, M. *Zur konformen Abbildung angenähert kreisformiger Gebiete.* Math. Z. 43(1938), 628-636. Zbl 17-408. [59]
949. Myrberg, L. *Über einige extremal Grössen in der Theorie der meromorphen Funktionen.* Ann. Acad. Sci. Fenn. Ser. A. I. No. 284 (1960), 9 pp. MR 23A-55. [19, 46]
950. Myrberg, P. J. *Über die Linearisierung der schlichten konformen Abbildungen.* Ann. Acad. Sci. Fennicae. Ser. A. I. Math.-Phys. No. 145, 8 pp. (1953). MR 14-861. [49]
951. Myrberg, P. J. *Über den Verzerrungssatz in der Theorie der konformen Abbildung.* Ofvers. AF Finska Vet. Soc. Forh 60, A. N. R. 7(1917-18), 12. FM 46-551. [9, 15, 19, 68]
952. Nabetani, K. *Some remarks on a theorem concerning star-shaped representation of an analytic function.* Proc. Imp. Acad. Jap. 10(1934), 537-540. Zbl 11-120. [3, 6, 10, 12, 16]
953. Nabetani, K. *Some inequalities on moduli of analytic functions.* Tohoku Math. J. 41(1935), 109-124. Zbl 12-212. [1, 3, 6, 10, 13, 15, 45, 63, 64, 66, 67]
954. Nabetani, K. *Remarks on some theorems concerning the sections of power series. I, II.* Tohoku Math. J. 41(1936), 329-336. FM 62-366. [3, 20]
955. Nabetani, K. *On multivalency of certain power series.* Tohoku Math. J. 41(1936), 402-405. Zbl 14-23. [14]
956. Nabetani, K. *On Study's theorem in the theory of conformal representation.* Tohoku Math. J. 41(1935-36), 406-410. FM 62-374. [1, 6, 10, 37]
957. Nagura, S. *Faber's polynomials.* Kodai Math. Sem. Rep. No. 5-6 (1949), 5-6. MR 11-718. [17, 53]

958. Nagura, S. *Faber's polynomials.* II. Kodai Math. Sem. Rep. 1950 (1950), 15-16. MR 12-327. [17, 53, 54]
959. Nagura, S.; Komatu, Y. *Distortion theorems in the theory of schlicht functions.* Nagoya-Math. J. 1(1950), 25-33. MR 12-490. [15, 24, 68]
960. Nakashima, K. *Note on subordination.* Mem. Fal. Sci. Eng. Wasedauniv. 16(1952), 119-122. MR 14-549. [1]
961. Nazim, T. A. *Über den Koebeschen Verzerrungssatz.* Rev. Fac. Sci. Univ. Istanbul (A) 15(1950), 113-118. MR 12-16. [19]
962. Nehari, Z. *Une propriété des valeurs moyennes d'une fonction analytique.* C. R. Acad. Sci. Paris 208(1939), 1785-1787. Zbl 21-142. [59]
963. Nehari, Z. *Sur la déformation de la frontière par les fonctions univalentes convexes.* C. R. Acad. Sci. Paris 209(1939), 781-783. MR 1-112. [4, 10, 15, 64]
964. Nehari, Z. *A generalization of Schwarz lemma.* Duke Math. J. 14(1947), 1035-1049. MR 9-340. [1, 19, 37]
965. Nehari, Z. *Une inégalité dans la théorie des fonctions bornées dans un anneau.* C. R. Acad. Sci. Paris 224(1947), 1093-1095. MR 8-508. [15, 22, 61]
966. Nehari, Z. *The elliptic modular function and a class of analytic functions first considered by Hurwitz.* Amer. J. of Math. 69(1947), 70-86. MR 8-454. [59]
967. Nehari, Z. *On analytic functions possessing certain properties of univalence.* Proc. London Math. Soc. (2) 50(1948), 120-136. MR 9-576. [1, 19, 22, 45]
968. Nehari, Z. *Analytic functions possessing a positive real part.* Duke Math. J. 15(1948), 1033-1042. MR 10-290. [2, 9, 37, 60]
969. Nehari, Z. *Sur un théorème de M. Montel.* C. R. Acad. Sci. Paris 228(1949), 1325-1327. MR 10-696. [59]
970. Nehari, Z. *The radius of univalence of an analytic function.* Amer. J. Math 71(1949), 845-852. MR 11-426. [12, 43]
971. Nehari, Z. *The Schwarzian derivative and schlicht functions.* Bull. Amer. Math. Soc. 55(1949), 545-551. MR 10-696. [4, 9, 31]
972. Nehari, Z. *On bounded analytic functions.* Proc. Amer. Math. Soc. 1(1950), 268-275. MR 11-590. [22, 37]
973. Nehari, Z. *A class of domain functions and some allied extremal problems.* Trans Amer. Math. Soc. 69(1950), 161-178. MR 12-251. [59]
974. Nehari, Z. *Note on positive harmonic functions.* J. London Math. Soc. 25(1950), 19-26. MR 11-435. [2]

975. Nehari, Z. *Extremal problems in the theory of bounded analytic functions.* Amer. J. Math. 73(1951), 78–106. MR 12–491. [11, 12, 22, 43]
976. Nehari, Z. *Sur la représentation conforme de deux domaines complementaires.* C. R. Acad. Sci. Paris 232(1951), 1532–1534. MR 13–25. [26, 49]
977. Nehari, Z. *Bounded analytic functions.* Bull. Amer. Math. Soc. 57(1951), 354–366. MR 13–222. [22, 60]
978. Nehari, Z. *Conformal Mapping.* McGraw-Hill, New York, 1952, 396 pp. MR 13–640. [44]
979. Nehari, Z. *Some inequalities in the theory of fuctions.* Trans. Amer. Math. Soc. 75(1953), 256–286. MR 15–115. [9, 17, 22, 25, 26, 29, 30]
980. Nehari, Z. *On the zeroes of solutions of second-order linear differential equations.* Amer. J. Math. 76(1954), 689–697. MR 16–131. [31]
981. Nehari, Z. *Some criteria of univalence.* Proc. Amer. Math. Soc. 5(1954), 700–704. MR 16–232. [4, 31]
982. Nehari, Z. *Univalent functions and linear differential equations.* Lect. of Funct. of a Complex Variable, 49–60. Univ. of Michigan Press, Ann Arbor, 1955. MR 16–1093. [4, 31]
983. Nehari, Z. *On the coefficients of R-univalent functions.* Duke Math. J. 22(1955), 223–227. MR 16–916. [9, 17, 25]
984. Nehari, Z. *On the coefficients of univalent functions.* Proc. Amer. Math. Soc. 8(1957), 291–293. MR 18–728. [4, 16, 17, 19, 42]
985. Nehari, Z.; Netanyahu, E. *On the coefficients of meromorphic schlicht functions.* Proc. Amer. Math. Soc. 8(1957), 15–23. MR 18–648. [2, 6, 9, 17, 21, 25, 36, 47]
986. Nehari, Z.; Schwarz, B. *On the coefficients of univalent Laurent series.* Proc. Amer. Math. Soc. 5(1954), 212–217. MR 15–786. [6, 9, 16, 17, 25, 36, 39]
987. Netanyahu, E. *The coefficient problem for schlicht functions In the exterior of the unit circle.* Tech. Report No. 39(1954), Dept. Math. Stanford U. Zbl 58–305 [9, 17, 25]
988. Nevanlinna, R. *Über beschrankte Funktione, die in gegebenen Punkten vorgeschriebene Werte annehmen.* Ann. Acad. Sci. Fenn. 13(1919), No. 1. FM 47–271. [2, 10, 22]
989. Nevanlinna, R. *Über die schlichten Abbildungen des Einheitkreises.* Finska Vetena Kaps-Soc. Forhandl (A), 62(1920), No. 7, 145. FM 47–324. [12, 15, 17, 68]
990. Nevanlinna, R. *Über die konforme Abbildung Sterngebieten.*

Oeversikt av Finska-Vetenskaps Societeten Forhandlingar 63(A), No. 6 (1921). FM 48-403. [6, 17, 36, 42]
991. Nevanlinna, R. *Eindeutige Analytische Funktionen*. Berlin, 1936. Zbl 14-163. [59]
992. Nevanlinna, R. *Eindeutige Analytische Funktionen*. J. W. Edwards, Ann Arbor, Mich., 1944. MR 6-59. [59]
993. Nevanlinna, R.; et al *Analytic Functions*. Princeton Univ. Press 1960, 197 pp. Zbl 100-287. [24]
994. Newman, D. J.; Shapiro, H. S. *The Taylor coefficients of inner functions*. Mich. Math. J. 9(1962), 249-255. [59]
995. Nishimiya, H. *On a coefficient problem for analytic functions typically-real in an annulus*. Kodai Math. Sem. Rep. 9(1957), 59-66. MR 20-18. [13, 17, 23, 51]
996. Nishimiya, H. *On the coefficients of functions starlike in the exterior of a circle*. Japan J. Math. 29(1959), 78-82. MR 23A-618. [2, 6, 9, 17, 25, 36]
997. Noshiro, K. *On the univalency of certain power series*. J. Fac. Sci. Hokkaido Univ. 1(1932), 157-161. Zbl 4-401. [4, 10, 12, 22, 39, 43]
998. Noshiro, K. *On the starshaped mapping by an analytic function*. Proc. Imp. Acad. Jap. 8(1932), 275-277. Zbl 5-251. [6, 20]
999. Noshiro, K. *On the theory of schlicht functions*. J. Fac. Sci. Hokkaido Univ. Jap (1), 2(1934-1935), 129-155. Zbl 10-263. [6, 11, 49]
1000. Noshiro, K. *On the univalency of certain analytic functions*. J. Fac. Sci. Hokkaido Univ. 2(1934), 89-101. Zbl 9-24. [2, 6, 43]
1001. Obrechkoff, N. *Sur les polynomes univalents*. C. R. Acad. Sci. Paris 198(1934), 2049-2050. FM 60-1033. [4, 6, 14, 32]
1002. Obrechkoff, N. *Sui polinomi univalenti*. Boll. Un. Mat. Ital. 14(1935), 246-247. FM 61-1154. [32, 49]
1003. Obrechkoff, N. *Sur les polynomes univalents ou multivalents*. Actes Congres Interbalkan Math. Athenes, (1934), 91-94. FM 61-1154. [4, 6, 14, 32]
1004. Obrechkoff, N. *Sur les polynomes univalents ou multivalents*. Bull. Sci. Math. (2) 60, (1935), 36-42. FM 62-377. [6, 14, 32]
1005. Ogawa, S. *A note on close-to-convex functions*. III. J. Nara Gakygei Univ. 9(1960), No. 2, 7-23. MR 25-427. [5, 6, 10, 20]
1006. Ogawa, S. *On some criteria for p-valence*. J. Math. Soc. Japon 13(1961), 431-441. MR 26-611. [4, 12, 14, 15, 68]
1007. Ogawa, S. *Some criteria for univalence*. J. Nara Gakugei Univ. 10(1961), No. 1, 7-12. MR 26-1210. [4, 5]
1008. Oikawa, K. *A distortion theorem on schlicht functions*. Kodai Math. Sem. Rep. 9(1957), 140-144. MR 19-1045. [15, 24, 46, 68]

1009. Ono, I. *On some properties of mean multivalent functions.* Sci. Rep. Tokyo Bunrika Daigaku Sect. A. 4(1951), 169–175. MR 13-453. [9, 17, 22, 27,, 33, 43]
1010. Onofri, L. *Sulle funzioni univalenti in una corona circolare.* Boll. Un. Mat. Ital. (2) 3(1940), 113–115. MR 3-201. [4]
1011. Onofri, L. *Contributo alla teoria delle funzioni univalenti.* Boll. Un. Mat. Ital. (2) 4(1942), 217–224. MR 7-424. [4, 19]
1012. Opitz, G. *Die Konvergenz des Verfahrens von Theodorsen zur konformen Abbildung kreisahnlicher Gebiete.* Arch. Math. 2(1950), 110–116. MR 11-341. [59]
1013. Ore, O. *On functions with bounded derivatives.* Trans. Amer. Math. Soc. 43(1938), 321–326. Zbl 18-395. [59]
1014. Oserovic, V. A. *On the conformal mapping of a circle onto a rectangular region.* Ukrain. Mat. Z. 13(1961), No. 1, 111–117. MR 25-611. [59]
1015. Osgood, W. F. *Lehrbuch der Funktiontheorie.* Berlin, 1920. [59]
1016. Osserman, R. *Koebe's general uniformization theorem: the parabolic case.* Ann. Acad. Sci. Fennicae Serie A. I. No. 258 (1958). [24]
1017. Ostrowski, A. *Über konforme Abbildungen annahernd kreisformiger Gebiete.* Jber. Deutschen Math. Verein. 39(1930), 78–81. FM 56-297. [59]
1018. Ozaki, S. *Remarks on some coefficients of a schlicht function.* Sci. Rep. Tokyo Bunrika Daig. A 1(1933), 283–287. Zbl 7-214. [17]
1019. Ozaki, S. *On the multivalency of functions.* Sci. Rep. Tokyo Bunrika Daig. A 2(1934), 99–102. Zbl 12-24. [4, 14, 17, 50]
1020. Ozaki, S. *Some remarks on the univalency and multivalency of functions.* Sci. Rep. Tokyo Bunrika Daig. A. No. 31-32(1934), 41–55. Zbl 12-23. [4, 9, 14, 17, 25, 39, 50]
1021. Ozaki, S. *On the theory of multivalent functions.* Sci. Rep. Tokyo Bunrika Daig. A 2(1935), 167–188. Zbl 12-24. [4, 6, 14]
1022. Ozaki, S. *On the theory of multivalent functions.* II. Sci. Rep. Tokyo Bunrika Daig. Sect. A (1941), 45–87. MR 14-34. [6, 13, 14, 17, 36, 40, 41, 45, 51, 52]
1023. Ozaki, S. *On the theory of multivalent functions in a multiply connected domain.* Sci. Rep. Tokyo Bunrika Daig. Sect. A 4(1944), 115–135. MR 14-35. [4, 14, 17]
1024. Ozaki, S.; Yosida, T. *On some properties of multivalent functions.* Sci. Rep. Tokyo Bunrika Daig. Sect. A 4(1949), 137–150. MR 13-453. [14, 17, 19, 43, 50]

1025. Ozaki. S.; Kashiwagi, S.; Tsuboi, T. *Some properties in matrix space*. Sci. Rep. Tokyo Bunrika Daig. Sect. A. 4(1952), 230–237. MR 14–368. [4]

1026. Ozaki, S.; Ono, I.; Ozawa, M. *On the function theoretic identities*. Sci. Rep. Tokyo Bunrika Daig. 4(1949), 157–160. MR 13–453. [59]

1027. Ozaki, S.; Ono, I.; Umezawa, T. *On a general second order derivative*. Sci. Rep. Tokyo Kyoiku Daig. Sect. A 5(1956), 111–114. MR 17–1195. [4]

1028. Ozaki, S.; Takatsuka, T.; Umezawa, T. *Analytic functions starlike of order p in one direction*. J. Dept. Educ. Shizuoka Univ. 1(1950), 81–87. [40]

1029. Ozawa, M. *On the conditions of univalency of conformal mapping*. Kodai Math. Sem. Rep. 1953(1953), 84–86. MR 15–414. [4, 53]

1030. Ozawa, M. *A distortion theorem on schlicht functions*. Kodai Math. Sem. Rep. 9(1957), 145–157. MR 19–1045. [15, 17, 24, 68]

1031. Ozegov, V. B. *On the convolutions of various classes of functions having a stieltjes integral representation*. Vestnik Leningrad Univ. 15(1960), No. 13, 32–40. (Russian. English summary) MR 23A–335. [13, 47]

1032. Paatero, V. *Uber die konforme Abbildung von Gebieten deren Rander von beschrankter Drehung sind*. Ann. Acad. Sci. Fenn. A. 33(1931), 1–78. Zbl 1–143. [10, 15]

1033. Paatero, V. *Uber Gebiete von beschrankter Randdrehung*. Ann. Acad. Sci. Fenn. Ser. A. 37(1933), No. 9. FM 59–1045. [10]

1034. Paydon, J. F., Wall, H. S. *The continued fraction as a sequence of linear transformations*. Duke Math. J. 9(1942), 360–372. MR 3–297. [59]

1035. Perron, O. *Über eine Schlichtheitsschranke von James S Thale*. Bayer Akad. Wiss. Math. -Nat. Kl. S. -B. (1956), 233–236(1957). MR 18–884. [16, 43, 55]

1036. Perry, R. L. *The univalent functions of a family*. J. London Math. Soc. 35(1960), 49–62. MR 22A(1)–291. [4, 32, 39]

1037. Peschl, E. *Über die Krummung von Niveaukurven bei der konformen Abbildung einfachzusammenhängender Gebiete auf das Innere eines Kreises*. Math. Ann. 106(1932), 574–594. Zbl 4–300. [10, 35]

1038. Peschl, E. *Zur Theorie der schlichten Funktionen*. J. Reine und Angewandte Math. 176(1936), 61–94. Zbl 16–35. [6, 17, 24, 29, 39, 54]

1039. Peschl, E. *Über die Bilder von Sternbereichen ein allgemeiner Ab-*

bildungsatz im Raume mehrerer komplexer Veränderlichen. Ber. Math. -Tagung Tubingen (1946), 112-112(1947). MR 9-25. [59]
1040. Peschel, E.; Erwe, F. *Über beschränkte Systeme Funktionen.* Math. Ann. 126(1953), 185-220. MR 15-520. [59]
1041. Petersen, G. M. *On functions with positive real part.* J. London Math. Soc. 36(1961), 49-51. MR 26-66. [2]
1042. Pflanz, E. *Über P-fach symmetrische schlichte Funktionen.* Tubingen, Dissertation, 1934. Zbl 10-307. [6, 10, 11, 42, 49]
1043. Pflanz, E. *Über P-fach symmetrische schlichte Funktionen.* Math. Z. 40(1935), 72-85. FM 61-350. [15, 49]
1044. Pfluger, A. *Extremallängen und Kapazität.* Comment. Math. Helv. 29(1955), 120-131. MR 16-810. [46, 62]
1045. Pfluger, A. *Theorie der Riemannschen Flachen.* Berlin, 1957. MR 18-796. [59]
1046. Pick, G. *Zur Theorie der konformen Abbildung kreisförmiger Bereiche.* Palermo Rend. 37(1914), 341-344. FM 45-671. [31]
1047. Pick, G. *Über die beschränkungen analytischer Funktionen, welche durch vorgegebene Funktionswerte bewirkt werden.* Math. Ann. 77(1916), 7-23. FM 46-474. [59]
1048. Pick, G. *Über den Koebeschen Verzerrungssatz.* Leipzig Ber. 68(1916), 58-64. FM 46-550. [15]
1049. Pick, G. *Über die konforme Abbildung eines Kreises auf ein schlichtes und zugleich beschranktes Gebiet.* Wien. Ber. 126(1917), 247-263. FM 46-553. [22, 49]
1050. Pick, G. *Zur schlichten konformen Abbildung.* Ber. Leipzig 81(1929), 3-8. FM 55-789. [60]
1051. Pick, G. *Über den Korbeschen Verzerrungssatz.* Ber. Sachs. Ges. Math. 176(1937), 61-94. [15, 22, 61, 68]
1052. Pinney, E. A. *On a note of Galbraith and Green.* Bull. Amer. Math. Soc. 54(1948), 527. MR 10-38. [59]
1053. Pir, L. E. *Über typisch reelle Funktionen im Kreisring.* Dokl. Akad Nauk SSSR (N.S.) 92(1953), 699-702. (Russian) Zbl 52-81. [13, 15, 48, 66]
1054. Pir, L. E. *Zur Theorie der schlichten Funktionen im Kreisring.* Dokl. Akad. Nauk SSSR (N.S.) 92(1953), 475-477. (Russian) Zbl 52-80. [48]
1055. Piranian, G. *An isolated schlicht function.* Abh. Math. Sem. Univ. Hamburg 24(1960), 236-238. MR 22A(2)-1629. [16]
1056. Pirl, Isotherme Kurvenscharen und zugehorige Extremalprobleme der konformen Abbildung. Wiss. Z. Martin-Luther-Univ. Halle-Wittenberg 4(1955), 1225-1252. MR 17-835. [59]

1057. Plemelj, J. *Über den Verzerrungssatz von P. Koebe*. Gesellschaft Deutscher Naturforscher und Aertze, Verhandlungen, 85(1913), II, 1, 163. [15]
1058. Pokornyi, V. V. *On some sufficient conditions for univalence*. Dokl. Akad. Nauk SSSR (N.S.) 79(1951), 743-746. MR 13-222. [4, 31]
1059. Polak, A. I. *On a property of locally univalent functions of a complex variable*. Dokl. Akad. Nauk SSSR (N.S.) 96(1954), 241-243. MR 16-25. [57]
1060. Pólya, G. *Über analytische Deformationen eines Rechtecks*. Ann. Math. 34(1933), 617-620. FM 59-348. [59]
1061. Pólya, G. *Sur la symmetrisation circulaire*. C. R. Acad. Sci. Paris 230(1950), 25-27. MR 11-435. [59]
1062. Pólya, G.; Schiffer, M. *Sur la représentation conforme de l'extérieur d'une courbe fermée convexe*. C. R. Acad. Sci. Paris 248(1959), 2837-2839. MR 21-783. [7, 49]
1063. Pólya, G.; Schoenberg, I. J. *Remarks on de la Vallée Poussin means and convex conformal maps of the circle*. Pacific J. Math. 8(1958), 295-334. MR 20-1176. [4, 6, 10, 16, 47]
1064. Pólya, G.; Szegö, G. *Aufgaben und Lehrsatze aus der Analysis*. I. II. Dover Publ., New York, 1945. MR 7-418. [44]
1065. Pólya, G.; Szegö, G. *Isoperimetric Inequalities in Mathematical Physics*. Princeton, 1951. MR 13-270. [59]
1066. Pommerenke, Ch. *On some problems by Erdös, Herzog, and Piranian*. Mich. Math. J. 6(1959), 221-225. MR 22A(1)-125. [59]
1067. Pommerenke, Ch. *On the derivative of a polynomial*. Mich. Math. J. 6(1959), 373-375. MR 22A(1)-16. [22, 32]
1068. Pommerenke, Ch. *On sequences of subordinate functions*. Mich. Math. J. 7(1960), 181-185. Zbl 91-252. MR 27-316. [1]
1069. Pommerenke, Ch. *Über die Mittlewerte und Koeffizienten multivalenter Funktionen*. Math. Ann. 145/146(1962), 285-296. MR 23A-611. [3, 14, 17, 33]
1070. Pommerenke, Ch. *Images of convex domains under convex conformal mappings*. Mich. Math. J. 9(1962), 257-269. [10]
1071. Pommerenke, Ch. *Über einige klassen meromorpher schlichter Funktionen*. Math. Z. 78(1962), 263-284. MR 28-258. [1, 3, 4, 5, 6, 7, 9, 10, 15, 17, 25, 36, 37, 38, 39, 49, 58, 63, 64]
1072. Pommerenke, Ch. *On the coefficients of close-to-convex functions*. Mich. Math. J. 9(1962), 259-269. MR 26-980. [1, 3, 4, 5, 6, 7, 9, 15, 17, 25, 36, 38, 49]
1073. Pommerenke, Ch. *On starlike and convex functions*. J. London

Math. Soc. 37(1962), 209-224. MR 25-254. [6, 7, 10, 15, 17, 18, 22, 23, 30, 36, 38, 63]
1074. Pommerenke, Ch. *On hyperbolic capacity and hyperbolic length.* Mich. Math. J. 10(1963), 53-63. MR 26-1208. [59]
1075. Pommerenke, Ch. *On starlike and close-to-convex functions.* Proc. London Math. soc. (3) 13(1963), 290-304. MR 26-499. [5, 6, 7, 8, 17, 36, 38, 54]
1076. Pommerenke, Ch. *On meromorphic starlike functions.* Pacific J. Math. 13(1963), 221-235. MR 27-59. [6, 9, 15, 17, 18, 23, 25, 36, 58, 63]
1077. Popov, B. S. *Quelques propriétés des fonctions d'une variable complexe.* Bull. Soc. Math. Phys. Macedoine 4(1953), 20-24(1954). MR 15-863. [59]
1078. Possel, R. de *Zum Parallelschlitztheorem unendlich vielfach zusammenhangender Gebiete.* Nachr. Ges. Wiss. Gottingen, Math. -Phys. Kl. (1931), 199-202. Zbl 3-314. [9, 60]
1079. Possel, R. *Quelques problèmes de représentation conforme.* J. École Polytechnique, Cahier 30(2), (1932), 1-98. Zbl 6-263. [6, 62]
1080. Possel, R. *Sur quelques propriétés de la représentation conforme des multiplement connexes, en relation avec le théorème des fentes parallèles.* Math. Ann. 107(1933), 496-504. Zbl 5-363. [60]
1081. Potugina, I. V. *On estimation of coefficients of odd univalent functions.* Dokl. Akad. Nauk SSSR (N.S.) 85(1952), 1215-1217. MR 14-260. [17, 45, 52]
1082. Prawitz, H. *Über Mittelwerte analytischer Funktionen.* Ark. Mat. Astron. och Fysik 20(1927/28), No. 6, 1-12. FM 53-307. [3, 27]
1083. Privalov, J. *Sur les fonctions qui donnent la représentation conforme biunivoque.* Rec. Math. D. I. Soc. Math. D. Moscou 31(1924), 350-365. FM 50-254. [6, 9, 10, 15, 58]
1084. Privalov, I. I. *Boundary Properties of Analytic Functions.* 2nd Ed. Gos. Iz. Tehn. -Teor. Lit., Moscow-Leningrad, 1950. (Russian) MR 13-926. [59]
1085. Rademacher, H. *On the Bloch-Landau constant.* Amer. J. Math. 65(1943), 387-393. MR 4-270. [19]
1086. Rado, T. *Zur Theorie der mehrdeutigen konfomen Abbildungen.* Acta Szeged 1(1922), 55-64. FM 48-1235. [59]
1087. Rado, T. *Sur la représentation conforme de domaines variables.* Acta Szeged 1(1923), 180-186. FM 49-247. [59]
1088. Rado, T. *Bemerkungen uber die konformen Abbildungen konvexer Gebiete.* Math. Ann. 102(1929), 428-429. FM 55-208. [1, 10, 37, 60]

1089. Rahmen, Q. I. *On extremal properties of the derivative of polynomials and rational functions.* J. London Math. Soc. 35(1960), 334–336. MR 23A–328. [59]
1090. Rahmanov, B. N. *On the theory of univalent functions.* Dokl. Akad. Nauk SSSR (N.S.) 78(1951), 209–211. MR 12-816. [4, 6, 10, 11, 12]
1091. Rahmanov, B. N. *On the theory of univalent functions.* Dokl. Akad. Nauk SSSR (N.S.) 82(1952), 341–344. MR 13-640. [6, 10]
1092. Rahmanov, B. N. *On the theory of univalent functions.* Dokl. Akad. Nauk SSSR (N.S.) 88(1953), 413–414. MR 14-740. [12]
1093. Rahmanov, B. N. *On the theory of univalent functions.* Dokl. Akad. Nauk SSSR (N.S.) 91(1953), 729–732. MR 15-413. [49]
1094. Rahmanov, B. N. *On the theory of univalent functions.* Dokl. Akad. Nauk SSSR (N.S.) 97(1954), 973–976. MR 16-122. [4, 6]
1095. Rahmanov, B. N. *On the theory of univalent functions.* Dokl. Akad. Nauk SSSR (N.S.) 103(1955), 369–371. MR 17-249. [4, 6, 10, 15, 20, 64]
1096. Rajagopal, C. T. *Caratheodory's inequality and allied results.* Math. Student 9(1941), 73–77. MR 3-201. [59]
1097. Rajagopal, C. T. *Caratheodory's inequality and allied results. II.* Math. Student 15(1947), 5–7(1948). MR 10-441. [15, 17, 22, 30, 61]
1098. Rajagopal, C. T. *On inequalities for analytic functions.* Amer. Math. Mon. 60(1953), 693–695. MR 15-412. [15, 17, 22, 30, 61]
1099. Rajagopal, C. T. *A note on power series.* Math. Student 20(1952), 99–106 (1953). MR 15-113. [2, 20, 22]
1100. Rajagopal, C. T. *On an absolute constant for a class of power series.* Math. Scand. 5(1957), 267–270. MR 20-874. [59]
1101. Rakovic, K. *Inequalities for absolute values and for coefficients of certain regular functions.* Acta Fac. Nat. Univ. Carol., Prague No. 172(1939), 28–31(1946). MR 9-232. [10, 14, 15, 17]
1102. Rao, K. *Two theorems on bounded functions.* J. London Math. Soc. 32(1957), 430–435. MR 19-736. [15, 22, 61]
1103. Rao, K. *A theorem on bounded functions.* J. London Math. Soc. 36(1961), 474–479. MR 26-615. [15, 22, 61]
1104. Rauch, S. E. *Mapping properties of Cesaro sums of order two of the geometric series.* Pacific J. Math. 4(1954), 109–121. MR 15-697. [20, 35]
1105. Reade, M. *Sur une classe de fonctions univalentes.* C. R. Acad. Sci. Paris 239(1954), 1758–1759. MR 16-579. [2, 4, 5, 6, 17, 36, 38, 41]
1106. Reade, M. *On close-to-convex univalent functions.* Mich. Math. J. 3(1955), 59–62. MR 17-25. [4, 5, 6, 17, 36, 38, 42]

1107. Reade, M. *The coefficients of close-to-convex functions.* Duke Math. J. 23(1956), 459–462. MR 17-1194. [4, 5, 6, 10, 17, 36, 38]
1108. Reade, M. *On the coefficients of certain univalent functions.* Ann. Acad. Sci. Fenn. Ser. A. I. No. 215(1956), 6 pp. MR 17-1069. [4, 5, 6, 15, 17, 36, 38, 40, 63, 64]
1109. Reade, M. *A radius of univalence for $\int_0^z e^{-t^2} dt$.* Preliminary Report Bull. Amer. Math. Soc. Abstract 63, 193(1957). [28]
1110. Reade, M. *On Umezawa's criteria for univalence.* J. Math. Soc. Japan 9(1957), 234–238. MR 19-642. [4, 5, 28]
1111. Reade, M. *On Umezawa's criteria for univalence. II.* J. Math Soc. Japan 10(1958), 255–259. MR 20-877. [4]
1112. Reade, M. *Two applications of close-to-convex functions.* Mich. Math. J. 5(1958), 91–94. MR 20-404. [4, 5, 28]
1113. Reade, M. *On the partial sums of certain Laurent expansions.* J. Math. Soc. Japan 19(1963), 66–68. MR 26-499. [9, 20, 43]
1114. Red'kov, M. I. *The domains of values of certain functions for certain classes of bounded univalent functions.* Dokl. Akad. Nauk SSSR 133(1960), 284–287. (Russian) Transl. Soviet Math. -Dokl. 1(1961), 484–851. MR 23A-330. [29]
1115. Red'kov, M. I. *The range of a certain functional in the class $S_1(\phi(w))$.* Izv. Vyss. Uc. Zaved. Mat. 1962, No. 2(27), 119–129. (Russian) MR 25-796. [22, 62]
1116. Red'kov, M. I. *The range of a certain functional in the class S_1.* Izv. Vyss. Uc. Zaved. Mat. 1962, No. 4(29), 134–142. (Russian) MR 25-796. [22, 62]
1117. Reich, E. *An inequality for subordinate analytic functions.* Pacific J. Math. 4(1954), 259–274. MR 15-862. [1, 3, 18]
1118. Reich, E. *An alternate proof of a theorem of Beckenbach.* Proc. Amer. Math. Soc. 5(1954), 578–579. MR 16-24. [3, 37]
1119. Reich, E. *On a Bloch-Landau constant.* Proc. Amer. Math. Soc. 7(1956), 75–76. MR 17-1066. [19]
1120. Reich, E. *Schlicht functions with real coefficients.* Duke Math. J. 23(1956), 421–427. Zbl 70-299. [13, 15, 22, 23, 24, 39]
1121. Reich, E. *On radial slit mappings.* Ann. Acad. Sci. Fenn. Ser. A. I. No. 296(1961), 12 pp. MR 23A-618. [59]
1122. Reich, E.; Warschawski, S. E. *On canonical conformal maps of regions of arbitrary connectivity.* Pacific J. Math. 10(1960), 965–985. MR 22A(2)-1376. [22, 24]
1123. Reinhardt, K. *Über schlichte konforme Abbildungen des Einheitskreises.* Jber. Deutsch. Ver. Math. 37(1928), 83–86. FM 54-378. [19]
1124. Remak, R. *Ueber eine spezielle Klasse schlichter konformer Ab-*

bildungen des Einheitskreises. Mathematica B 11, 175–192; 12, 43–49(1943). MR 8–22. [4, 6, 32, 62]

1125. Remizova, M. P. *Extremal problems in the class of typically-real functions.* Izv. Vyss. Uc. Zaved. Mat. (1963), No. 1(32), 135–144. MR 26–982. [9, 13, 15, 23, 29, 58, 66]

1126. Rengel, E. *Einige Schlitztheoreme der konformen Abbildung.* Schr. Math. Sem. Inst. Math. Univ. Berlin 1(1933), 139–162. FM 59–351. [59]

1127. Rengel, E. *Existenzbeweise fur schlichte Abbildungen mehrfach zusammenhängender Bereiche auf gewisse Normalbereiche.* Jber. Deutsch. Math. -Ver. 44(1934), 51–55. FM 60–286. [60]

1128. Rengel, E. *Verzerrung des Randes bei schlichter konformer Abbildungen.* I, II. Deutsche Math. 6(1942), 370–393. MR 5–37. [9, 19]

1129. Rényi, A. *On the coefficients of schlicht functions.* Publ. Math. Debrecen 1(1949), 18–23. MR. 11–92. [10, 17, 38, 41, 42]

1130. Rényi, A. *On the geometry of conformal mapping.* Acta Sci. Math. Szeged 12B(1950), 215–222. MR 11–649. [7, 49]

1131. Rényi, A. *Some remarks on univalent functions II.* Ann. Acad. Sci. Fenn. Ser. A. I. No. 250/29 (1958), 7 pp. MR 22A(1)–128. [5, 12, 15, 16, 17, 38, 64]

1132. Rényi, A. *Some remarks on univalent functions.* Bulgar. Akad. Nauk Izv. Mat. Inst. 3, No. 2, 111–121(1959). Bulgarian, Russian summaries) MR 22A(1)–128. [5, 17, 38]

1133. Ricci, G. *Su un problema di massimo per le funzioni maggioanti delle serie di potenze.* Atti. Accad. Naz. Nincei. Rend. Cl. Sci. Fis. Mat. Nat. (8) 18(1955), 609–613. MR 17–1070. [39]

1134. Riesz, F. *Uber die Fourierkoeffizienten einer stetigen Funktion von beschrankter Schwankung.* Math. Z. 2(1918), 312–315. FM 46–452. [59]

1135. Riesz, M. *Eine trigonometrische Interpolationsformel und einige Ungleichungen fur Polynome.* Jber. Deutsch. Math. Ver. 23(1915), 354–368. FM 45–405. [59]

1136. Riesz, M. *Sur les fonctions conjugées.* Math. Z. 27(1928), 218–244. [59]

1137. Riesz, M. *Remarque sur les fonctions holomorphes.* Acta Sci. Math. Szeged 12A(1950), 53–56. Zbl 38–229. [59]

1138. Ringleb, F. *Über das Verhalten der Krummung ebener Kurven bei konformer Abbildung.* Jber. D. M. V. 45(1935), 57–60. FM 61–1156. [35]

1139. Robertson, M. S. *A note on schlicht polynomials.* Trans. Royal Soc. Canada (3) 26(1932), 43–48. Zbl 6–147. [32, 43]

1140. Robertson, M. S. *On the coefficients of a typically-real functions.* Bull. Amer. Math. Soc. 41(1935), 565-572. Zbl 12-212. [2, 13, 17, 23, 45, 51, 52]
1141. Robertson, M. S. *A remark on the odd schlicht functions.* Bull. Amer. Math. Soc. 42(1936), 366-370. FM 62-373. [16, 17, 24, 42, 45, 52]
1142. Robertson, M. S. *Analytic functions starlike in one direction.* Amer. J. Math. 58(1936), 465-472. Zbl 14-120. [2, 6, 10, 13, 17, 23, 36, 38, 40, 41, 42]
1143. Robertson, M. S. *On the order of the coefficients of a univalent function.* Amer. J. Math. 59(1937), 205-210. Zbl 16-126. [3, 7, 15, 17, 42, 54, 62, 68]
1144. Robertson, M. S. *On the theory of univalent functions.* Ann. Math. 37(1936), 374-408. Zbl 14-165. [1, 2, 3, 4, 6, 10, 13, 15, 16, 17, 20, 21, 22, 23, 30, 36, 38, 41, 51, 54, 56, 61, 63, 64, 66, 68]
1145. Robertson, M. S. *On the univalency of Cesàro sums of univalent functions.* Bull. Amer. Math. Soc. 42(1936), 241-243. FM 62-367. [4, 20, 32]
1146. Robertson, M. S. *On the order of the coefficients of a univalent function.* Amer. J. Math. 59(1936), 205-210. Zbl 16-126. [3, 7, 15, 17, 42, 54, 62, 68]
1147. Robertson, M. S. *A representation of all analytic functions in terms of functions with positive real part.* Ann. Math. 38(1937), 770-783. Zbl 17-407. [2, 13, 15, 17, 21, 23, 28, 36, 40, 45, 51, 52, 56, 63, 66, 67]
1148. Robertson, M. S. *Multivalent functions of order p.* Bull. Amer. Math. Soc. 44(1938), 282-285. Zbl 18-315. [14, 17]
1149. Robertson, M. S. *On certain power series having infinitely many zero coefficients.* Ann. Math. 40(1939), 339-352. Zbl 21-143. [13, 15, 17, 39, 45, 51, 66]
1150. Robertson, M. S. *Piecemeal univalency of analytic functions.* Ann. Math. 40(1939), 120-128. Zbl 20-141. [3, 14, 15, 17, 50, 54, 57, 65, 68]
1151. Robertson, M. S. *The variation of the sign of V for an analytic function $U + iV$.* Duke Math. J. 5(1939), 512-519. MR 1-9. [2, 3, 14, 15, 17, 21, 49, 50, 65]
1152. Robertson, M. S. *Typically-real functions with $a_n = 0$ for $n = 0$ (mod 4).* Bull. Amer. Math. Soc. 46(1940), 136-141. MR 1-214. [13, 17, 39, 51, 52]
1153. Robertson, M. S. *The partial sums of multivalently starlike functions.* Ann. Math (2) 42(1941), 829-838. MR 3-79. [6, 14, 20, 43]
1154. Robertson, M. S. *Star center points of multivalent functions.*

Duke Math. J. 12(1945), 669–684. MR 7-379. [2, 6, 14, 15, 17, 19, 21, 36, 50, 54, 56, 63, 65, 68]

1155. Robertson, M. S. *The coefficients of univalent functions.* Bull. Amer. Math. Soc. 51(1945), 733–738. MR 7-150. [13, 17, 39, 45, 54]

1156. Robertson, M. S. *Univalent power series with multiply monotonic sequences of coefficients.* Ann. Math. (2) 46(1945), 533–555. MR 7-201. [4, 6, 13, 20, 23, 39, 41]

1157. Robertson, M. S. *Applications of a lemma of Fejer to typically-real functions.* Proc. Amer. Math. Soc. 1(1950), 555–561. MR 12-248. [13, 39, 41, 47]

1158. Robertson, M. S. *A coefficient problem for functions regular in an annulus.* Canad. J. Math. 4(1952), 407–423. MR 14-460. [13, 14, 17, 49, 50, 51]

1159. Robertson, M. S. *Multivalently starlike functions.* Duke Math. J. 20(1953), 539–549. MR 15-613. [6, 13, 14, 16, 17, 36, 39, 50, 51]

1160. Robertson, M. S. *Schlicht solutions of* $w'' + pw = 0$. Trans. Amer. Math. Soc. 76(1954), 254–274. MR 15-786. [4, 6, 10, 13, 31, 41]

1161. Robertson, M. S. *Schlicht Dirichlet series.* Canad. J. Math. 10(1958) 161–176. MR 20-664. [28]

1162. Robertson, M. S. *Cesàro partial sums of harmonic series expansions.* Pacific J. Math. 8(1958), 829–846. MR 21-413. [4, 13, 20, 23, 41]

1163. Robertson, M. S. *Applications of the subordination principle to univalent functions.* Pacific J. Math. 11(1961), 315–324. MR 23A-329. [1, 6, 10]

1164. Robertson, M. S. *Convolutions of schlicht functions.* Proc. Amer. Math. Soc. 13(1962), 585–589. MR 25-254. [6, 9, 47]

1165. Robertson, M. S. *Variational methods for functions with positive real part.* Trans. Amer. Math. Soc. 102(1962), 82–93. MR 24A-612. [1, 23, 24]

1166. Robertson, M. S. *Extremal problems for analytic functions with positive real part and applications.* Trans. Amer. Math. Soc. 106(1963), 236–253. MR 26-65. [2, 4, 6, 9, 10,, 12, 40]

1167. Robertson, M. S. *Some radius of convexity problems.* Mich. Math. J. 10(1963), 231–236. MR 27-313. [2, 6, 9, 12, 16]

1168. Robinson, R. M. *The Bloch constant U for a schlicht function.* Bull. Amer. Math. Soc. 41(1935), 535–540. Zbl 12-171. [19]

1169. Robinson, R. M. *A note on Chen's paper: "On the theory of schlicht functions."* Tohoku Math. J. 41(1936), 327–328. Zbl 14-70. [16]

1170. Robinson, R. M. *Bloch functions.* Duke Math. J. 2(1936), 453-459. Zbl 15-30. [19]
1171. Robinson, R. M. *On the mean values of an analytic function.* Bull. Amer. Math. Soc. 46(1940), 489-851. MR 2-79. [3]
1172. Robinson, R. M. *Bounded univalent functions.* Trans. Amer. Math. Soc. 52(1942), 426-449. MR 4-77. [15, 22, 24, 61, 68]
1173. Robinson, R. M. *Bounded analytic functions.* Univ. Calif. Publ. Math. (N.S.) 1(1944), 131-146. MR 5-259. [15, 22]
1174. Robinson, R. M. *Univalent majorants.* Trans. Amer. Math. Soc. 61(1947), 1-35. MR 8-370. [1]
1175. Robinson, R. M. *Extremal problems for star mappings.* Proc. Amer. Math. Soc. 6(1955), 364-377. MR 16-1096. [6, 15, 68]
1176. Rogozin, V. S. *Two sufficient conditions for univalence of a mapping.* Rostov. Gos. Univ. Uc. Zap. Fiz. Mat. Fak. 32(1955), 135-137. MR 17-724. [4, 9, 28]
1177. Rogosinski, W. *Über Bildschranken bei Potenzreihen und ihren Abschnitten.* Math. Z. 17(1923), 260-276. FM 49-231. [2, 13, 20, 23]
1178. Rogosinski, W. *Über positive harmonische Sinusentwicklungen.* Jber. Dtsch. Math. Ver. 40(1931), 2. Abt. 33-35. [2]
1179. Rogosinski, W. *Über den Wertevorrat einer analytischen Funktion.* Schrift. Konigsberg, Gel. Ges., Naturw. Klasse, 8(1931), 1-31. Zbl 2-272. [1, 9, 19, 37]
1180. Rogosinski, W. *Über positive harmonische Entwicklungen und typischreele Potenzreihen.* Math. Zeit. 35(1932), 93-121. Zbl 3-393. [2, 13, 23]
1181. Rogosinski, W. *Zum Majorantenprinzip der Funktiontheorie.* Math. Z. 37(1933), 210-236. Zbl 7-167. [1, 2]
1182. Rogosinski, W. *Zum Schwarzschen Lemma.* Jber. Deutsch. Math. Ver. 44(1934), 258-261. Zbl 10-307. [22, 37]
1183. Rogosinski, W. *Nachtrag zu meiner Arbeit: Zum Schwarzschen Lemma.* Jber. D. M. V. 45(1935), 48, 243. FM 61-348. [37]
1184. Rogosinski, W. *Über den Wertevorrat einer analytischer Funktion von der zwei Werte vorgegeben sind.* Compositio Math. 3(1936), 199-226. Zbl 14-353. [9, 19, 22]
1185. Rogosinski, W. *On subordinate functions.* Proc. Camb. Phil. Soc. 35(1939), 1-26. Zbl 20-140. [1]
1186. Rogosinski, W. *On a theorem of Bieberbach-Eilenberg.* J. London Math. Soc. 14(1939), 4-11. Zbl 20-376. (1, 3, 16, 17, 19, 22, 34]
1187. Rogosinski, W. *On the coefficients of subordinate functions.* Proc. London Math. Soc. (2) 48(1943), 48-82. MR 5-36. [1, 2, 3,

6, 10, 17, 19, 27, 39, 42]
1188. Rogosinski, W.; Shapiro, H. S. *On certain extremum problems for analytic functions.* Acta Math. 90(1953), 287-318. MR 15-516. [59]
1189. Rosenblatt, A. *Sur la représentation conforme du cercle de convergence d'une série de puissances.* Krak. Bull. (1917), 575-585. FM 46-554. [9, 15, 17, 18, 19, 54, 58, 68]
1190. Rosenblatt, A. *Sur les séries de puissances univalentes dans le cercle unité.* Cir. Acad. Sci. Paris 207(1938), 442-444. Zbl 19-272. [17, 45, 52, 54]
1191. Rosenblatt, A. *Sur les séries bornée dans le cercle unité.* Bull. soc. Sc. Liege 4(1935), 124-127. FM 61-346. [15, 20, 22, 61]
1192. Rosenblatt, A. *Über die regularen schlichten Funktionen im Einheitskreis.* Rev. Ci. Lima 40(1938), 165-176. (Spanish) Zbl 20-141. [17, 27, 45, 52, 54]
1193. Rosenblatt, A. *Über die Koeffizienten der schlichten Reihen im Einheitskreis.* Rev. Ci. Lima 40(1938), 177-179. Zbl 20-141. [17, 27, 45, 52, 54]
1194. Rosenblatt, A. *Bemerkung zur vorstehenden Note.* Rev. Ci. Lima 40(1938), 181-182. (Spanish) Zbl 20-141. [17, 27, 45, 52, 54]
1195. Rosenblatt, A. *Zusatz zu meinen Noten uber die schlichten Funktionen.* Rev. Ci. Lima 40(1938), 183-184. Zbl 20-141. [17, 27, 45, 52, 54]
1196. Rosenblatt, A. *Sur les coefficients des fonctions univalente dans le cercle unité.* Rev. Ci. Lima 40(1938), 541-545. Zbl 21-143. [17, 54]
1197. Rosenblatt, A. *Nouvelles recherches sur les coefficients des séries univalentes.* Rev. Ci. Lima 40(1938), 547-553. Zbl 21-143. [17, 54]
1198. Rosenblatt, A. *Sulle funzioni univalenti dispari nel cerchio unitario.* Atti. Accad. Naz. Lincei Rend. VIs 28(1938), 144-146. Zbl 20-141. [17, 27, 45, 52, 54]
1199. Rosenblatt, A. *Sur les coefficients des séries univalentes dans le cercle unité.* Bull. Soc. Math. Grece. 19(1939), 127-128. Zbl 21-143. [17, 54]
1200. Rosenblatt, A. *On the coefficients of univalent series.* Actas Acad. Ci. Lima 4(1941), 145-155. MR 4-7. [17, 39, 45, 52, 54]
1201. Rosenblatt, A. *On power series in the unit circle.* Revista Ci. Lima 45(1943), 195-225. (Spanish) MR 5-176. [17, 22, 30]
1202. Rosenblatt, A. *On the modulus of functions analytic in the unit circle.* Actas Acad. Ci. Lima 8(1945), 27-44. (Spanish) MR 8-19. [59]

1203. Rosenblatt, A.; Turski, S. *Sur les coefficients des séries de puissances univalentes dans le cercle unité.* C. R. Acad. Sci. Paris 200(1935), 1270-1272. Zbl 11-261. [17, 54]
1204. Rosenbloom, P. C. *Some properties of absolutely monotonic functions.* Bull. Amer. Math. Soc. 52(1946), 458-462. MR 8-65. [59]
1205. Royden, H. L. *The coefficient problem for bounded schlicht functions.* Proc. Nat. Acad. Sci. U.S.A. 35(1949), 657-662. MR 11-426. [22, 24, 29]
1206. Royden, H. L. *The conformal rigidity of certain subdomains on a Riemann surface.* Trans. Amer. Math. Soc. 76(1954), 14-25. MR 15-519. [59]
1207. Royden, H. L. *The interpolation problem for schlicht functions.* Ann. of Math. (2) 60(1954), 326-344. MR 16-232. [24, 29]
1208. Royden, H. L. *Conformal deformation.* Reprint from: Lectures on Functions of a Complex Variable; Univ. of Michigan Press, 1955. MR 16-1096. [24]
1209. Royster, W. C. *Convexity and starlikeness of analytic functions.* Duke Math. J. 19(1952), 447-457. MR 14-261. [6, 10]
1210. Royster, W. C. *Note on values omitted by p-valent functions.* Duke Math. J. 22(1955), 153-156. MR 16-685. [14, 16, 19, 22]
1211. Royster, W. C. *Rational univalent functions.* Amer. Math. Monthly 63(1956), 326-328. MR 17-1069. [39]
1212. Royster, W. C. *Coefficient problem for functions regular in an ellipse.* Duke Math. J. 26(1959), 361-371. MR 21-1064. [6, 13, 17, 40, 41, 57]
1213. Royster, W. C. *Functions having positive real part in an ellipse.* Proc. Amer. Math. Soc. 10(1959), 266-269. MR 21-783. [2, 17, 57]
1214. Royster, W. C. *Extremal problems for functions starlike in the exterior of the unit circle.* Canad. J. Math. 14(1962), 540-551. MR 26-499. [6, 9, 16, 17, 24, 25]
1215. Royster, W. C. *Meromorphic starlike multivalent functions.* Trans. Amer. Math. Soc. 107(1963), 300-308. MR 25-1211. [6, 9, 14, 17, 25, 36, 50]
1216. Saginyan, A. L. *On the theory of univalent functions.* Akad. Nauk Armyan, SSR, Izv. Fiz. -Mat. Estest. Tehn. Nauk; 9(1956), No. 7, 29-35. (Russian. Armenian summary) MR 18-728. [19, 49]
1217. Sakaguchi, K. *On functions star like in one direction.* J. Math. Soc. Japan 10(1958), 260-271. MR 20-771. [6, 10, 11, 12, 20,, 40]
1218. Sakaguchi, K. *On a certain univalent mapping.* J. Math. Soc. Japan 11(1959), 72-75. MR 21-1063. [4, 5, 6]

1219. Sakaguchi, K. *Some classes of multivalent functions.* Sci. Rep. Tokyo Kyoiku Daigaku. Sec. A 6(1959), 205-222. MR 22A(1)-18. [6, 9, 10, 11, 12, 14, 15, 17, 20, 49]
1220. Sakaguchi, K. *A note on p-valent functions.* J. Math. Soc. Japan 14(1962), 312-321. MR 26-1209. [4, 14]
1221. Sakai, E. *On the multivalency of analytic functions.* J. Math. Soc. Japan 2(1950), 105-113. MR 12-601. [6, 14, 22, 43]
1222. Sakashita, H. *Some subclasses of the convex functions.* Bull. Kyoto Gakugei Univ. Ser B 14(1959), 4-7. MR 21-1350. [4, 10, 15, 17, 38]
1223. Salem, R. *Power series with integral coefficients.* Duke Math. J. 12(1945), 153-172. MR 6-206. [59]
1224. Salem, R. *Algebraic numbers and Fourier analysis.* D. C. Heath and Co., Boston 1963, 66 pp. [39]
1225. Salinas, B. R. *Note on the region of value of a schlicht function.* Revista Mat. His. -Amer. (4) 12(1952), 223-228. MR 14-460. [15, 24, 29, 68]
1226. Sansone, G.; Gerretsen, J. *Lectures on the Theory of Functions of a Complex Variable I:* Holomorphic Functions, P. Noordhoff (1960), 488 pp. Zbl 93-268. [59]
1227. Santalo, L. A. *A theorem on conformal mapping.* Math. Notae 5(1945), 29-40. (Spanish) MR 6-261. [35]
1228. Sario, L. *On univalent functions.* Treizième Congrès des Mathematiciens Scandinaves, Tenu à Helsinki 18-23 Août (1957), 202-208. Mercators Tryckeri, Helsinki, 1958. 209 pp. MR 21-783. [9, 60]
1229. Sasaki, Y. *Theorems on the convexity of bounded functions.* Proc. Japan Acad. 27(1951), 122-129. MR 13-733. [10, 12, 22]
1230. Sasaki, Y. *On the Hauptsehne of the regions to which the unit circle is mapped by the bounded function.* Proc. Japan Acad. 27(1951), 216-218. MR 13-642. [7, 22]
1231. Sasaki, Y. *On some family of multivalent functions.* Kodai Math. Sem. Rep. 1952(1952), 89-92. MR 14-649. [4, 6, 10, 14, 35]
1232. Schaeffer, A. C. *An extremal boundary value problem.* Contrib. to Theory of Riemann Surfaces (1953), pp. 41-47. Annals of Math. Studies No. 30. Princeton Univ. Press, Princeton, N. J. MR 15-24. [62]
1233. Schaeffer, A. C.; Spencer, D. C. *The coefficients of schlicht functions.* I. Duke Math. J. 10(1943), 611-635. MR 5-175. [9, 17, 24, 25, 45, 52, 54]
1234. Schaeffer, A. C.; Spencer, D. C. *The coefficients of schlicht functions.* II. Duke Math. J. 12(1945), 107-125. MR 6-206. [17, 22, 24, 42, 54]

1235. Schaeffer, A. C.; Spencer, D. C. *The coefficients of schlicht functions*. III. Proc. Nat. Acad. Sci. U.S.A. 32(1946), 111–116. MR 7-424. [24, 29]
1236. Schaeffer, A. C.; Spencer, D. C. *A variational method in conformal mapping*. Duke Math. J. 14(1946), 949–966. MR 9-341. [24]
1237. Schaeffer, A. C.; Spencer, D. C. *A general class of problems in conformal mapping*. Proc. Nat. Acad. Sci. U.S.A. 33(1947), 185–189. MR 8-575. [24]
1238. Schaeffer, A. C.; Spencer, D. C. *The coefficients of schlicht functions*. IV. Proc. Nat. Acad. Sci. U.S.A. 35(1949), 143–150. MR 10-523. [24]
1239. Schaeffer, A. C., Spencer, D. C. *Models illustrating the third coefficient region for schlicht functions*. Scripta. Math. 16(1950), 67–71. MR 12-89. [29]
1240. Schaeffer, A. C.; Spencer. D. C. *Coefficient regions for schlicht functions*. Amer. Math. Soc. Coll. Pub. 35(1950). MR 12-326. [44]
1241. Schaeffer, A. C.; Spencer, D. C. *Coefficient regions for schlicht functions*. Proc. of Int. Cong. of Math., Cambridge, Mass. 2(1950), 224–232. MR 13-546. [44]
1242. Schaeffer, A. C.; Spencer, D. C. *A variational method for simply connected domains*. Proc. of a Symp. 189–191. Nat. Bur. of Standards, Appl. Math. Ser. No. 18, U.S. Govt. Printing Office, Wash., D. C. (1952). MR 14-743. [24]
1243. Schaeffer, A. C.; Spencer, D. C.; Schiffer, M. *The coefficient regions of schlicht functions*. Duke Math. J. 16(1949), 493–527. MR 11-91. [24]
1244. Schiffer, M. *Sur un principe nouveau pour l'evaluation des fonctions holomorphes*. Bull. Soc. Math. France 64(1936), 231–240. Zbl 15-359. [1, 15, 17]
1245. Schiffer, M. *Un calcul de variation pour une famille des fonctions univalentes*. C. R. Acad. Sci. Paris 205(1937), 709–711. Zbl 17-270. [24, 42, 49]
1246. Schiffer, M. *Sur un problème d'extremum de la représentation conforme*. Bull. de. la Societe Mathematique de France, 66(1938), 48–55. Zbl 18-409. [9, 17, 24, 25]
1247. Schiffer, M. *A method of variation within the family of simple functions*. Proc. London Math. Soc. (2) 44(1938), 432–449. Zbl 19-222. [24, 42, 53]
1248. Schiffer, M. *On the coefficients of simple functions*. Proc. London Math. Soc. (2) 44(1938), 450–452. Zbl 19-222. [24, 42, 53]
1249. Schiffer, M. *The span of multiply connected domains*. Duke Math. J. 10(1943), 209–216. MR 4-271. [9, 17]

1250. Schiffer, M. *Variation of the Green function and theory of the p-valued functions.* Amer. J. Math. 65(1943), 341–360. MR 4-215. [9, 14, 17, 24, 42, 53, 54]
1251. Schiffer, M. *Sur l'équation différentielle de M. Löwner.* C. R. Acad. Sci. Paris 221(1945), 369–371. MR 7-515. [24]
1252. Schiffer, M. *On the modulus of double-connected domains.* Quart. J. of Math. 17(1946), 197–213. MR 8-325. [59]
1253. Schiffer, M. *Hadamard's formula and variation of domain-functions.* Amer. J. Math. 68(1946), 417–448. MR 8-325. [7, 9, 24]
1254. Schiffer, M. *An application of orthonormal functions in the theory of conformal mapping.* Amer. J. Math. 70(1948), 147–156. MR 9-341. [9, 60]
1255. Schiffer, M. *Faber polynomials in the theory of univalent functions.* Bull. Amer. Math. Soc. 54(1948), 503–517. MR 10-26. [9, 15, 42, 53, 58]
1256. Schiffer, M. *Variational methods in the theory of conformal mapping.* Proc. Int. Cong. Math. 1950 V-2. MR 13-547. [24]
1257. Schiffer, M. *Variation of domain functionals.* Bull. Amer. Math. Soc. 60(1954), 303–328. MR 16-233. [24]
1258. Schiffer, M. *Applications of variational methods in the theory of conformal mapping.* Proc. of the Symposia in Applied Math. 8(1956), 93–113. MR 20-157. [24]
1259. Schiffer, M. *Extremum problems and variational methods in conformal mapping.* Proc. Internat. Congress Math. 1958, pp. 211–231. Cambridge Univ. Press, New York, 1960. MR 24A-612. [9, 17, 24, 25, 53]
1260. Schiffer, M. *Applications of variational methods in the theory of conformal mapping.* McGraw-Hill, 1958, 153 pp. MR 20-157. [24]
1261. Schiffer, M.; Spencer, D. C. *Lectures on Conformal Mapping and Extremal Methods.* Princeton Lectures, 1949–1950. [59]
1262. Schiffer, M.; Spencer, D. C. *The coefficient problem for multiply-connected domains.* Ann. of Math. 52 (2) (1950), 362–402. MR 12-171. [59]
1263. Schiffer, M.; Spencer, D. C. *Some remarks on variational methods applicable to multiply connected domains.* Proc. of a Symp. 193–198. Nat. Bur. of Stand., Appl. Math. Ser. No. 18, U.S. Govt. Print. Office, Wash., D. C. (1952). MR 14-743. [24]
1264. Schiffer, M.; Spencer, D. C. *Functionals of Finite Riemann Surfaces.* Princeton, 1954. MR 16-461. [59]
1265. Schild, A. *On a problem in conformal mapping of schlicht functions.* Proc. Amer. Math. Soc., 4(1953), 43–51. MR 14-861. [6, 10, 12, 19, 31, 39]

1266. Schild, A. *On a class of functions schlicht in the unit circle.* Proc. Amer. Math. Soc. 5(1954), 115-120. MR 15-694. [4, 6, 12, 15, 19, 32]
1267. Schild, A. *On a class of univalent, star shaped mappings.* Proc. Amer. Math. Soc. 9(1958), 751-757. MR 20-404. [4, 6, 10, 11, 15, 17, 27, 29, 31, 32, 36, 63]
1268. Schmittroth, L. *The conformal mapping of annuli.* Thesis. Stanford, Univ., 1954. [59]
1269. Scholz, D. R. *Some minimum problems in the theory of functions.* Pacific J. Math. 4(1954), 275-299. MR 15-862. [59]
1270. Schottlaender, S. *Der Hadamardsche Multiplikationssatz und weitere Kompositionssatze der Funktionentheorie.* Math. Nachr. 11(1954), 239-294. MR 16-346. [47]
1271. Schur, I. *Über Potenzreihen die im Innern des Einheitskreises beschränkt sind.* Jour. fur reine und angewandte Math., 147(1917), 205-232; 148(1918), 122-145. FM 46-475. [59]
1272. Schur, I. *On Faber polynomials.* Amer. J. of Math. 67(1945), 33-41. MR 6-210. [53]
1273. Schwarz, B. *Complex non-oscillation theorems and criteria of univalence.* Trans. Amer. Math.. Soc. 80(1955), 159-186. MR 17-370. [4, 31]
1274. Schweitzer, M. *The partial sums of second order geometric series.* Duke Math. J. 18(1951), 527-533. MR 13-23. [59]
1275. Scott, W. T. *A covering theorem for univalent functions.* Amer. Math. Monthly 64(1957), 90-94. MR 20-877. [19, 39]
1276. Scott, W. T. *Comment on a paper of C. Uluçay.* Proc. Amer. Math. Soc. 10(1959), p. 395. MR 21-662. [19]
1277. Seda, V. *A note to a paper of Clunie.* Acta. Fac. Nat. Univ. Comenian. 4(1959), 255-260. (Czech and Russian summaries) MR 23A-56. [59]
1278. Seidel, W. *Über die Ränderzuordnumg bei konformen Abbildungen.* Mat. Ann. 104(1931), 182-213. Zbl 1-19. [1, 2, 3, 6, 10, 22, 37, 46]
1279. Seidel, W. *On the order of growth of univalent functions.* Bull. Amer. Math. Soc. 42(1936), 335. FM 62-379. [62]
1280. Seidel, W.; Walsh, J. L. *On the derivatives of functions analytic in the unit circle and their radii of univalence and of p-valence.* Trans. Amer. Math. Soc. 52(1942), 128-216. MR 4-215. [3, 15, 18, 19, 22, 42, 43, 62, 68]
1281. Seshu, S.; Seshu, L. *Bounds and stieltjes transform representations for positive real functions.* J. Math. Analysis and Appl. 3(1961), 592-604. MR 25-800. [59]

1282. Sewell, W. E. *Generalized derivatives and approximation by polynomials.* Amer. Math. Soc. Trans. 41(1937), 84–128. Zbl 16-107. [59]
1283. Shah, T. *On the distortion of schlicht functions.* Acad. Sinica Sci. Record 4(1951), 209–212. MR 15-948. [9]
1284. Shah, T. *On the coefficients of schlicht functions.* J. Chinese Math. Soc. (N.S.) 1(1951), 98–107. MR 17-141. [17, 39, 54]
1285. Shah, T. *On the product of mapping radii for a system of non-overlapping domains.* Acta Math. Sinica 3(1953), 1–7. MR 17-141. [24, 26]
1286. Shah, T. *The principle of area in the theory of univalent functions.* Acta Math. Sinica 3(1953), 208–212. MR 17-141. [4, 27, 53]
1287. Shah, T. *The product of the mapping radii of non-overlapping domains.* Acta Math. Sinica 5(1955), 27–36. MR 17-142. [59]
1288. Shah, T. *On the moduli of some classes of analytic functions.* Acta. Math. Sinica 5(1955), 439–454. (Chinese. English summary) MR 17-724. [15, 34]
1289. Shah, T. *Some covering properties of convex domains in the theory of conformal mapping.* Acta Math. Sinica 7(1957), 421–432. (Chinese. English summary) MR 21-389. [10, 19, 24]
1290. Shah, T. *Goluzin's number $(3 - \sqrt{5})/2$ is the radius of superiority in subordination.* Sci. Record (N.S.) 1(1957), 219–222. MR 20-1074. [1]
1291. Shah, T. *On the radius of superiority in subordination.* Sci. Record (N.S.) 1(1957), 329–333. MR 20-1074. [1]
1292. Shah, T. *Some covering properties of convex domains in the theory of conformal mapping.* Sci. Sinica 7(1958), 816–828. MR 21-389. [10, 19, 24]
1293. Shah, T.; Chang, K. *Some inequalities in the theory of subordination.* Acta. Math. Sinica 8(1958), 408–412. (Chinese. English summary) MR 21-1350. [1, 17, 19]
1294. Shapiro, H. S. *Applications of normed linear spaces to function-theoretic extremal problems.* Lectures on functions of a complex variable, pp. 399–404. The Univ. of Mich. Press, Ann Arbor, 1955. MR 17-25. [59]
1295. Shi, S. *A covering theorem on bounded schlicht functions.* Advancement in Math. 2(1956), 675–677. MR 20-1074. [19, 22]
1296. Shich, S. *On the coefficients of schlicht functions.* Advancement in Math. 3(1957), 597–601. (Chinese) MR 20-968. [17]
1297. Shiffman, M. *On the isoperimetric inequality for saddle surfaces with singularities.* Studies and Essays presented to R. Courant, (1948), 383–394. MR 9-303. [59]

1298. Shlionsky, H. G. *On extremal problems for differentiable functionals in the theory of univalent functions.* Dokl. Akad. Nauk SSSR (N.S.) 113(1957), 280-282. (Russian) MR 19-738. [24, 60]. See Slionskii.
1299. Shniad, H. *On analytic maps of circles into convex regions.* Amer. Math. Mon. 57(1950), 473-474. MR 12-401. [10, 15]
1300. Shniad, H. *Convexity properties of integral means of analytic functions.* Pacific J. Math. 3(1953), 657-666. MR 15-112. [3]
1301. Sidon, S. *Über Potenzreihen mit monotoner Koeffizientenfolge.* Acta Litt. Sci. Szeged 9(1940), 244-246. MR 1-213. [4, 32, 39]
1302. Siewierski, L. *Sur les fonctions univalentes, algébriques dans le demi-plan.* Bull. Soc. Sci. Lett. Lodz. Cl. III Math. Natur. 7(1956), No. 4. 17 p. Zbl 90-291. [9, 57]
1303. Siewierski, L. *Sur la variation locale des fonctions univalentes, algebriques dans le demi-plan.* Bull. Soc. Sci. Lett. Lodz. Cl. III Sci. Math. Natur. 8(1957), No. 3, 16 p. Zbl 90-291. [9, 57]
1304. Siewierski, L. *Sur les fonctions extremales dans les familles des fonctions univalentes, algébriques dans le demi-plan.* Bull Soc. Sci. Lett. Lodz. Cl. III Sci. Math. Natur. 8(1957), 30 p. Zbl 90-292. [9, 57]
1305. Singh, S. K. *Univalent and multivalent functions.* Math. Student 30(1962), 79-90. MR 26-743. [44]
1306. Singh, S. K.; Shah, S. M. *On the maximum function of a meromorphic function.* Math. Student 22(1954), 121-128. MR 16-459. [9, 16]
1307. Singh, V. *Interior variations and some extremal problems for certain classes of univalent functions.* Pacific J. Math. 7(1957), 1485-1504. MR 20-20. [22, 24]
1308. Singh, V. *Some extremal problems for a new class of univalent functions.* J. Math. Mech. 7(1948), 811-821. MR 20-771. [15, 17, 24]
1309. Singh, V. *Extremum problems for the coefficient a_3 of the bounded univalent functions.* Proc. London Math. Soc. (3) 9(1959), 397-416. MR 22A(1)-18. [17, 22, 24, 30, 39]
1310. Singh, V. *Grunsky inequalities and coefficients of bounded schlicht functions.* Ann. Acad. Sci. Fenn. AI No. 310(1962), 22 pp. MR 26-277. [4, 22]
1311. Siryk, G. V. *On conformal mapping of nearby regions.* Uspehi Mat. Nauk (N.S.) 11(1956), No. 5(71), 57-60. (Russian) MR 19-258. [59]
1312. Šlionskii, G. *On finite sums of bounded univalent functions.* Dokl. Akad. Nauk SSSR (N.S.) 93(1953), 707-709 MR 15-516.

[16, 20, 22, 43]
1313. Šlionskii, G. *On the theory of bounded univalent functions.* Dokl. Akad. Nauk SSSR (N.S.) 111(1956), 962–964. (Russian) MR 18-798. [9, 15, 22, 58, 61]
1314. Šlionskii, G. *On the extremal problems for differentiable functionals in the theory of univalent functions.* Vestnik Leningrad. Univ. 13(1958), No. 13, 64–83. (Russian. English summary) MR 20-771. [24]
1315. Šlionskii, G. *On the theory of bounded schlicht functions.* Vestnik Leningrad Univ. 14(1959), No. 13, 42–51. (Russian. English summary) MR 22A(1)-455. [22]
1316. Slobodeckii, L. N. *On a problem of the theory of univalent functions.* Dokl. Akad. Nauk SSSR (N.S.) 92(1953), 235–238. MR 15-413. [9, 15, 58, 60]
1317. Špaček, L. *Contribution à la theorie des fonctions univalentes.* Casopis Pest. Mat. 62(1932), 12–19. Zbl 6-64. [4, 6, 10, 17, 36, 38, 42]
1318. Špak, G. S. *On some estimates for the argument of an analytic function.* Dokl. Akad. Nauk SSSR (N.S.) 92(1953), 711–713. (Russian) MR 15-613. [59]
1319. Špak, G. S. *A covering theorem in function theory.* Izv. Vyss. Ucebn. Zaved. Matematika (1959), No. 1(8), 218–223. (Russian) MR 23A-731. [1, 19, 22]
1320. Špak, G. S. *Some estimates for the modulus and real aprt of pseudo-positive functions.* Izv. Vyss. Ucebn. Zaved. Mat. 1962, Izo. 6(31), 148–154. [2, 15, 56]
1321. Specht, E. J. *Estimates of the mapping function and its derivatives in conformal mapping of nearly circular regions.* Trans. Amer. Math. Soc. 71(1951), 183–196. MR 13-337. [59]
1322. Spencer, D. C. *On finitely mean valent functions.* II. Trans. Amer. Math. Soc. 48(1940), 418–435. MR 2-82. [3, 17, 33]
1323. Spencer, D. C. *Note on some function-theoretic identities.* J. London Math. Soc. 15(1940), 84–86. MR 2-82. [3, 27]
1324. Spencer. D. C. *On an inequality of Grunsky.* Proc. Nat. Acad. Sci. U.S.A. 26(1940), 616–621. MR 2-79. [18, 19]
1325. Spencer, D. C. *On finitely mean valent functions.* Proc. London Math. Soc. (2) 47(1942), 201–211. MR 3-79. [3, 16, 17, 18, 33, 45, 52]
1326. Spencer, D. C. *On distortion in analytic transformations.* J. Math. Phys. Mass. Inst. Tech. 20(1941), 124–126. MR 2-186. [16, 19, 33]

1327. Spencer, D. C. *On mean one-valent functions.* Ann. of Math. (2) 42(1941), 614–633. MR 3-78. [15, 16, 17, 19, 27, 33]
1328. Spencer, D. C. *Note on mean one-valent functions.* J. Math. Phys. Mass. Inst. Tech. 21(1942), 178–188. MR 4-138. [33]
1329. Spencer, D. C. *Some remarks concerning the coefficients of schlicht functions.* J. Math. Phys. Mass. Inst. Tech. 21(1942), 63–68. MR 4-76. [16, 17, 42, 54]
1330. Spencer, D. C. *A function-theoretic identity.* Amer. J. Math. 65(1943), 147–160. MR 4-137. [33, 46]
1331. Spencer, D. C. *Some problems in conformal mapping.* Bull. Amer. Math. Soc. 53(1947), 417–439. MR 8-575. [44]
1332. Springer, G. *The coefficient problem for schlicht mappings of the exterior of the unit circle.* Trans. Amer. Math. Soc. 70(1951), 421–450. MR 13-24. [9, 15, 16, 17, 24, 25, 27, 53, 58]
1333. Springer, G. *Extreme Punkte der konvexen Hülle schlichter Funktionen.* Math. Ann. 129(1955), 230–232. MR 16-1011. [9, 15, 17, 25, 27, 58]
1334. Srivastav, R. P. *On the mean value of integral functions and their derivatives.* Riv. Mat. Univ. Parma 8(1957), 361–369. MR 21-1063. [59]
1335. Stein, P. *On a theorem of M. Riesz.* J. London Math. Soc. 8(1933), 242–247. FM 59-325. [3]
1336. Stone, M. H. *Hilbert space methods in conformal mapping.* Proc. Internat. Sympos. Linear Spaces (Jerusalem 1960), pp. 409–425. Jerusalem Academic Press, Jerusalem; Pergamon, Oxford; 1961. MR 25-39. [59]
1337. Strelic, S. I. *On a connection between typically-real and univalent functions.* Uspehi Mat. Nauk (N.S.) 12(1957) No. 3(75) pp. 211–220. (Russian) MR 19-643. [4, 13, 41]
1338. Strelicas, S. *Some properties of functions analytic in a circle.* Vilniaus Valst. Univ. Mokslo Darbai. Mat. Fiz. Chem. Mokslu Ser. 4(1955), 67–71. (Lithuanian. Russian summary) MR 22A(2)-2093. [3, 4, 17, 54]
1339. Stroganoff, W. G. *Über den Arc $f'(z)$ unter der Bedinbung dass $f(z)$ die konforme Abbildung eines sternartigen Gebietes auf das innere des Einheitskreises der z-Ebene liefert.* Trudy Mat. Inst. Steklova 5(1934), 247–258. Zbl 9-173. [6, 15, 68]
1340. Strohhäcker, E. *Beitrage zur Theorie der schlichten Funktionen.* Math. Z. 37(1933), 356–380. Zbl 7-214. [2, 6, 10, 15, 17, 19, 29, 38, 45, 63, 64, 67]
1341. Study, E. *Konforme Abbildung Einfachzusammenhangender*

Bereiche. B. C. Teubner, Leipzig and Berlin, 1913. FM 44-755. [1, 10, 37]
1342. Suetin, P. K. *Faber polynomials for regions with non-analytic boundaries.* Dokl. Akad. Nauk SSSR (N.S.) 88(1953), 25-28. MR 14-740. [53]
1343. Suvorov, G. D. *On the order of equicontinuity of a class univalent mappings in closed domains.* Dokl. Akad. Nauk SSSR (N.S.) 107(1956), 22-23. MR 17-956. [22, 60, 62]
1344. Suvorov, G. D. *On the continuity of univalent mappings of arbitrary closed domains.* Dokl. Acad. Nauk SSSR (N.S.) 108(1956), 777-779. (Russian) MR 18-724. [59]
1345. Suvorov, G. D. *On distortion of distances in univalent mappings of closed regions.* Mat. Sb. N.S. 45(87) (1958), 159-180. (Russian) MR 20-968. [49]
1346. Suvorov, G. D. *Correction to the article "On distortion of distances in univalent mappings of closed regions".* Mat. Sb. (N.S.) 48(90) (1959), 251-252. (Russian) MR 23A-51. [16]
1347. Suvorov, G. D. *The length and area principles for Q-quasiconformal mappings.* Dokl. Akad. Nauk SSSR 140(1961), 1267-1269. (Russian) MR 24A-372. [7, 15, 24]
1348. Szász, O. *Ungleichtheits beziehungen fur die Ableitungen einer Potenzreihe die eine im Einheitskreise beschrankte Funktion darstellt.* Math. Z. 8(1920), 303-309. FM 47-274. [15, 22, 61]
1349. Szász, O. *Uber Funktionen die den Einheitskreis schlicht abbilden.* Jber, Deutsch. Math. Ver. 42(1932), 73-75. Zbl 5-108. [17, 39, 42]
1350. Szász, O. *On the Cesàro and Riesz means of Fourier series.* Composito Mathematica 7(1939), 112-122. MR 1-138. [59]
1351. Szász, O. *On the partial sums of harmonic developments and of power series.* Trans. Amer. Math. Soc. 52(1942), 12-21. MR 4-37. [59]
1352. Szczepankiewicz, E.; Zamorski, J. *Close-to-convex and almost starlike functions.* Prace Mat. 6(1961), 141-148. (Polish. Russian and English summaries) MR 26-980. [5, 6]
1353. Szegö, G. *Über orthogonale Polynome die zu einer gegebenen Kurve der komplexen Ebene gehören.* Math. Z. 9(1921), 218-270. FM 48-374. [59]
1354. Szegö, G. *Bemerkungen zu einem Staz von J. H. Grace uber die Wurzeln algebraischer Gleichungen.* Math. Z. 13(1922), 28-55. FM 48-82. [59]
1355. Szegö, G. *Relative Minimalflachen.* Jber. Deutsch. Math. -Ver. 31(1922), 41-42. [59]

1356. Szegö, G. *Über das Maximum einer quadratischen Form von unendlichvielen Veranderlichen.* Jber. Deutsch. Math. -Ver. 31(1922), 85–88. FM 48–490. [3]
1357. Szegö, G. *Über Potenzreihen mit endlich vielen verschiedenen Koeffizienten.* S. -B. Preuss. Acad. Wiss. (1922), 88–91. FM 48–330. [59]
1358. Szegö, G. *Über eine Extremalaufgabe aus der Theorie der schlichten Abbildungen.* Sitzber. Berl. Math. Ges. 22(1923), 38–47. FM 49–248. [19]
1359. Szegö, G. *Über einen Satz von A. Markoff.* Math. Z. 23(1925), 45–61. FM 51–97. [59]
1360. Szegö, G. *Zur Theorie der schlichten Abbildungen.* Math. Ann. 100(1928), 188–211. FM 54–336. [6, 20, 43]
1361. Szegö, G. *Some recent investigations concerning the sections of power series and related developments.* Bull. Amer. Math. Soc. 42(1936), 505–522. Zbl 14–352. [2, 20, 22]
1362. Szegö, G. *Power series with multiply monotonic sequences of coefficients.* Duke Math. J. 8(1941), 559–564. MR 3–76. [4, 39]
1363. Szegö, G. *Conformal mapping of the interior of an ellipse onto a circle.* Amer. Math. Mon. 57(1950), 474–478. MR 12–401. [57]
1364. Taam, Choy-Tak. *Schlicht functions and linear differential equations of second order.* J. Rat. Mech. Analysis 4(1955), 467–480. Zbl 64–321. [31]
1365. Takahashi, S. *Über schlichte Potenzreihen und ihre Abschnitte.* Tohoku Math. J. 33(1930), 55–60. FM 56–985. [4, 6, 10, 20]
1366. Takahashi, S. *Bemerkung zu einer Arbeit von Herrn Rado.* Jap. J. Math. 7(1930), 161–162. FM 56–986. [1, 6, 10, 37]
1367. Takahashi, S. *Über die notwendige und hinreichende Bedingung fur die schlichte Abbildung des Einheitskreises.* Proc. Imp. Acad. Jap. 8(1932), 344–347. Zbl 6–65. [4, 6, 11, 20, 43]
1368. Takahashi, S. *Über die Koeffizientenabschatzungen von ungeraden schlichten Potenzreihen.* Proc. Imp. Acad. Jap. 9(1933), 1–2. Zbl 6–262. [8, 17, 39, 42, 45, 52]
1369. Takahashi, S. *Zwei Bemerkungen über schlichte Funktionen.* Proc. Imp. Acad. Jap. 9(1933), 461–464. Zbl 8–168. [42, 43]
1370. Takahashi, S. *Über die Abschnitte einer ungeraden schlichten Potenzreihe.* Proc. Phys. -Math. Soc. Jap. (3) 16(1934), 7–15. FM 60–283. [6, 10, 11, 12, 20, 45]
1371. Takahashi, S. *On the necessary and sufficient condition for the multivalency of an analytic function regular in the unit circle.* Proc. Phys. -Math. Soc. Jap. (3) 18(1936), 353–355. FM 62–376. [4, 14]

1372. Takahashi, S. *On the multivalency of an analytic function*. Proc. Imp. Acad. Jap. 12(1936), 59–60. Zbl 15–71. [4, 14]

1373. Takahashi, S. *Univalent mappings in several complex variables*. Ann. of Math. (2) 53(1951), 464–471. MR 12–818. [59]

1374. Tammi, O. *On the maximalization of the coefficients of schlicht and related functions*. Ann. Acad. Sci. Fennicae Ser. A. I. Math. -Phys. No. 114, 51 pp. (1952). MR 14–366. [6, 15, 17, 23, 24, 29, 42, 54, 68]

1375. Tammi, O. *On certain combinations of the coefficients of schlicht functions*. Ann. Acad. Sci. Fennicae Ser. A. E. Math. -Phys. No. 140 13 pp. (1952). MR 14–740. [12, 15, 17, 23, 54, 68]

1376. Tammi, O. *On the coefficients of the solutions of Lowner's differential equation*. Ann. Acad. Sci. Fennicae. Ser. A. I. Math. -Phys. No. 148, 7 pp. (1953). MR 15–302. [24]

1377. Tammi, O. *On the maximalization of the coefficient a_3 of bounded schlicht functions*. Ann. Acad. Sci. Fennicae Ser. A. I. Math. -Phys. No. 149. 14 pp. (1953). MR 15–302. [17, 22, 23, 24, 54]

1378. Tammi, O. *On the extremal domains belonging to the coefficient a_3 of bounded schlicht functions*. Ann. Acad. Sci. Fennicae. Ser. A. I. Math. -Phys. No. 162, 12 pp. (1953). MR 15–516. [22, 24]

1379. Tammi, O. *On the coefficients of bounded schlicht functions*. Tolfte Skandinaviska Matematikerkongressen, Lund, (1953), 297–301 (1954). MR 16–347. [22, 42]

1380. Tammi, O. *On the conformal mapping of symmetric schlicht domains*. Ann. Acad. Sci. Fenn. Ser. A. I. Math. -Phys. No. 173, 12 pp. (1954). MR 16–233. [22, 24, 39, 42]

1381. Tammi, O. *Note on symmetric schlicht domains of bounded boundary rotation*. Ann. Acad. Sci. Fenn. Ser. A. I. No. 198(1955) 10 pp. MR 17–599. [17, 23, 39, 49, 54]

1382. Tammi, O. *Note on Gutzmer's coefficient theorem*. Rev. Fac. Sci. Univ. Istanbul Ser. A22(1957), 9–12. (Turkish summary) MR 20–877. [1, 17, 49]

1383. Tammi, O. *On bounded univalent functions*. Bull. Acad. Polon. Sci. Ser. Sci. Math. Astr. Phys. 7(1959), 413–417. (Russian summary) Zbl 94–49. [22, 29]

1384. Tammi, O. *On a method of symmetrization in the theory of univalent functions*. Ann. Acad. Sci. Fenn. Ser. A. I. No. 302, (1961), 7 pp. MR 24A–38. [24]

1385. Tammi, O. *On a method of extremalization in the class S_k of univalent functions*. Ann. Acad. Sci. Fenn. Ser. A. I. No. 320(1962), 6 pp. MR 25–426. [24]

1386. Tamrazov, P. M. *Relative boundary distortion under a schlicht conformal mapping of a doubly connected domain.* Dopovidi Akad. Nauk Ukrain. RST (1962), 338-340. (Ukrainian. Russian and English summaries) MR 26-743. [60]

1387. Tamrazov, P. M. *On the theory of schlicht conformal mappings of doubly connected domains.* Dopovidi Akad. Nauk Ukrain. RSR 1962, 563-566. (Ukrainian. Russian and English summaries) MR 26-743. [60]

1388. Tanaka, C. *On the class H_p of functions analytic in the unit circle.* Yokohama Math. J. 4(1956), 47-53. MR 19-130. [59]

1389. Tang, S. *The sections of schlicht functions.* Advancement in Math. 3(1957), 468-477. (Chinese. English summary) MR 20-1074. [20]

1390. Tang, S. *On the distortions of a class of schlicht functions.* Advancement in Math. 3(1957), 478-484. (Chinese. English summary) MR 21-26. [9, 15, 16, 22, 24, 58, 61]

1391. Taylor, A. E. *New proofs of some theorems of Hardy by Banach space methods.* Math. Mag. 23(1950), 115-124. MR 11-507. [3]

1392. Teichmüller, O. *Ungleichungen zwischen den Koeffizienten schlichter Funktionen.* Stizgsber. Preuss. Akad. Wiss., Phys.-Math. Kl. (1938), 363-375. [9, 15, 17, 24, 25]

1393. Teichmüller, O. *Extremale quasikonforme Abbildungen und quadratische Diffrentiale.* Abh. Preuss. Akad. Wiss., Math.-Naturwiss. Kl. (1939), No. 22. MR 2-187. [59]

1394. Teichmüller, O. *Über Extremalprobleme der konformen Geometrie.* Dtsch. Math. 6(1941), 50-77. MR 3-202. [17, 27]

1395. Terzioğlu, A. N. *Über den Verzerrungssatz.* Rev. Fac. Sci. Univ. Istanbul A 15(1950), 113-118. Zbl 38-52. [15]

1396. Terzioğlu, A. N.; Kahramaner, S. *Ein Verzerrungssatz des Argumentes der schlichten Funktionen.* Rev. Fac. Sci. Univ. Istanbul Ser. A. 20(1955), 81-90. MR 17-836. [15, 17, 54, 68]

1397. Terzioğlu, A. N.; Kahramaner, S. *Über das Argument der analytischen Funktionen.* Rev. Fac. Sci. Univ. Istanbul Ser. A. 21(1956), 145-153(1957). (Turkish summary) MR 19-846. [9, 15, 58]

1398. Thale, J. S. *Univalence of continued fractions and stieltjes transforms.* Proc. Amer. Math. Soc. 7(1956), 232-244. MR 17-1063. [4, 6, 11, 28, 57]

1399. Tims, S. R. *A theorem on functions schlicht in convex domains.* Proc. London Math. Soc. (3) 1(1951), 200-205. MR 13-336. [4]

1400. Titchmarsh, E. C. *The Theory of Functions.* Oxford, 1932, 454 pp. Zbl 5-210. [44]

1401. Toeplitz, O. *Über die Fourier'sche Entwicklung positiver Funk-*

tionen. Rend. Cir. Mat. Palermo 32(1911), 191–192. FM 42–428. [2]
1402. Tong, K. C. *On a covering theorem in the theory of schlicht functions.* Sci. Record 4(1951), 37–40. Zbl 42–315. [19]
1403. Townes, S. B. *A theorem on schlicht functions.* Proc. Amer. Math. Soc. 5(1954), 585–588. MR 16–25. [39]
1404. Tsuji, M. *On a regular function whose real part is positive in a unit circle.* Proc. Japan Acad. 21(1945), 321–329(1949). MR 11–338. [2]
1405. Tsuji, M. *On Lowner's differential equation in the theory of univalent functions.* Japan J. Math. 19(1947), 321–341. MR 10–440. [24]
1406. Tsuji, M. *A deformation theorem on conformal mapping.* J. Math. Soc. Japan 2(1951), 213–215. MR 13–224. [49]
1407. Tsuji, M. *A remark on Rengel's theorem concerning Szego's conjecture.* Kodai Math. Sem. Rep. 1953(1953), 117–118. MR 15–613. [19]
1408. Tsuji, M. *On the radial order at a certain regular function in a unit circle.* J. Math. Soc. Japan 6(1954), 336–342. MR 16–809. [62]
1409. Tsuji, M. *On a Riemann surface, which is conformally equivalent to a Riemann surface with a finite spherical area.* Comment. Math. Univ. St. Paul 6(1957), 1–7. MR 19–1043. [46]
1410. Tsuji, M. *A theorem on the boundary behaviour of a meromorphic function in $|z| < 1$.* Comment. Math. Univ. St. Paul 8(1960), 53–55. MR 22A(2)–1899. [7, 16, 46, 62]
1411. Turán, P. *Über die monotone Konvergenz der Cesàro-mittel bei Fourier und Potenzreihen.* Proc. Cambridge Phil. Soc. 34(1938), 134–143. Zbl 19–17. [59]
1412. Turkovskii, V. A. *On a theorem in the theory of univalent functions.* Izv. Vyss. Uc. Zaved. Mat. (1959), No. 6, (13), 189–191. MR 24A–38. [4, 10, 16, 17]
1413. Ulina, G. V. *On the domains of values of certain systems of functionals in the classes of univalent functions.* Vestnik Leningrad Univ. 15(1960), No. 1, 34–54. (Russian. English summary) MR 22A(2)–965. [24, 29, 39]
1414. Ullman, J. L. *Two mapping properties of schlicht functions.* Proc. Amer. Math. Soc. 2(1951), 654–657. MR 13–223. [6, 9, 23]
1415. Ullman, J. L. *Studies in Faber polynomials.* Trans. Amer. Math. Soc. 94(1960), 515–528. Zbl 90–47. [53]
1416. Ullrich, E. *Über eine Anwendung des Verzerrungssätzes auf meromorphe Funktionen.* J. Reine Angewandte Math. 166(1931/32), 220–234. [9, 15]

1417. Ullrich, E. *Betragflachen mit ausgezeichnetem Krummungsverhalten*. Math. Z. 54(1951), 297-328. MR 13-124. [2, 35]
1418. Uluçay, C. *Sur les fonctions de Bloch de la troisième espèce.* Comm. Fac. Sci. Univ. Ankara Ser. A. 6(1954), 5-10. MR 16-1011. [19]
1419. Ullrich, E. *Sur les fonctions de Bloch de la première et de la seconde espèce*. Comm. Fac. Sci. Univ. Ankara Ser. A. 6(1954), 11-16. MR 16-1011 [19]
1420. Ullrich, E. *On the constant C*. Comm. Fac. Sci. Univ. Ankara Ser. A. 6(1954), 77-88. MR 16-1011. [19]
1421. Ullrich, E. *On the Bloch-Landau constants*. Comm. Fac. Sci. Univ. Ankara Ser. A. 7(1955), 233-252. (Turkish summary) MR 17-472. [19]
1422. Ullrich, E. *Note on a generalization of Schwarz's lemma*. Comm. Fac. Sci. Univ. Ankara Ser. A. 8(1956), 1-5. (Turkish summary) MR 19-736. [59]
1423. Ullrich, E. *Bloch functions of the third kind and the constant U*.Proc. Amer. Math. Soc. 8(1957), 923-925. MR 18-736. [16, 19]
1424. Ullrich, E. *The exact values of the Bloch-Landau constants B, L*. J. Reine Angew. Math. 199(1958), 188-191. MR 20-877. [19]
1425. Ullrich, E. *On the coefficients of schlicht functions*. J. Reine Angew. Math. 202(1959), 1-5. MR 21-1195. [3, 15, 16, 17, 42, 54, 68]
1426. Ullrich, E. *On inequalities in the theory of schlicht functions*. J. Reine Angew. Math. 202(1959), 6-8. MR 21-1195. [6, 15, 16, 17, 45, 52, 54, 67, 68]
1427. Umezawa, T. *A class of multivalent functions with assigned zeros*. Proc. Amer. Math. Soc. 3(1952), 813-820. MR 14-260. [6, 13, 14, 15, 17, 36, 40, 50, 63, 65, 66]
1428. Umezawa, T. *Analytic functions convex in one direction*. J. Math. Soc. Japan 4(1952), 194-202. MR 14-461. [4, 12, 41]
1429. Umezawa, T. *On the multivalency of analytic functions*. J. Math. Soc. Japan 4(1952), 279-285. MR 14-859. [4, 6, 10, 14]
1430. Umezawa, T. *Analytic functions starlike of order p in one direction*. Tohoku Math. J. 4(1952), 264-277. MR 14-859. [15, 17, 39, 40, 49]
1431. Umezawa, T. *On the coefficients of multivalent functions*. J. Math. Soc. Japan 5(1953), 137-144. MR 15-302. [13, 17, 39, 49, 51]
1432. Umezawa, T. *Star-like theorems and convex-like theorems in an annulus*. J. Math. Soc. Japan 6(1954), 68-75. MR 15-947. [4, 48]
1433. Umezawa, T. *The coefficients of meromorphic functions*. Duke Math. J. 21(1954), 355-361. MR 15-947. [16, 17, 36, 40]

1434. Umezawa, T. *On the theory of univalent functions.* Tohoku Math. J. 7(1955), 212–228. MR 17–1068. [4, 9, 10, 14, 60]
1435. Umezawa, T. *Multivalently close-to-convex functions.* Proc. Amer. Math. Soc. 8(1957), 869–874. MR 19–846. [4, 5, 7, 14, 15, 17, 38, 64, 65]
1436. Unkelbach, H. *Über die Randvrzerrung bei konformer Abbildung.* Math. Z. 43(1938), 739–742. Zbl 18–225. [46]
1437. Unkelbach, H. *Über den Vorrat analytischer Funktionen an grossen Funktionswerten.* Jber. Deutsch. Math. Ver. 49(1939), 38–49. Zbl 21–143. [6, 10, 19]
1438. Unkelbach, H. *Über die Randverzerrung bei schlichter konformer Abbildung.* Math. Z. 46(1940), 329–336. MR 2–83. [6, 10, 46]
1439. Unkelbach, H. *Über die Approzimation schlichter konformer Abbildungen durch kongruente Iterationen spezieller sterniger oder konvexer konformer Abbildungen.* Math. Z. 58(1953), 63–70. MR 14–861. [15, 22]
1440. Urazbaev, B. M. *On the argument of the derivative of a univalent star-shaped function.* Izv. Akad. Nauk Kazah SSR 56(1948), Ser. Mat. Meh. 2, 102–121. MR 13–546. [6, 15, 63]
1441. Valiron, G. *Recherches sur le théorème de M. Borel dans la theorie des fonctions meromorphes.* Acta Math. 52(1928), 67–92. FM 54–348. [59]
1442. Valiron, G. *Sur les domaines d'univalence de fonctions entières d'ordre nul.* Composito Math. 3(1936), 129–135. Zbl 14–266. [28]
1443. Valiron, G. *Division en feuillets de la surface de Riemann definie par $w = (e^z - 1)/z + h$.* J. Math. Pures Appl. 9(19) (1940), 339–358. MR 2–358. [59]
1444. Verblunsky, S. *On the curvature of level curves.* Math. Ann. 125(1953), 472–476. MR 15–206. [35]
1445. Visser, C. *Sur la dérivée angulaire des fonctions univalentes.* C. R. Acad. Sci. Paris 199(1934), 924–925. Zbl 10–120. [46]
1446. Visser, C. *Sur la dérivée angulaire des fonctions univalentes.* Akad. Wetensch. Amsterdam Proc. 38(1935), 402–411. Zbl 12–25. [46]
1447. Visser, C. *Sur la derivee angulaire des fonctions univalentes.* II. Akad. Wetensch. Amsterdam Proc. 38(1935), 618–627. Zbl 11–407. [46]
1448. Waadeland, H. *Über einen spezialfall k-fach symmetrischer schlichter Funktionen.* Norske Vid. Selsk. Forh., Trondheim 27(1954), 151–155(1955), MR 16–916. [17, 19, 39]
1449. Waadeland, H. *Über k-fach symmetrische sternförmige schlichte*

Abbildungen des Einheitskreises. Math. Scand. 3(1955), 150–154. MR 17–25. [6, 17, 36, 49]

1450. Waadeland, H. *Bemerkung zu einer Arbeit von Golusin.* Norske Vid. Selsk. Forh., Trondheim 29(1956), 29–32. MR 18–201. [17, 42, 45, 52, 54]

1451. Waadeland, H. *Ein Golusinscher Satz über schlichte Abbildungen von $|\zeta| > 1$.* Norske Vid. Selsk. Forh. Trondheim 30(1957), 165–168. MR 20–664. [9, 17, 25]

1452. Waadeland, H. *Über ein koeffizienten Problem für schlichte Abbildungen des $|\zeta| > 1$.* Norske Vid. Selsk. Forh. Trondheim, 30(1957), 168–170. MR 20–664. [9, 17, 25]

1453. Waadeland, H. *Bemerkung zu einem Bedeckungssatz von Scott.* Norske Vid. Selsk. Forh. Trondheim 32(1959), 112–116. MR 24A–370. [17, 54]

1454. Waadeland, H. *Untersuchungen über Äquivalenzklassen normierter, schlichter Funktionen.* Norske Vid. Selsk. Skr. Trondheim (1959), No. 3, 47 pp. MR 24A–254. [17, 29, 54]

1455. Waadeland, H. *Über beschrankte schlichte Funktionen.* I, II. Norske Vid. Selsk. Forh. Trondheim 32(1959), 84–91, 92–94. MR 24A–254. [17, 22, 24, 30]

1456. Waadeland, H. *Zur Theorie der beschränkten Funktionen.* Norske Vid. Selsk. Forh. Trondheim 35(1962), 82–85. MR 26–748. [15, 22, 61]

1457. Wall, H. S. *Analytic Theory of Continued Fractions.* New York, Van Nostrand, 1948. MR 10–32. [59]

1458. Walsh, J. L. *Lemminscates and equipotential curves of Green's function.* Amer. Math. Monthly 42(1935), 1–17. [59]

1459. Walsh, J. L. *Note on the curvature of level curves of Green's function.* Proc. Nat. Acad. Sci. U.S.A. 23(1937), 84–89. FM 63–297. [9, 10]

1460. Walsh, J. L. *On the shape of level curves of Green's functions.* Amer. Math. Monthly 44(1937), 202–213. FM 63–296. [59]

1461. Walsh, J. L. *Note on the curvature of orthogonal trajectories of level curves of Green's functions.* Bull. Amer. Math. Soc. 44(1938), 520–523. Zbl 19–271. [59]

1462. Walsh, J. L. *On the circles of curvature of the images of circles undr a conformal map.* Amer. Math. Monthly 46(1939), 472–485. MR 1–111. [19, 35]

1463. Walsh, J. L. *Note on the derivatives of functions analytic in the unit circle.* Bull. Amer. Math. Soc. 53(1947), 515–523. MR 9–23. [15, 17, 18, 19]

1464. Walsh, J. L. *On the conformal mapping of multiply connected regions.* Trans. Amer. Math. Soc. 82(1956), 128–146. MR 18-290. [60]
1465. Walsh, J. L. *Interpolation and approximation by rational functions in the complex domain.* Amer. Math. Soc. Colloq. Publ. 20, 2nd Ed. (1956). [59]
1466. Walsh, J. L.; Landau, H. J. *On canonical maps of multiply-connected regions.* Trans. Amer. Math. Soc. 93(1959), 81–96. Zbl 86-281. [59]
1467. Warschawski, S. *Über das Randverhalten der Ableitung der Abbildungsfunktion bei konformer Abbildung.* Math. Z. 35(1932), 321–456. Zbl 4-404. [19, 46]
1468. Warschawski, S. *On the higher derivatives at the boundary in conformal mapping.* Trans. Amer. Math. Soc. 38(1935), 310–340. Zbl 14-267. [59]
1469. Warschawski, S. *Über die Winkelderivierten schlichter Funktionen.* Compositio Math. 4(1937), 346–366. Zbl 16-407. [46]
1470. Warschawski, S. *On Theodorsen's method of conformal mapping of nearly circular regions.* Quart. of Applied Math. 3(1945), 12–28. MR 6-207. [59]
1471. Warschawski, S. *On conformal mapping of nearly circular regions.* Proc. Amer. Math. Soc. 1(1950), 562–574. MR 12-170. [3]
1472. Warschawski, S. *On differentiability of the boundary in conformal mapping.* Proc. Amer. Math. Soc. 12(1961), 614–620. MR 24A-253. [59]
1473. Whittaker, J. M. *A note on the series $\Sigma a_n f(nz)$.* Duke Math. J. 21(1954), 571–573. MR 16-232. [59]
1474. Wiatrowski, P. *Sur le théorème de Koebe pour la classe de fonctions univalentes bornees.* Bull. Soc. Sci. Lettres Łódź 10(1959), No. 14, 13 pp. MR 24A-37. [19]
1475. Wiener, N. *The quadratic variation of a function and its Fourier coefficients.* Mass. J. Math. 3(1924), 72–74. FM 50-203. [59]
1476. Wigner, E. P. *On a class of analytic functions from the quantum theory of collisions.* Ann. of Math. (2) 53(1951), 36–67. MR 12-490. [9, 13]
1477. Wilf, H. S. *Subordinating factor sequences for convex maps of the unit circle.* Proc. Amer. Math. Soc. 12(1961), 689–693. MR 23A-475. [1, 2, 10, 16, 47]
1478. Wilf, H. S. *The radius of univalence of certain entire functions.* Illinois J. Math. 6(1962), 242–244. MR 25-426. [28]
1479. Wing, G. M. *Averages of the coefficients of schlicht functions.*

Proc. Amer. Math. Soc. 2(1951), 658-661 MR 13-123. [17, 20, 42, 54]
1480. Wintner, A. *On the principle of subordination in the theory of analytic differential equations.* Acta Math. 96(1956), 143-156. MR 18-798. [1]
1481. Wirtinger, W. *Note zur Theorie der schlichten Funktionen.* Studia Math. 4(1933), 66-69. Zbl 8-319. [9, 10, 15, 58, 64]
1482. Wittich, H. *Bemerkung zur Modulgrösse eines Schlitz-gebietes.* Arch. Math. 2(1950), 303-305. MR 12-491. [49]
1483. Wittich, H. *Zur konformen Abbildung schlichter Gebiete.* Math. Nachr. 18(1958), 226-234. MR 20-401. [19, 24]
1484. Wittich, H. *Konforme Abbildung schlichter Gebiete.* Ann. Acad. Sci. Fenn. Ser. A. I. No. 249/6 (1958), 12 pp. MR 20-539. [49]
1485. Wolff, J. *L'intégrale d'une fonction holomorphe et à partie réelle positive dans un demi plan est univalente.* C. R. Acad. des Sci. (Paris) 198(1934), 1209-1210. Zbl 8-363. [2, 4]
1486. Wolff, J. *Sur les fonctions holomorphes univalentes.* C. R. Acad. Sci. Paris 213(1941), 158-160. MR 5-36. [22, 49, 62]
1487. Wolff, J. *Inégalités remplies par les fonctions univalentes.* Nederl. Akad. Wetensch., Proc. 44, 956-1963 (1941); Errata p. 1163. MR 7-379. [7, 18, 22, 57]
1488. Wolff, J. *Deux théorèmes dur la dérivée d'une fonction holomorphe univalente et bornée dans un demi-plan au voisinage de la frontière.* Nederl. Akad. Wetensch., Proc. 45(1942), 574-577. MR 5-259. [57]
1489. Wolff, J. *Inégalités remplies par les dérivées des fonctions holomorphes, univalentes et bornées dans un demi-plan.* Comment. Math. Helv. 15(1943), 296-298. MR 5-234. [22, 57]
1490. Wolibner, W. *Sur les coefficients des fonctions analytique univalentes a l'extérieur d'un cercle.* Studia Math. 11(1949), 126-132. MR 12-16. [9, 16, 17, 25, 42]
1491. Wolibner, W. *Sur certaines conditions nécessaires et suffisantes pour qu'une fonction analytique soit univalente.* Coll. Math. 2(1951) (1952), 249-253. MR 14-35. [4, 9, 53]
1492. Wolontis, V. *Porperties of conformal invariants.* Amer. J. Math. 74(1952), 587-606. MR 14-36. [59]
1493. Woolf, W. B. *The boundary behavior of meromorphic functions.* Ann. Acad. Sci. Fenn. Ser. A. I. No. 305(1961), 11 pp. MR 26-748. [59]
1494. Wu, Z. *Some classes of functions of starlikeness.* Acta Math. Sinica 7(1957), 167-182. (Chinese. English summary) MR 21-22.

[1, 6, 10, 15, 23, 49, 63, 64]
1495. Wu, Z. *A class of functions with star shaped images*. Acta Math. Sinica 7(1957), 433–438. (Chinese. English summary) MR 23A-177. [6, 20, 43]
1496. Yamaguchi, K. *On a property of schlicht polynomials*. Sugaku 11(1959/60), 98–99. (Japanese) MR 26-744. [15, 32]
1497. Yan-sin, Cin *On the arguments of the coefficients of the expansion of a univalent function*. Acta Math. Sinica 4(1954), 81–86. (Chinese. Russian summary) MR 17-142. [4, 39]
1498. Yoshikawa, H. *On the conformal mapping of nearly circular domains*. J. Math. Soc. Japan 12(1960), 174–186. MR 23A-329. [6, 49]
1499. Yosida, Tokunosuke *Bemerkungen über die p-wertigen Funktionen*. Proc. Imp. Acad. Tokyo 20(1944), 16–19. MR 7-288. [9, 14, 15, 17, 25, 43, 58]
1500. Yosida, Tokunosuke *Ein Satz uber die p-wertigen Funktionen*. Proc. Imp. Acad. Tokyo 20(1944), 409. MR 7-288. [14, 15]
1501. Yurchenko, A. K.; Dundučenko, L. E. *On the boundary values of functions regular and univalent in the circle $|z| < 1$ belonging to certain special classes*. Ukrain. Mat. Z. 9(1957), 455–460. (Russian) MR 19-846. [6, 23]
1502. Zamorski, J. *Equations satisfied by the extremal schlicht functions with a pole*. Ann. Polon. Math. 3(1956), 41–45. MR 18-568. [9, 17, 24, 25, 29]
1503. Zamorski, J. *Equations satisfied by the extremal star-like functions*. Ann. Polon. Math. 5(1958/59), 285–291. MR 21-928. [6, 9, 17, 25]
1504. Zamorski, J. *Remarks on a class of analytic functions*. Bull. Acad. Polon. Sci. Ser. Sci. Math. Astronom. Phys. 8(1960), 277–380. (Russian summary, unbound inserts) MR 24A-40. [2, 6, 9, 16, 17, 25, 36]
1505. Zamorski, J. *Differential equations for the extremal starlike functions*. Ann. Polon. Math. 7(1960), 279–283. MR 22A(1)-295. [6, 9, 24]
1506. Zamorski, J. *The estimation of the third coefficient of the starlike function with a pole*. Ann. Polon. Math. 8(1960), 185–191. MR 22A(2)-1377. [6, 9, 17, 25, 36]
1507. Zamorski, J. *About the extremal spiral schlicht functions*. Ann. Polon. Math. 9(1960/61), 265–273. MR 24A-37. [6]
1508. Zamorski, J. *Estimation of the coefficients of functions belonging to two classes of k-symmetric schlicht functions*. Prace Mat.

5(1961), 101-105. (Polish. Russian and English summaries) MR 24A-37. [4]

1509. Zamorski, J. *On Bazilevič schlicht functions.* Ann. Polon. Math. 12(1962), 83-90. MR 26-278. [4, 17]

1510. Zasko, V. N. *On certain classes of schlicht functions in domains of connectivity h.* Izv. Vyss. Ucebn. Zaved. Matematika (1960), No. 1(14), 116-122. (Russian) MR 24A-254. [4, 5, 15, 23, 64]

1511. Zawadzki, R. *Sur les fonctions univalentes algebriques.* Bull. Sci. Lett. Lodz. Cl. III Sci. Math. Natur. 8(1957), No. 1, 21 pp. Zbl 91-69. [28]

1512. Zawadzki, R. *Les equations des fonctions extremales dans les familles des fonctions univalentes algebriques non bornees.* Bull. Soc. Sci. Lettres Lodz. Cl. III Sci. Math. Natur. 8, No. 10, 21 pp. (1957). Zbl 91-69. [28]

1513. Zawadzki, R. *Sur les modules des coefficients B_0 et B_1 des fonctions holomorphes univalentes bornees inferieurement.* Bull. Soc. Sci. Lettres Lodz. 12(1961), No. 5. 9 pp. MR 24A-611. [9, 15, 17, 24, 25, 58]

1514. Zawadzki, R. *Sur la fontionelle $\|B_1\| - |B_0\|$ definie dans la famille des fonctions univalentes holomorphes bornees inferieurement.* Bull. Soc. Sci. Lettres Lodz 12(1961), No. 6, 15 pp. MR 24A-612. [9, 17, 24]

1515. Zhang, M. *Ein Uberdeckungssatz fur konvexe Gebiete.* Acad. Sinica Sci. Rec. 5(1952), 17-21. MR 15-413. [10, 16, 19]

1516. Zmorovič, V. A. *On the structure formulas of some classes of univalent functions.* Dokl. Akad. Nauk SSSR (N.S.) 72(1950), 833-836. MR 15-207. [2, 22]

1517. Zmorovič, V. A. *On certain variational problems of the theory of univalent functions.* Ukrain. Mat. Zurnal. 4(1952), 276-298. MR 15-301. [7, 9, 10, 15, 23, 35, 49]

1518. Zmorovič, V. A. *On some classes of analytic functions univalent in a circular ring.* Mat. Sbornik N.S. 32(74) (1953), 623-652. MR 14-1075. [2, 6, 9, 10, 48]

1519. Zmorovič, V. A. *On some special classes of analytic functions univalent in a circle.* Uspehi. Mat. Nauk (N.S.) 9(1954), No. 4(62), 175-182. MR 16-459. [4, 6, 10, 22]

1520. Zmorovič, V. A. *On the theory of special classes of univalent functions.* Dopovidi Akad Nauk Ukrain. RSR (1959), 5-9. (Ukrainian. Russian, English summaries) MR 21-662. [4, 5, 10, 23]

1521. Zmorovič, V. A. *On bounds for the variation of the curvature of*

the image of a plane curve under a univalent conformal mapping. Dopovidi Akad. Nauk Ukrain. RSR (1959), 351-354. (Ukrainian. Russian and English summaries) MR 21-662. [35]
1522. Zmorovič, V. A. *Theory of special classes of univalent functions.* I. Uspehi Mat. Nauk 14(1959), No. 3(87), 137-143. (Russian) MR 22A(1)-645. [4, 5, 16, 41]
1523. Zmorovič, V. A. *Theory of special classes of univalent functions.* II. Uspehi Mat. Nauk 14(1959), No. 4(88), 169-172. (Russian) MR 22A(1)-645. [4, 5, 10]
1524. Zmorovič, V. A. *On structure formulas of certain classes of analytic functions univalent in a circular ring.* Dokl. Akad. Nauk SSSR (N.S.) 86(1952), 465-468. (Russian) MR 5-207. [2, 6, 10, 23]
1525. Zygmund, A. *Sur les fonctions conjuguees.* Fund. Math. 13(1929), 284-303. FM 55-751. [59]
1526. Zygmund, A. *Trigonometrical Series.* Cambridge Univ. Press 1959. MR 21-1208. [59]

SUPPLEMENTARY BIBLIOGRAPHY

1. Aleksandrov, I. A. *A variational method of solving extremal problems in certian classes of analytic functions.* (Russian) Dokl. Akad. Nauk SSSR 151(1963), 999–1002. MR 27–312. [2, 10, 13, 22, 24]
2. Aleksandrov, I. A. *Variational formulae for univalent functions in doubly connected domains.* (Russian) Sibirsk. Mat. Z. 4(1963), 961–976. MR 27–1124. [6, 24, 48]
3. Aleksandrov, I. A. *Extremal properties of the class $S(\omega_0)$.* (Russian) Trudy Tomsk. Gos. Univ. Ser. Meh. -Mat. 169(1963), 24–58. MR 29–694. [29]
4. Aleksandrov, I. A.; Cernikov, V. V. *Extremal properties of univalent star-like mappings.* (Russian) Sibirsk. Mat. Z. 4(1963), 1201–1207. MR 28–258. [6, 9, 17, 25]
5. Alenicyn, Ju. E. *Conformal mappings of multiply connected domains onto multivalent canonical surfaces.* (Russian) Dokl. Akad.

Nauk SSSR 150(1963), 711-714. MR 29-258. [15, 18]
6. Alenicyn, Ju. E. *Conformal mappings of a multiply connected domain onto many-sheeted canonical surfaces.* (Russian) Izv. Akad. Nauk SSSR Ser. Mat. 28(1964), 607-644. MR 29-258. [15, 18, 60]
7. Artemiadis. N. *On a class of holomorphic functions.* Proc. Amer. Math. Soc. 16(1965), 879-885. [13, 16, 17, 42, 51]
8. Bagemihl, F. *Ambiguous points and ambiguous prime ends of functions in simply connected regions, and boundary multiplicities of schlicht functions.* Math. Ann. 156(1964), 198-204. MR 30-51. [62]
9. Basilewitsch, J. *Complément a mes notes "Zum Koeffizientenproblem der schlichten Funktionen" et "Sur les théorèmes de Koebe-Bieberbach".* Rec. math Moscou, N.S. 2, 689-697 u. franz. Zusammenfassung 697-698 (1937). [Russisch] [29]
10. Bazilevič, I. E. *Generalization of an integral formula for a subclass of univalent functions.* (Russian) Mat. Sb. (N.S.) 64(106) (1964), 628-630. MR 29-694. [4, 23]
11. Beckenbach, E. F. *Isoperimetric inequalities for related conformal maps.* Mich. Math. J. 11(1964), 321-326. MR 30-48. [6, 7, 18, 47]
12. Beckenbach, E. F.; Cootz, T. A. *Extensions of the convexity theorem of Study.* Notices A.M.S. vol. 12, no. 3, issue no. 81, April 1965, p. 351 (Abstract). [6]
13. Betz, A. *Konforme Abbildung.* Springer-Verlag, Berlin, 1964. xi + 407 pp. MR 29-258. [59]
14. Bielecki, A. *Quelques résultats récents sur les majorantes dans la théorie des fonctions holomorphes.* Colloq. Math. 11(1963/64), 141-145. MR 29-259. [1]
15. Bielecki, A.; Lewandowski, Z. *Sur certaines majorantes des fonctions holomorphes dans le cercle unité.* Colloq. Math. 9(1962), 299-303. MR 27-510. [1, 6, 10]
16. Bilimovitch, A. *Sur les transformations des fonctions non analytiques.* C. R. Acad. Sci. Paris 247(1958), 1954-1956. [59]
17. Bombieri, E. *Sul problema di Bieberbach per le funzioni univalenti.* Atti Accad. Naz. Lincei Rend. Cl. Sci. Fis. Mat. Natur. (8) 35(1963), 469-471. MR 29-693. [42]
18. Bombieri, E. *On functions which are regular and univalent in a half plane.* Proc. London Math. Soc. (series 3), vol. 14A(1965), 47-50. [57]
19. Cantor, D. G. *Power series with integral coefficients.* Bull. Amer. Math. Soc. 69(1963), 362-366. MR 27-311. [59]
20. Carathéodory, C. *Über die Winkelderivierten von beschränkten analytischen Funktionen.* Sitzungber. Preuss. Akad. (1929), 1-18. [46]

21. Černikov, V. V. *Extremal properties of univalent functions with real coefficients.* I, II. (Russian) Trudy Tomsk. Gos. Univ. Ser. Meh. -Mat. 169(1963), 69-85; ibid 169(1963), 86-95. MR 29-260. [39]
22. Černikov, V. V. *On univalent functions with real coefficients.* (Russian. Armenian summary) Izv. Akad. Nauk Armjan. SSR Ser. Fiz. -Mat. Nauk16(1963), no. 3, 39-47. MR 27-314. [39]
23. Charzyński, Z.; Smialkowna, H. *The general equation of extremal functions with respect to any differentiated functional.* Bull. Soc. Sci. Lettres Lódź 12(1961), no. 13, 16 pp. MR 29-1123. [22]
24. Chen Jian-gong [Ch'en Chien-kung] *Schlicht function theory in China.* Advancement in Math. 1(1955), 748-774. (Chinese) [44]
25. Chu, Soon-hu *Functions with images of bounded measure.* Advancement in Math. 2(1956), 667-674. (Chinese. English summary) [7, 15, 16, 17, 19, 22, 30, 61]
26. Clunie, J.; Keogh, F. R. *Addendum to a note on schlicht functions.* J. London Math. Soc. 39(1964), 63-64. MR 28-797. [6, 10, 22]
27. Curtiss, J. H. *Harmonic interpolation in Fejér points with the Faber polynomials as a basis.* Math. Z. 86(1964), 75-92. MR 29-1128. [53]
28. Curtiss, J. H. *On the coefficients of the Faber polynomials.* Amer. Math. Soc. Notices, Vol. 11, no. 5, issue no. 76, Aug. 1964, p. 557 (Abstract). [53]
29. Das, K. M. *A note on measure of curvature of level curves and orthogonal trajectories of functions finite-valent in the unit circle.* Ganita 13(1962), 107-109. MR 29-46. [31]
30. Distler, R. J. *The domain of univalence of certain classes of meromorphic functions.* Proc. Amer. Math. Soc. (6), 15(1964), 923-928. [4, 9]
31. Dunducenko, L. E.; Kas'janjuk, S. A. *Some properties of analytic functions with positive real part in a circular annulus.* (Ukrainian. Russian and English summaries) Dopovidi Akad. Nauk Ukrain. RSR 1960, 878-883. MR 27-65. [2, 15, 21, 56]
32. Duren, P. L. *On the Marx conjecture for starlike functions.* Amer. Math. Soc. Notices, vol. 10, no. 6, issue no. 70, Oct. 1963. (Abstract) [6, 11, 12, 16, 29]
33. Duren, P. L. *An arclength problem for close-to-convex functions.* J. London Math. Soc. 39(1964), 757-761. MR 30-48. [2, 5, 7, 16, 18, 22, 23, 42]
34. Duren, P. L. *Corrections to the paper Distortion in Certain Conformal Mappings of an Annulus.* Mich. Math. J. 11(1964), 95. [48]
35. Ezrohi, T. G. *On a class of functions univalent in the domain* $1 < |z| < \infty$. (Russian) Izv. Vyss. Ucebn. Zaved. Matematika

1964, no. 1(38), 166–172. MR 28–615. [4, 6, 9, 10, 15, 58]
36. Èzrohi, T. G. *Analytic functions of the generalized Schild classes.* (Russian. Polish and French summaries) Ann. Uiv. Mariae Curie-Sklodowska Sect. A 16(1962), 151–161 (1964). MR 29–260. [19, 32]
37. Èzrohi, T. G. *On certain classes of p-valent functions.* (Russian. Polish and French summaries) Ann. Univ. Mariae Curie-Sklodowska Sect. A 16(1962), 137–149(1964). MR 29–260. [4, 6, 10, 14, 15, 63, 65]
38. Èzrohi, T. G. *On certain classes of p-valent functions.* (Russian) Ukrain. Mat. Ž. 16(1964), 558–568. MR 29–461. [4, 6, 10, 14, 15, 63, 65]
39. Fox, W. C. *A new inequality for the Green's function.* Proc. Amer. Math. Soc. 10(1959), 562–569. [19]
40. Gehring, F. W.; af Hällström, Gunnar *A distortion theorem for functions univalent in an annulus.* Ann. Acad. Sci. Fenn. Ser. AI No. 325 (1963), 16 pp. MR 27–60. [7, 48, 57, 60]
41. Gel'fer, S. A. *Typically real functions.* (Russian) Mat. Sb. (N.S.) 64 (106) (1964), 171–184. MR 29–259. [11, 13, 15, 66]
42. Goodman, A. W. *A note on bivalent functions.* J. London Math. Soc. 39(1964), 215–219. MR 29–45. [57]
43. Goodman, A. W. *A partial differential equation and parallel plane curves.* Amer. Math. Monthly, March 1964, pp. 257–264. [59]
44. Goodman, A. W. *On a characterization of analytic functions.* Amer. Math. Monthly, March 1964, pp. 265–267. [59]
45. Goodman, A. W.; Hummel, J. A. *A note on multivalent starlike functions.* Proc. Amer. Math. Soc. 16, no. 2, April 1965, 284–288. [6, 14]
46. Górski, J. *On the coefficients of univalent functions in the unit circle.* Colloq. Math. 11(1963/64), 181–185. MR 29–259. [17, 54]
47. Gray, E.; Schild, A. *A new proof of a conjecture of Schild.* Proc. Amer. Math. Soc. 16(1965), 76–77. MR 30–417. [6, 12, 39]
48. Hayden, T. L. *Chain sequences and univalence.* Amer. Math. Soc. Notices, vol. 11, no. 1, part 1, issue no. 72, Jan. 1964, p. 83. [6, 11, 28, 43]
49. Hayden, T. L.; Merkes, E. P. *On classes of univalent continued fractions.* Proc. Amer. Math. Soc. 16, no. 2, April 1965, 252–257. [55]
50. Hayman, W. K. *Meromorphic Functions.* Oxford Mathematical Monographs, Oxford Univ. Press, New York (1964). [9, 59]
51. Hayman, W. K. *Coefficient problems for univalent functions and related function classes.* J. London Math. Soc. 40(1965), 385–406. [44]

52. Hille, Einar *Oscillation theorems in the complex domain.* Trans. Amer. Math. Soc. vol. 23(1922), 350–385. [31]
53. Hille, Einar *On the zeroes of Sturm-Liouville functions.* Arkiv för Matematik, Astronomi och Fysik, 16(1922), 1–20. [31]
54. Hille, Einar *Convex distribution of the zeroes of Sturm Liouville functions.* Bull. Amer. Math. Soc. 28(1922), 261–265. [31]
55. Hille, Einar *Non-oscillation theorems.* Trans. Amer. Math. Soc. 64(1948), 234–252. MR 10–376. [31]
56. Ibragimov, G. I. *On the completeness of subsystems of Faber polynomials on curves in the complex plane.* (Russian) Mat. Sb. (N.S.) 65 (107) (1964), 3–17. MR 29–1129. [53]
57. Ionin, V. K. *On a disk embedded in a multiply connected domain.* (Russian) Trudy Tomsk. Gos. Univ. Ser. Meh. -Mat. 169(1963), 8–12. MR 29–460. [59]
58. Janowski, W. *Sur le borne supérieure d'une fonctionelle dans la famille des fonctions univalentes bornées.* Bull. Soc. Sci. Lettres Lódź 13(1962), no. 6, 10 pp. MR 27–314. [17, 22, 24, 30]
59. Janowski, W. *Évaluation de la fonctionelle $|\log(\phi(z_1) - \phi(z_2))/(z_1 - z_2)|$ dans la famille des fonctions univalentes et bornées inférieurement dans le cercle $K(\infty, 1)$.* Colloq. Math. 11(1963/64), 191–194. MR 29–259. [9, 22]
60. Janowski, W. *Sur une évaluation des coefficents des fonctions holomorphes univalentes et bornées inférieurement dans le cercle $K(\infty, 1)$.* Colloq. Math. 11(1963/64), 187–190. MR 29–259. [9, 22]
60a. Jen, Fu-yao *On the functions of Bieberbach and of Lebedev-Milin.* Sci. Record (N.S.) 2(1958), 121–125. MR 27–314. [34]
61. Jenkins, J. A. *On some span theorems.* Ill. J. Math. 7(1963), 104–117. MR 27–59. [59, 60]
62. Jenkins, J. A. *Some area theorems and a special coefficient theorem.* Ill. J. Math. 8(1964), 80–99. MR 28–797. [9, 27, 53]
63. Juve, Yrjö *Über gewisse Verzerrungseigenschaften konformer und quasikonformer Abbildungen.* Ann. Acad. Sci. Fennicae. Ser. A. I. Math. -Phys. no. 174, (1954), 40 pp. MR 16–26. [59]
64. Kazdan, J. L. *A boundary value problem arising in the theory of univalent functions.* J. Math. Mech. 13(1964), 283–303. MR 28–797. [42]
65. Kir'jackiĭ, È. G. [Kirjackis, E.] *Some extremal problems in the classes $K_n(E)$ and $P(E)$.* (Rusian. Lithuanian and German summaries.) Litovsk. Mat. Sb. 3(1963), 83–90, no. 2. MR 29–1123. [10, 15, 17, 38, 64]
66. Kir'jackiĭ, È. G. [Kirjackis, E.] *An extension of some theorems of*

Aksent'ev and Čakalov to the class $K_n(D)$. (Russian. Lithuanian and German summaries) Litovsk. Mat. Sb. 3(1963), 91–96, no. 2. MR 29-1123. [32]

67. Kir'jackiĭ, E. G. [Kirjackis, E.] *Functions with non-zero divided difference.* (Russian. Lithuanian and German summaries) Litovsk. Mat. Sb. 3(1963), 157–168. no. 1. MR 29-260. [10, 15, 64]

68. Kirwan, W. E. *Extremal problems for the class of typically real functions.* Notices A.M.S., vol. 12, no. 5, issue no. 83, August 1965, p. 549 (Abstract). [11, 13, 43]

69. Kiselev, P. Ja. *On the approximation of analytic functions by Faber-Walsh polynomials.* (Russian) Ukrain. Mat. Ž. 15(1963), 193–199. MR 29-264. [53]

70. Kobori, A. *Über die Schlichtheit der Potenzreihen mit beschränktem Realteil.* Tohoku Math. J. 37(1933), 368–373. Zbl 7-351. [22]

71. Kocur, M. F. *Some special classes of analytic functions in a circular annulus.* I, II. (Russian) Izv. Vyss. Ucebn. Zaved. Matematika, no. 4(41) (1964), 79–85. no. 5(42), 30–40. MR 29-1123. [2, 6, 10, 15, 63]

72. Komatu, Y. *On angular derivative.* Kodai Math. Sem. Rep. no. 3, 13(1961), 167–179. [2, 46]

73. Komatu, Y. *On fractional angular derivative.* Kodai Math. Sem. Rep. no. 4, 13(1961), 249–254. [2, 13, 22, 46]

74. Komatu, Y.; Nishimiya, H. *Conformal mapping onto polygons bounded by spiral arcs.* Kodai Math. Sem. Rep. 16(1964), 243–248. MR 30-241. [49]

75. Koseki, K. *Über die Integralgleichungen, welche durch die schlichten Funktionen genügt werden.* Math. J. Okayama Univ. 11 (1963), 119–124. MR 27-314. [23]

76. Koseki, K. *Über die Koeffizienten der schlichten Funktionen.* IV. Math. J. Okayama Univ. 11(1963), 125–157. MR 27-314. [24, 42]

77. Kresnjakova, L. V. *On regular functions with bounded mean modulus.* (Russian) Dokl. Akad. Nauk SSSR 147(1962), 290–293. MR 27-317. [3, 11, 15, 43]

78. Krzyz, J. *Some remarks on close-to-convex functions.* Bull. Acad. Polon. Sci. Sér. Sci. Math. Astronom. Phys. 12(1964), 25–28. MR 28-1000. [2, 5, 6, 15, 29, 64]

79. Krzyz, J. *Some remarks concerning my paper "On univalent functions with two preassigned values."* (Polish and Russian summaries) Ann. Univ. Mariae Curie-Sklodowska Sect. A 16(1962), 129–136 (1964). MR 29-461. [16]

80. Krzyz, J.; Lewandowski, Z. *On the integral of univalent functions.* Bull. Acad. Polon. Sci. Sér. Sci. Math. Astronom. Phys. 11(1963), 447–448. MR 27-734. [4, 16]

81. Krzyz, J.; Reade, M. O. *The radius of univalence of certain analytic functions*. Mich. Math. J. 11(1964), 157-159. MR 29-45. [4, 6]
82. Kubo, T. *On a theorem of analytic functions in an annulus*. Math. Jap. no. 1, 5(1958), 17-20. [59]
83. Kubo, T. *Some theorems on analytic functions in an annulus*. Japanese J. of Math. 29(1959), 43-47. [19]
84. Kühnau, R. *Berechnung einer Extremalfunktion der konformen Abbildung*. Wiss. Z. Martin-Luther-Univ. Halle-Wittenberg. Math. -Nat. Reihe 9(1960), 285-287. MR 27-312. [59]
85. Kuran, Ü. *Two theorems on univalent functions*. J. London Math. Soc. 40(1965), 96-98. MR 30-417. [42]
86. Kuz'mina, G. V. *Covering theorems for functions holomorphic and univalent within a disk*. Soviet Math. 6(1965), 21-25. [6, 17, 19, 39]
87. Lai, Wan-tsai [Lai, Wan-tzei] *On starlike typically-real functions*. Acta Math. Sinica 13(1963), 389-404. (Chinese) Translated as Chinese Math. 4(1964), 423-439. MR 29-693. [6,13]
88. Lavrent'ev, M. A. *Boundary problems in the theory of univalent functions*. Amer. Math. Soc. Transl. (2) 32(1963), 1-35. MR 27-1124. [7]
89. Lewandowski, Z.; Zlotkiewicz, E. *Variational formulae for functions meromorphic and univalent in the unit disc*. Bull. Acad. Polon. Sci. Sér. Sci. Math. Astronom. Phys. 12(1964), 253-254. MR 29-694. [9, 24]
90. Li Ji-Min [Li, Chi-min] *The effect of the amplitudes of coefficients of holomorphic functions in the unit circle upon the schlicht-ness of the functions*. Acta Math. Sinica 14(1964), 367-378. (Chinese) Translated as Chinese Math.-Acta 5(1964), 396-408. MR 30-48. [4]
91. Libera, R. J. *Some radius of convexity problems*. Duke Math. J. 31(1964), 143-158. MR 28-797. [2, 4, 5, 6, 10, 12, 13, 15, 17, 19, 23, 36, 42, 63]
92. Libera, R. J. *Meromorphic close-to-convex functions*. Notices, Amer. Math. Soc. Vol. 11, no. 5, issue no. 76, Aug. 1964, p. 593 (Abstract). Duke Math. J. 32, no. 1 (1965), 121-128. [2, 5, 6, 9, 10, 12, 17, 25, 38, 43]
93. Libera, R. J. *Some classes of regular univalent functions*. Proc. Amer. Math. Soc. no. 4, 16(1965), 755-758. [4, 5, 6, 10]
94. Mac Gregor, T. H. *A class of univalent functions*. Proc. Amer. Math. Soc. 15(1964), 311-317. MR 28-434. [2, 7, 8, 10, 11, 12, 15, 17, 18, 39, 56]
95. Mac Gregor, T. H. *A covering theorem for convex mappings*. Proc. Amer. Math. Soc. 15(1964), 310. MR 28-434. [10, 19]
96. Mac Gregor, T. H. *Length and area estimates for analytic functions*.

Mich. Math. J. 11(1964), 317-320. [7, 18]
97. Marcus, M. *Transformations of domains in the plane and applications in the theory of functions.* Pacific J. Math. 14(1964), 613-626. MR 29-461. [24]
98. McGregor, M. T. *On three classes of univalent functions with real coefficients.* J. London Math. Soc. 39(1964), 43-50. MR 29-45. [6, 10, 19, 39]
99. Merkes, E. P. *Covering theorems for starlike and convex functions.* Notices A.M.S. vol. 12, no. 3, issue no. 81, April 1965, pp. 316-317 (Abstract). [19]
100. Milin, I. M. *The area method in the theory of univalent functions.* Dokl. Akad. Nauk SSSR 154(1964), 264-267; Soviet Math. 5(1964), 78-81. no. 1. MR 28-258. [4, 27, 53]
101. Milin, I. M. *Estimates of coefficients of univalent functions.* Soviet Math. 6(1965), 196-198. [9, 17, 42]
102. Minatani, T. *On a property of some class of univalent function.* Sci. Rep. Kagoshima Univ. No. 11(1962), 7-8. MR 27-59. [6, 19]
103. Mitjuk, I. P. *The symmetrization principle for multiply connected domains.* (Russian) Dokl. Akad. Nauk SSSR 157(1964), 268-270. MR 29-460. [24]
104. Mocanu, P. T. *Sur le rayon d'étoilement et le rayon de convexité de fonctions holomorphes.* Mathematica (Cluj) 4 (27) (1962), 57-63. MR 27-734. [5, 6, 11, 12]
105. Nehari, Z. *Some examples in the theory of univalent functions.* Amer. Math. Monthly, no. 6, 72(1965), 695. (Abstract) [42]
106. Netanyahu, Elisha Extremal problems for schlicht functions in the exterior of the unit circle. Amer. Math. Soc. Notices, no. 2, vol. 11, issue no. 73, Feb. 1964 (Abstract). [6, 9, 16, 17, 24, 25]
107. Nosenko, O. S. *On the domain of values of the derivative of a function convex and univalent outside the unit circle.* (Ukrainian. Russian and English summaries) Dopovidi Akad. Nauk Ukrain. RSR 1963, 1001-1005. MR 29-460. [9, 10]
108. Nosenko, O. S. *On the range of certain functionals.* (Ukrainian. Russian and English summaries) Dopovidi Akad. Nauk Ukrain. RSR 1963, 1303-1307. MR 29-460. [10, 29]
109. Obrock, A. E. *The extremal functions for certain problems concerning schlicht functions.* Bull. Amer. Math. Soc. (whole no. 697), vol. 71, no. 4, July 1965, pp. 626-628. [24]
110. Oikawa, K.; Suita, N. *On parallel slit mappings.* Kodai Math. Sem. Rep. 16(1964), 249-254. MR 30-241. [24]
111. Ozawa, M. *On certain coefficient inequalities of univalent func-

tions. Kodai Math. Sem. Rep. 16(1964), 183-188. MR 29-1124. [17, 54]
112. Pederson, R. N. *On an inequality of Opial, Beesack and Levinson.* Proc. A.M.S., vol 16, no. 1, Feb. 1965, p. 174. [59]
113. Pfaltzgraff, J. A. *Extremal problems and coefficient regions for analytic functions represented by a Stieltjes integral.* Notices, Amer. Math. Soc., vol. 11, no. 3, issue no. 74, April 1964, pp. 363-364. [29, 59]
114. Pommerenke, Ch. *Über die Faberschen Polynome schlichter Funktionen.* Math. Z. 85(1964), 197-208. MR 29-1129. [53]
115. Pommerenke, Ch. *Linear-invariante Familien analytischer Funktionen.* I, II. Math. Ann. 155(1964), 108-154; ibid. 156(1964), 226-262. MR 29-1124. [44]
116. Pommerenke, Ch. *Lacunary power series and univalent functions.* Mich. Math. J. 11(1964), 219-223. MR 29-919. [39]
117. Pommerenke, Ch. *On close-to-convex analytic functions.* Trans. Amer. Math. Soc. 114(1964), 176-186. [3, 4, 5, 6, 9, 10, 12, 15, 17, 22, 30, 38, 41, 58, 62, 64]
118. Poole, J. T. *A note on variational methods.* Proc. Amer. Math. Soc. no. 6, 15(1964), 929-932. [24]
119. Poole, J. T. *Coefficient extremal problems for schlicht functions.* Notices Amer. Math. Soc., vol. 12, no. 5, issue no. 83, August 1965, p. 586. [9, 17, 25, 54]
120. Popov, V. I. *Starlikeness of arcs of level lines under univalent conformal mappings.* Soviet Math., March-April 1064, vol. 5, no. 2, pp. 510-513. [6, 9, 10]
121. Privalova, G. K.; Saran, L. A. *Some extremal problems for a class of functions which is star-shaped in the direction of the real axis.* (Russian) Izv. Vyss. Ucebn. Zaved. Matematika no. 4, 41(1964), 126-130. MR 29-461. [6, 15, 40, 63]
122. Reade, M. O. *On the partial sums of certain Laurent expansions.* Colloq. Math. 11(1963/64), 173-179. MR 29-260. [9, 20, 43]
123. Reade, M. O. *On Ogawa's criterion for univalence.* Publ. Math. Debrecen 11(1964), 39-43. MR 30-417. [4, 5, 10]
124. Red'kov, M. I. *The range of values of the functional $I(f(\omega), \overline{f(\omega)}, f'(\omega), f'(0))$ in the class $S_1[|f(\omega)|]$.* (Russian) Trudy Tomsk. Gos. Univ. Ser. Meh. -Mat. 169(1963), 59-68. MR 29-260. [29]
125. Reza, F. M. *On the schlicht behavior of certain impedance functions.* IRE Trans. Circuit Theory, CT-9(1962), 231-232. [28]
126. Robertson, M. S. *An extremal problem for functions with positive real part.* Notices, A.M.S. Vol. 11, no. 6, issue no. 77, Oct. 1964,

p. 673. [1, 2, 4, 6, 10]
127. Robertson, M. S. *Radii of starlikeness and close-to-convexity.* Proc. Amer. Math. Soc. 16(1965), 847–852. [2, 4, 5, 6, 11, 12]
128. Rogosinski, W. *Über beschränkte Potenzreihen.* I, II. Composito Math. 5(1937), 67–106; (1938), 442–476. Zbl 17-269. [1, 22]
129. Rogosinski, W. *Fourier Series.* Chelsea Publ. Co., New York, 1950, 176 pp. MR 11–347. [59]
130. Rosenblatt, A. *Sur la représentation conforme de domaines limités sur le cercle de rayon un.* Krak. Anz. 1917. [15, 22, 61]
131. Royster, W. C. *A Poisson integral formula for the ellipse and some applications.* Proc. Amer. Math. Soc. no. 4, 15(1964), 661–670. [6, 7, 18, 57]
132. Rubinstein, Z. *Extension of two results in the theory of rational functions to certain classes of analytic functions.* Amer. Math. Soc. Notices, vol. 11, no. 2, issue no. 73, Feb. 1964 (Abstract). [59]
133. Rubinstein, Z. *Some inequalities for polynomials and their zeroes.* Proc. Amer. Math. Soc., no. 1, 16(1965), 72–75. [4, 6]
134. Sakaguchi, K. *Meromorphic functions convex of order at most p in one direction.* J. Nara Gakugei Univ. Natur. Sci. 19(1961/62), 109–112. MR 27–313. [10, 14, 41]
135. Sakaguchi, K. *A representation theorem for a certain class of mregular functions.* J. Math. Soc. Japan 15(1963), 202–209. MR 27–313. [4, 5, 6, 17, 23]
136. Sakaguchi, K.; Watanabe, S. *Meromorphic functions multivalently starlike in a wide sense.* J. Nara Gakugei Univ. Natur. Sci. 11(1963), 3–7. MR 27–314. [14, 23]
137. Sato, T. *Ein Satz über Schlichtheit von einer meromorphen Funktion.* Proc. Imp. Acad. Jap. 11(1935), 212–213. Zbl. 12-171. [10]
138. Schild, A. *On starlike functions of order a.* Amer. J. Math. 87(1965), 65–70. [6, 11, 15, 17]
139. Schneider, W. J. *Some notes on positive functions.* Notices, A.M.S., vol. 12, no. 5, issue no. 83, August 1965, p. 586 (Abstract). [2]
140. Sone, N. *Univalent functions and non-convex domains.* J. Math. Soc. Japan 15(1963), 191–201. MR 27–314. [4, 5, 16, 17, 38]
141. Stepanova, O. V. *Certain properties of level curves for univalent conformal mappings.* (Russian) Mat. Sb. (N.S.) 61 (103) (1963), 350–361. MR 27–313. [6, 7, 10, 24, 25]
142. Suetin, P. K. *The basic properties of Faber polynomials.* (Russian) Uspehi Mat. Nauk 19(1964), no. 4 (118), 125–154. MR 29–1129. [53]
143. Suffridge, T. J. *Convolutions of convex functions.* Notices, A.M.S., vol. 12, no. 2, issue no. 80, February 1965, p. 244 (Abstract). [4, 5, 10, 47]

144. Suffridge, T. J. *Convolutions of convex functions.* Notices, A.M.S., vol. 12, no. 5, issue no. 83, Augus 1965, pp. 579–580 (Abstract). [10, 47]
145. Suita, N. *On the coefficient theorem of Jenkins.* (Japanese) Sûgaku 14(1962/63), 129–137. MR 29–45. [24]
146. Suita, N. *A distortion theorem of univalent functions related to symmetric three points.* Kodai Math. Sem. Rep. 14(1962), 26–30. MR 27–59. [9, 15, 17, 25, 58]
147. Suvorov, G. D. *Univalent mappings of plane domains and sets of prime ends of a domain of generalized measure zero.* Dokl. Akad. Nauk SSSR 152(1963), 296–298. MR 27–737. [49]
148. Taam, Choy-tak *Oscillation theorems.* Amer. J. Math. vol. 74(1952), 317–324. MR 14–50. [31]
149. Taam, Choy-tak *On the complex zeroes of functions of Sturm-Liouville type.* Pacific J. Mathematics 3(1953), 837–843. MR 15–625. [31]
150. Taam, Choy-tak *Non-oscillation and comparison theorems of linear differential equations with complex-valued coefficients.* Portugaliae Mathematica 12(1953), 57–72. MR 14–873. [31]
151. Tamrazov, P. M. *On univalent conformal mapping of doubly connected regions.* (Ukrainian. Russian and English summaries) Dopovidi Akad. Nauk Ukrain. RSR 1962, 1142–1145. MR 27–59. [60]
152. Tamrazov, P. M. *Some estimates in the theory of univalent conformal mappings of doubly connected domains.* (Ukrainian. Russian and English summaries) Dopovidi Akad. Nauk Ukrain. RSR 1963, 1160–1163. MR 29–1125. [60]
153. Walsh, J. L. *On the location of the roots of the derivative of a polynomial.* Comptes Rendus du Congrès International des Mathématiciens. Strasbourg, 22–30 Septembre 1920. [59]
154. Walsh, J. L. *Note on the curvature of orthogonal trajectories of the level curves of Green's function III.* Bull. Amer. Math. Soc. no. 2, 46(1940), 101–108. MR 1–210. [59]
155. Walsh, J. L. *On the location of the zeroes of the derivatives of a polynomial symmetric in the origin.* Bull. Amer. Math. Soc. No. 10, 54(1948), 942–945. MR 10–250. [59]
156. Walsh, J. L. *Note on the shape of level curves of Green's function.* Amer. Math. Monthly, Dec. 1953, 671–674. MR 15–424. [59]
157. Walsh, J. L. *Note on the shape of level curves of Green's function.* Duke Math. J. no. 4, 20(1953), 611–616. MR 15–310. [59]
158. Walsh, J. L. *On the convexity of the ovals of lemniscates.* Studies in mathematical analysis and related topics, pp. 419–423. Stanford

Univ. Press, Stanford, Calif., 1962. MR 27-317. [10, 49]
159. Waszkiewicz, J. *Sur certains problèmes extrêmaux dans la famille des fonctions univalentes bornées inférieurement dans le cercle $K(\infty, 1)$.* Bull. Soc. Sci. Lettres Łódź 14(1963), no. 3, 30 pp. MR 29-694. [9, 15, 58]
160. Whiteley, J. N. *Notes on a theorem on convolutions.* Proc. Amer. Math. Soc. no. 1, 16(1965), 1-7. [47]
161. Widder, D. V. *The Laplace Transform.* Princeton Univ. Press, London, 1946, 406 pp. MR 3-232. [59]
162. Wigner, E. P. *On the connection between the distribution of poles and residues for an R function and its invariant derivative.* Ann. of Math. (2) 55(1952), 7-18. MR 13-733. [46]
163. Wilf, H. S. *Mathematics for the Physical Sciences.* John Wiley and Sons, Inc., New York, 1962. 284 pp. [44]
164. Wilf, H. S. *Calculations relating to a conjecture of Pólya and Schoenberg.* Math. Comp. 17(1963), 200-201. MR 28-798. [10, 47]
165. Wintner, A. *A criterion for non-oscillatory differential equations.* Quarterly of Applied Mathematics, 7(1949), 115-117. MR 10-456. [31]
166. Wintner, A. *On the non-existence of conjugate points.* Amer. J. Math. 73(1951), 368-380. MR 13-37. [31]
167. Wolff, J. *Sur une généralisation d'une théorème de Schwarz.* C. R. Paris 183(1926), 500-502. [46]
168. Złotkiewicz, E. *On a variational formula for starlike functions.* (Polish and Russian summaries) Ann. Univ. Mariae Curie-Sklodowska Sect. A 15(1961), 111-113. MR 28-434. [6, 24]

MATHEMATICAL JOURNALS

Following is a list of twenty-two mathematical journals and papers on schlicht functions which they contain. Numbers within brackets refer to the main bibliography; numbers preceded by the letter "A" indicate references to the Supplementary Bibliography.

Acta Mathematica, vol. 1–113 (1882–1965) [18, 293, 508, 632, 648, 692, 840, 895, 945, 1188, 1441, 1480].

American Journal of Mathematics, vol. 22–86 (1900–1964) [595, 598, 599, 966, 970, 975, 980, 1085, 1142, 1146, 1250, 1253, 1254, 1272, 1330, 1492, A148, A166].

Annals of Mathematics, vol. 8–81 (1906–1965, series 2) [52, 59, 215, 224, 360, 364, 452, 456, 457, 545, 593, 603, 609, 614, 615, 616, 621, 633, 634, 944, 1060, 1144, 1147, 1149, 1150, 1153, 1156, 1207, 1262, 1327, 1373, 1476, A162].

Archive for Rational Mechanics and Analysis, vol. 1–19

(1957-1965) [205, 266, 344, 346].

Bulletin of the American Mathematical Society, vol. 7-71 (1900-1965) [61, 84, 88, 166, 191, 277, 278, 333, 347, 429, 446, 536, 628, 635, 750, 869, 877, 879, 971, 977, 1052, 1109, 1140, 1141, 1143, 1145, 1148, 1152, 1155, 1168, 1171, 1204, 1255, 1257, 1279, 1331, 1361, 1461, 1463, A19, A54, A154, A155].

Canadian Journal of Mathematics, vol. 1-17 (1949-1965) [91, 169, 172, 444, 626, 1158, 1161, 1214].

Duke Mathematical Journal, vol. 1-32 (1935-1965) [11, 90, 92, 103, 104, 105, 108, 109, 110, 111, 161, 219, 248, 301, 320, 329, 336, 358, 380, 529, 531, 604, 755, 875, 904, 911, 964, 968, 983, 1034, 1107, 1120, 1151, 1154, 1159, 1170, 1209, 1210, 1212, 1223, 1233, 1234, 1236, 1243, 1249, 1274, 1362, 1433, 1473, A91, A92, A157].

Illinois Journal of Mathematics, vol. 1-9 (1957-1965) [620, 622, 753, 1478, A48, A61].

Journal de Mathematiques Pures et Appliquées, vol. 2-44 (1923-1965) [366, 941, 1443].

Journal für die reine und angewandte Mathematik (formerly Crelle Journal), vol. 122-217 (1900-1965) [581, 1038, 1416, 1424, 1425, 1426].

Journal of Mathematical Analysis and Applications, vol. 1-11 (1960-1965) [449, 1281].

Journal of the London Mathematical Society, vol. 1-40 (1926-1965) [85, 113, 214, 216, 217, 228, 295, 304, 316, 325, 326, 337, 340, 489, 519, 521, 522, 594, 608, 612, 662, 663, 664, 754, 817, 865, 892, 974, 1036, 1041, 1073, 1089, 1102, 1103, 1186, 1323, 1335, A26, A33, A42, A51, A85, A98].

Mathematische Annalen, vol. 53-158 (1900-1965) [94, 116, 123, 173, 174, 185, 187, 189, 197, 223, 294, 483, 487, 541, 690, 882, 909, 1037, 1040, 1047, 1069, 1080, 1088, 1278, 1333, 1360, 1444].

Mathematische Zeitschrift, vol. 1-88 (1930-1965) [20, 62, 118, 162, 163, 274, 280, 300, 302, 303, 324, 341, 342, 459, 485, 486, 523, 528, 532, 549, 666, 669, 693, 694, 804, 809, 810, 812, 844, 845, 881, 893, 916, 919, 948, 1043, 1071, 1134, 1136, 1177, 1180, 1181, 1340, 1353, 1354, 1359, 1417, 1436, 1438, 1439, 1467, A114].

Michigan Journal of Mathematics, vol. 1-12 (1952-1965) [233, 265, 629, 656, 657, 859, 872, 888, 914, 994, 1066, 1067, 1068, 1070, 1072, 1074, 1106, 1112, 1167, A11, A34, A81, A96, A116, A126].

Pacific Journal of Mathematics, vol. 1-14 (1951-1964) [279, 345, 551, 752, 876, 915, 1063, 1076, 1104, 1117, 1122, 1162, 1163, 1269, 1300, 1307, A149].

Proceedings of the American Mathematical Society, vol. 1-16

(1950-1965) [157, 160, 171, 181, 264, 271, 288, 335, 432, 433, 434, 437, 443, 492, 509, 530, 552, 553, 580, 596, 597, 601, 602, 607, 815, 873, 889, 890, 891, 913, 917, 918, 972, 981, 984, 985, 986, 1118, 1119, 1157, 1164, 1175, 1213, 1265, 1266, 1267, 1276, 1398, 1403, 1414, 1423, 1427, 1435, 1471, 1472, 1477, 1479, A30, A118, A131].

Proceedings of the London Mathematical Society, vol. 1-54 series 2 (1903-1951); vol. 1-15, series 3 (1951-1965) [334, 338, 339, 490, 495, 497, 504, 512, 665, 846, 862, 967, 1075, 1187, 1247, 1248, 1309, 1325, 1399].

Quarterly Journal of Mathematics, vol. 33-50, series 1 (1900-1927); vol. 1-20, Oxford series 1 (1930-1949); vol. 1-16, Oxford series 2 (1950-1965) [195, 196, 317, 494, 498, 863].

Soviet Mathematics-Doklady, vol. 1-6 (1960-1965) [81, 179, 831, 1114, A120].

Transactions of the American Mathematical Society, vol. 1-118 (1900-1965) [10, 89, 93, 234, 249, 299, 348, 357, 361, 428, 431, 435, 436, 438, 439, 440, 441, 445, 451, 579, 592, 605, 606, 610, 611, 613, 617, 624, 625, 630, 814, 878, 887, 973, 979, 1013, 1160, 1165, 1166, 1172, 1174, 1206, 1215, 1273, 1280, 1282, 1321, 1322, 1332, 1415, 1464, 1466, 1468, A52, A55].

Translations of the American Mathematical Society, vol. 1-31. series 2 (1955-1963) [371, 373, 375, 376, 423, 426, 826, 828, A88].

EXPOSITORY PAPERS

Following is a list of publications which are devoted wholly or in part to a general survey of the development of the theory of univalent and p-valent functions. Numbers in brackets refer to the main bibliography. Numbers preceded by the letter "A" refer to the Supplementary Bibliography. An asterisk indicates a particularly comprehensive source.

[239 (Chapter 8); 1400 (pp. 198–212); 864 (pp. 163–185, 205–226); 125 (pp. 71–83); 1064 (vol. 2, Chapter 2); 403*; 1331*; 1240*; 104; 978 (Chapter 5); 1241; 126 (Chapter 5); 514; 515*; 618*; 537 (vol. 2, Chapters 17, 18); 1305; 411; 886; 939; A24; A51; A115; A163 (Chapter 6)]

TOPIC REFERENCES

Following is a list of sixty-eight topics dealing directly with or closely related to the theory of univalent and p-valent functions. Following each topic is a list of references which contain results pertaining to that topic. Numbers within brackets refer to the main bibliography; numbers preceded by the letter "A" refer to the Supplementary Bibliography. The topics are numbered T1, T2, T3, ... , T68 for purposes of reference.

Topic T15 (Distortion Theorems) contains so many references that a breakdown of T15 is given by Topics T56, T58, T61, T63, T64, T65, T66, T67 and T68. Similarly, a breakdown of Topic T17 (Coefficient Bounds) is given by T21, T25, T30, T36, T38, T50, T51, T52, and T54. The references given under Topic T44 (Survey Articles) should be consulted in connection with any of the other topics.

T1. *The Principle of Subordination.*

[1, 35, 43, 87, 100, 102, 111, 130, 132, 134, 139, 140, 212, 231, 320, 367, 382, 392, 414, 417, 420, 443, 446, 503, 598, 605, 611, 677, 685, 743, 802, 836, 838, 855, 856, 863, 870, 887, 910, 953, 956, 960, 964, 967, 1068, 1071, 1072, 1088, 1117, 1144, 1163, 1165, 1174, 1179, 1181, 1185, 1186, 1187, 1244, 1278, 1290, 1291, 1293, 1319, 1341, 1366, 1382, 1477, 1480, 1494, A14, A15, A126, A128]

T2. *Functions of Positive Real Part.*

[4, 34, 46, 60, 108, 186, 196, 226, 233, 268, 272, 315, 350, 380, 412, 426, 456, 519, 526, 528, 530, 577, 703, 717, 718, 719, 720, 721, 887, 889, 890, 891, 913, 968, 974, 985, 988, 996, 1000, 1041, 1099, 1105, 1140, 1142, 1144, 1147, 1151, 1154, 1166, 1167, 1177, 1178, 1180, 1181, 1187, 1213, 1278, 1320, 1340, 1361, 1401, 1404, 1417, 1477, 1485, 1504, 1516, 1518, 1524, A1, A31, A33, A71, A72, A73, A78, A91, A92, A94, A126, A127, A139]

T3. *Integral Means.*

[79, 86, 88, 135, 139, 145, 147, 148, 152, 181, 184, 196, 230, 294, 300, 305, 328, 337, 338, 339, 378, 407, 409, 410, 413, 417, 423, 489, 490, 495, 498, 511, 519, 747, 748, 749, 792, 831, 832, 833, 834, 835, 846, 862, 894, 901, 952, 953, 954, 1069, 1071, 1072, 1082, 1117, 1118, 1143, 1144, 1146, 1150, 1151, 1171, 1186, 1187, 1278, 1280, 1300, 1322, 1323, 1325, 1335, 1338, 1356, 1391, 1425, 1471, A77, A117]

T4. *Sufficient Conditions for Univalency (p-valency).*

[22, 23, 24, 25, 26, 52, 77, 108, 121, 151, 171, 172, 175, 176, 179, 180, 182, 183, 205, 211, 217, 227, 241, 256, 259, 289, 295, 296, 297, 327, 335, 336, 350, 354, 421, 424, 428, 437, 485, 492, 493, 500, 530, 542, 554, 652, 655, 656, 668, 672, 674, 687, 719, 743, 746, 848, 850, 859, 874, 876, 877, 883, 887, 890, 891, 911, 912, 917, 918, 925, 940, 963, 971, 981, 982, 984, 997, 1001, 1006, 1007, 1010, 1011, 1019, 1020, 1021, 1023, 1025, 1027, 1029, 1036, 1058, 1063, 1071, 1072, 1090, 1094, 1095, 1105, 1106, 1107, 1108, 1110, 1111, 1112, 1124, 1144, 1145, 1156, 1160, 1162, 1166, 1176, 1218, 1220, 1222, 1231, 1266, 1267, 1273, 1286, 1301, 1310, 1317, 1337, 1338, 1362, 1365, 1367, 1371, 1372, 1398, 1399, 1412, 1428, 1429, 1432, 1434, 1435, 1485, 1491, 1497, 1508, 1509, 1510, 1519, 1520, 1522, 1523, A10,

A30, A35, A37, A38, A80, A81, A90, A91, A93, A100, A117, A123, A126, A127, A133, A135, A140, A143]

T5. *Close-to Convex Functions* (and generalizations).

[129, 282, 656, 687, 765, 766, 851, 854, 859, 887, 896, 897, 898, 922, 1005, 1007, 1071, 1072, 1075, 1105, 1106, 1107, 1108, 1110, 1112, 1131, 1132, 1218, 1352, 1435, 1510, 1520, 1522, 1523, A33, A78, A91, A92, A93, A104, A117, A123, A127, A135, A140, A143]

T6. *Starlike Functions* (and generalizations).

[4, 23, 28, 30, 31, 32, 34, 52, 55, 60, 77, 80, 81, 82, 87, 90, 92, 97, 98, 99, 108, 109, 111, 119, 128, 132, 134, 137, 142, 156, 157, 216, 217, 235, 242, 243, 247, 255, 258, 260, 261, 274, 282, 287, 301, 336, 352, 353, 381, 385, 400, 417, 422, 426, 430, 431, 433, 434, 437, 442, 445, 481, 500, 519, 551, 552, 553, 574, 642, 653, 659, 661, 663, 665, 673, 674, 675, 676, 677, 678, 716, 719, 722, 723, 733, 734, 846, 853, 856, 888, 889, 890, 891, 896, 897, 898, 908, 909, 917, 918, 932, 934, 952, 953, 956, 985, 986, 990, 996, 998, 999, 1000, 1001, 1003, 1004, 1005, 1021, 1022, 1038, 1042, 1063, 1071, 1072, 1073, 1075, 1076, 1079, 1083, 1090, 1091, 1094, 1095, 1105, 1106, 1107, 1108, 1124, 1142, 1144, 1153, 1154, 1156, 1159, 1160, 1163, 1164, 1166, 1167, 1175, 1187, 1209, 1212, 1214, 1215, 1217, 1218, 1219, 1221, 1231, 1265, 1266, 1267, 1278, 1317, 1339, 1340, 1352, 1360, 1365, 1366, 1367, 1370, 1374, 1398, 1414, 1426, 1427, 1429, 1437, 1438, 1440, 1449, 1494, 1495, 1498, 1501, 1503, 1504, 1505, 1506, 1507, 1518, 1519, 1524, A2, A4, A11, A12, A15, A26, A32, A35, A37, A38, A45, A47, A48, A71, A78, A81, A86, A87, A91, A92, A93, A98, A102, A104, A106 A117, A120, A121, A126, A127, A131, A133, A135, A138, A141, A168]

T7. *Relations Involving Arc-Length.*

[2, 80, 81, 82, 86, 88, 145, 152, 184, 294, 300, 341, 346, 350, 366, 444, 519, 545, 610, 612, 613, 662, 663, 664, 665, 700, 717, 718, 720, 767, 796, 820, 887, 902, 909, 932, 1062, 1071, 1072, 1073, 1075, 1130, 1143, 1146, 1230, 1253, 1347, 1410, 1435, 1487, 1517, A11, A25, A33, A40, A88, A94, A96, A131, A141]

T8. *Bounds for* $\|a_{n+1}| - |a_n\|$.

[56, 146, 149, 150, 241, 521, 1075, 1368, A94]

T9. *Meromorphic Univalent* (*p*-valent) *Functions*.

[3, 5, 36, 73, 74, 75, 76, 81, 82, 116, 117, 141, 155, 176, 183, 209, 215, 216, 234, 235, 264, 277, 280, 285, 307, 335, 363, 364, 368, 372, 381, 383, 387, 390, 393, 395, 397, 400, 404, 407, 408, 416, 418, 426, 434, 436, 452, 455, 481, 482, 485, 492, 507, 508, 544, 579; 595, 601, 619, 623, 624, 625, 629, 682, 697, 705, 709, 712, 715, 721, 722, 731, 732, 734, 741, 782, 791, 804, 807, 859, 866, 881, 888, 921, 924, 925, 951, 971, 979, 983, 985, 986, 987, 996, 1009, 1020, 1071, 1072, 1076, 1078, 1083, 1113, 1125, 1128, 1164, 1166, 1167, 1176, 1179, 1184, 1189, 1214, 1215, 1219, 1228, 1246, 1249, 1250, 1253, 1254, 1255, 1259, 1283, 1302, 1303, 1304, 1306, 1313, 1316, 1332, 1333, 1390, 1392, 1397, 1414, 1416, 1434, 1451, 1452, 1459, 1476, 1481, 1490, 1491, 1499, 1502, 1503, 1504, 1505, 1506, 1513, 1514, 1517, 1518, A4, A30, A35, A50, A59, A60, A62, A89, A92, A101, A106, A107, A117, A119, A120, A122, A146, A159]

T10. *Convex Functions* (and generalizations)

[2, 4, 23, 28, 29, 31, 34, 52, 63, 80, 81, 82, 87, 90, 108, 109, 111, 118, 132, 134, 142, 189, 217, 243, 247, 255, 261, 262, 274, 295, 296, 301, 335, 336, 338, 339, 349, 355, 381, 383, 412, 430, 431, 433, 437, 443, 445, 453, 454, 492, 546, 550, 552, 564, 570, 636, 642, 659, 660, 661, 663, 673, 674, 675, 676, 677, 703, 716, 719, 722, 723, 725, 731, 751, 761, 762, 767, 846, 859, 870, 881, 884, 887, 888, 889, 890, 891, 896, 897, 898, 909, 941, 952, 953, 956, 963, 988, 997, 1005, 1032, 1033, 1037, 1042, 1063, 1070, 1071, 1073, 1083, 1088, 1090, 1091, 1095, 1101, 1107, 1129, 1142, 1144, 1160, 1163, 1166, 1187, 1209, 1217, 1219, 1222, 1229, 1231, 1265, 1267, 1278, 1289, 1292, 1299, 1317, 1340, 1341, 1365, 1366, 1370, 1412, 1429, 1434, 1437, 1438, 1459, 1477, 1481, 1494, 1515, 1517, 1518, 1519, 1520, 1523, 1524, A1, A15, A26, A35, A37, A38, A65, A67, A71, A91, A92, A93, A94, A95, A98, A107, A108, A117, A120, A123, A126, A134, A137, A141, A143, A144, A158, A164]

T11. *Radii of Starlikeness*.

[28, 173, 174, 242, 243, 368, 369, 381, 396, 484, 500, 502, 673, 675, 676, 677, 682, 747, 748, 749, 854, 889, 890, 891, 897, 908, 909, 914, 915, 928, 934, 975, 999, 1042, 1090, 1217, 1219, 1267, 1367, 1370, 1398, A32, A41, A48, A68, A77, A94, A104, A127, A138]

T12. *Radii of Convexity*.

[28, 29, 31, 173, 174, 243, 368, 502, 673, 676, 677, 747, 765, 848, 849, 850, 887, 889, 890, 891, 897, 914, 915, 928, 952, 970, 975, 989, 997, 1006, 1090, 1092, 1131, 1166, 1167, 1217, 1219, 1229, 1265, 1266, 1370, 1375, 1428, A32, A47, A91, A92, A94, A104, A117, A127]

T13. *Typically-Real Functions*

[4, 5, 46, 55, 57, 58, 60, 109, 111, 131, 167, 168, 169, 170, 335, 372, 374, 375, 413, 416, 422, 426, 427, 432, 436, 445, 580, 607, 626, 629, 715, 751, 804, 858, 901, 913, 953, 995, 1022, 1031, 1053, 1120, 1125, 1140, 1142, 1144, 1147, 1149, 1152, 1155, 1156, 1157, 1158, 1159, 1160, 1162, 1177, 1180, 1212, 1337, 1427, 1431, 1476, A1, A7, A41, A68, A73, A87, A91]

T14. *p-valent Functions*

[3, 36, 45, 48, 89, 90, 92, 135, 136, 140, 141, 144, 146, 157, 197, 208, 242, 246, 255, 258, 261, 282, 288, 305, 325, 330, 336, 348, 353, 356, 367, 368, 373, 376, 388, 389, 396, 406, 410, 428, 430, 431, 438, 441, 445, 505, 507, 508, 512, 574, 575, 576, 614, 644, 668, 678, 679, 680, 681, 682, 683, 684, 685, 698, 737, 759, 760, 816, 877, 879, 923, 925, 941, 942, 955, 1001, 1003, 1004, 1006, 1019, 1020, 1021, 1022, 1023, 1024, 1069, 1101, 1148, 1150, 1151, 1153, 1154, 1158, 1159, 1210, 1215, 1219, 1220, 1221, 1231, 1250, 1371, 1372, 1427, 1428, 1429, 1434, 1435, 1499, 1500, A37, A38, A45, A134, A136]

T15. *Distortion Theorems*

[4, 5, 39, 40, 41, 48, 49, 61, 64, 65, 68, 70, 71, 72, 76, 92, 118, 121, 131, 141, 142, 148, 155, 156, 157, 170, 191, 197, 209, 231, 234, 242, 246, 247, 255, 256, 257, 258, 259, 260, 261, 262, 263, 281, 282, 306, 325, 353, 355, 356, 366, 367, 369, 375, 376, 381, 383, 385, 386, 387, 388, 389, 390, 392, 394, 395, 396, 397, 399, 400, 404, 407, 408, 412, 413, 415, 416, 418, 426, 427, 431,, 434, 435, 436, 453, 454, 455, 460, 461, 474, 481, 482, 484, 500, 505, 507, 519, 532, 544, 557, 558, 559, 560, 570, 573, 577, 585, 590, 591, 595, 599, 605, 621, 626, 627, 641, 644, 659, 661, 677, 679, 680, 682, 683, 684, 687, 697, 698, 709, 711, 712, 716, 723, 724, 725, 733, 737, 742, 749, 758, 759, 760, 761, 762, 763, 764, 766, 788, 793, 804, 808, 826, 828, 833, 834, 835, 846, 848, 850, 852, 853, 858, 867, 870, 880, 881, 882, 884, 887, 892, 896, 897, 906, 908, 909, 916, 921, 932, 940, 943, 951, 953, 959, 963, 965, 989, 1006, 1008, 1030, 1032, 1043, 1048, 1051, 1053, 1057, 1071,

1072, 1073, 1076, 1083, 1095, 1097, 1098, 1101, 1102, 1103, 1108, 1120, 1125, 1131, 1143, 1144, 1146, 1147, 1149, 1150, 1151, 1154, 1172, 1173, 1175, 1189, 1191, 1219, 1222, 1244, 1255, 1266, 1267, 1280, 1288, 1299, 1308, 1313, 1316, 1320, 1327, 1332, 1333, 1340, 1347, 1348, 1374, 1375, 1390, 1392, 1395, 1396, 1397, 1416, 1425, 1426, 1427, 1435, 1439, 1440, 1456, 1463, 1481, 1494, 1496, 1499, 1500, 1510, 1513, 1517, A5, A6, A25, A31, A35, A37, A38, A41, A65, A67, A71, A78, A79, A94, A117, A121, A130, A138, A146, A159]

T16. *Conjectures, Problems, Open Questions.*

[3, 24, 110, 138, 146, 151, 152, 161, 172, 184, 278, 301, 304, 333, 339, 346, 360, 364, 392, 428, 431, 432, 438, 441, 445, 467, 512, 516, 518, 519, 523, 595, 598, 614, 629, 639, 640, 642, 643, 645, 673, 674, 682, 684, 714, 715, 736, 749, 767, 781, 791, 812, 834, 835, 849, 851, 855, 863, 869, 903, 906, 934, 946, 952, 984, 986, 1035, 1055, 1063, 1131, 1141, 1144, 1159, 1167, 1169, 1186, 1210, 1214, 1306, 1312, 1325, 1326, 1327, 1329, 1332, 1346, 1390, 1410, 1412, 1425, 1426, 1433, 1477, 1490, 1504, 1515, 1522, A7, A25, A32, A33, A80, A106, A140]

T17. *Coefficient Bounds*

[3, 5, 32, 37, 45, 48, 55, 56, 67, 69, 70, 72, 75, 92, 103, 107, 117, 123, 135, 138, 147, 150, 170, 184, 194, 205, 206, 207, 208, 215, 216, 217, 226, 235, 241, 242, 255, 256, 257, 259, 261, 263, 264, 269, 271, 277, 294, 304, 305, 328, 329, 349, 355, 363, 364, 365, 370, 372, 374, 375, 376, 379, 388, 389, 393, 394, 396, 407, 408, 409, 412, 413, 421, 428, 430, 431, 432, 435, 436, 445, 493, 505, 509, 511, 512, 519, 521, 552, 554, 562, 566, 567, 568, 576, 577, 580, 583, 586, 587, 599, 601, 605, 610, 619, 623, 624, 625, 629, 638, 639, 640, 645, 651, 654, 660, 661, 677, 679, 682, 684, 685, 686, 687, 705, 709, 714, 715, 727, 728, 737, 740, 741, 742, 749, 755, 788, 789, 790, 793, 795, 804, 812, 833, 834, 835, 842, 843, 844, 846, 859, 863, 865, 866, 868, 870, 882, 883, 884, 887, 888, 892, 898, 901, 906, 907, 924, 925, 957, 958, 983, 984, 985, 986, 987, 989, 990, 995, 996, 1009, 1018, 1019, 1020, 1022, 1023, 1024, 1030, 1038, 1069, 1071, 1072, 1073, 1075, 1076, 1081, 1097, 1098, 1101, 1105, 1106, 1107, 1108, 1129, 1131, 1132, 1140, 1141, 1142, 1143, 1144, 1146, 1147, 1148, 1149, 1150, 1151, 1152, 1154, 1155, 1158, 1159, 1186, 1187, 1189, 1190, 1192, 1193, 1194, 1195, 1196, 1197, 1198, 1199, 1200, 1201, 1203, 1212, 1213, 1214,

1215, 1219, 1222, 1233, 1234, 1244, 1246, 1249, 1250, 1259, 1267, 1284, 1293, 1296, 1308, 1309, 1317, 1322, 1325, 1327, 1329, 1332, 1333, 1338, 1340, 1349, 1368, 1374, 1375, 1377, 1381, 1382, 1392, 1394, 1396, 1412, 1425, 1426, 1427, 1430, 1431, 1433, 1435, 1448, 1449, 1450, 1451, 1452, 1453, 1454, 1455, 1463, 1479, 1490, 1499, 1502, 1503, 1504, 1506, 1509, 1513, 1514, A4, A7, A25, A46, A58, A65, A86, A92, A94, A101, A106, A111, A117, A119, A135, A138, A140, A146]

T18. *Relations Involving Area.*
[2, 75, 97, 137, 152, 184, 217, 231, 236, 237, 250, 277, 285, 350, 417, 429, 435, 444, 452, 496, 687, 717, 718, 720, 758, 767, 887, 1073, 1076, 1117, 1189, 1280, 1324, 1325, 1463, 1487, A5, A6, A11, A33, A94, A96, A131]

T19. *Covering Theorems.*
[3, 10, 20, 39, 53, 73, 74, 75, 79, 97, 98, 99, 100, 101, 110, 116, 120, 122, 124, 137, 139, 140, 141, 158, 184, 193, 218, 234, 270, 271, 307, 309, 310, 348, 360, 384, 391, 393, 404, 406, 429, 444, 446, 452, 468, 470, 471, 479, 488, 493, 505, 507, 509, 524, 538, 550, 598, 614, 621, 627, 642, 647, 685, 705, 709, 761, 762, 768, 772, 774, 801, 804, 807, 810, 813, 822, 830, 846, 849, 852, 853, 868, 877, 881, 887, 892, 893, 916, 933, 937, 949, 951, 961, 964, 967, 984, 1011, 1024, 1085, 1119, 1123, 1128, 1154, 1168, 1170, 1179, 1184, 1186, 1187, 1189, 1210, 1216, 1265, 1266, 1277, 1276, 1280, 1289, 1292, 1293, 1295, 1319, 1324, 1326, 1327, 1340, 1358, 1402, 1407, 1418, 1419, 1420, 1421, 1423, 1424, 1437, 1448, 1462, 1463, 1467, 1474, 1483, 1515, A25, A36, A39, A83, A86, A91, A94, A95, A98, A99, A102]

T20. *Partial Sums, Cesàro Sums.*
[25, 111, 274, 293, 295, 296, 301, 350, 351, 354, 412, 532, 557, 558, 559, 560, 561, 564, 565, 567, 568, 570, 572, 575, 595, 637, 638, 641, 653, 672, 673, 675, 676, 677, 733, 791, 883, 887, 922, 954, 998, 1005, 1095, 1099, 1104, 1113, 1144, 1145, 1153, 1156, 1162, 1177, 1191, 1217, 1219, 1312, 1360, 1361, 1365, 1367, 1370, 1389, 1479, 1495, A122]

T21. *Coefficient Bounds for Functions of Positive Real Part.*
[226, 412, 519, 887, 985, 1144, 1147, 1151, 1154, A31]

124 BIBLIOGRAPHY OF SCHLICHT FUNCTIONS

T22. *Bounded Functions.*

[38, 39, 78, 79, 116, 173, 174, 191, 193, 194, 198, 201, 203, 204, 242, 254, 255, 261, 272, 290, 293, 327, 330, 369, 378, 379, 396, 412, 452, 482, 505, 519, 532, 538, 543, 573, 574, 583, 585, 586, 587, 588, 589, 590, 591, 612, 613, 627, 639, 641, 662, 664, 665, 700, 743, 758, 759, 760, 761, 762, 768, 769, 772, 797, 821, 826, 866, 867, 877, 879, 925, 937, 965, 967, 972, 975, 977, 979, 988, 997, 1009, 1049, 1051, 1067, 1073, 1097, 1098, 1099, 1102, 1103, 1115, 1116, 1120, 1122, 1144, 1172, 1173, 1182, 1184, 1186, 1191, 1201, 1205, 1210, 1221, 1229, 1230, 1234, 1278, 1280, 1295, 1307, 1309, 1310, 1312, 1313, 1315, 1319, 1343, 1348, 1361, 1377, 1378, 1379, 1380, 1383, 1390, 1439, 1455, 1456, 1486, 1487, 1489, 1516, 1519, A1, A23, A25, A26, A33, A58, A59, A60, A70, A73, A117, A128, A130]

T23. *Representation Theorems, Structural Formulas.*

[24, 25, 46, 60, 92, 170, 256, 257, 258, 262, 317, 349, 353, 355, 372, 419, 422, 426, 436, 519, 540, 577, 626, 651, 659, 660, 665, 686, 703, 716, 723, 858, 869, 882, 883, 884, 889, 896, 897, 899, 995, 1073, 1076, 1120, 1225, 1140, 1142, 1144, 1147, 1156, 1162, 1165, 1177, 1180, 1374, 1375, 1377, 1381, 1414, 1494, 1501, 1510, 1517, 1520, 1524, A10, A33, A75, A91, A135, A136]

T24. *Variational Methods*

[16, 34, 69, 71, 75, 76, 77, 78, 151, 198, 199, 202, 204, 205, 206, 238, 267, 272, 278, 352, 360, 363, 364, 366, 371, 373, 376, 377, 390, 392, 401, 404, 405, 408, 415, 416, 418, 419, 422, 426, 438, 439, 440, 441, 442, 463, 465, 473, 475, 476, 487, 505, 551, 552, 553, 554, 592, 598, 600, 601, 602, 603, 606, 614, 617, 619, 620, 621, 623, 624, 625, 628, 630, 639, 640, 641, 643, 651, 685, 696, 700, 707, 711, 713, 722, 724, 735, 736, 739, 754, 755, 759, 760, 763, 764, 773, 777, 778, 779, 784, 787, 824, 825, 826, 827, 828, 830, 857, 867, 882, 907, 928, 929, 931, 935, 936, 959, 993, 1008, 1030, 1038, 1120, 1122, 1141, 1165, 1172, 1205, 1207, 1208, 1214, 1225, 1233, 1234, 1235, 1236, 1237, 1238, 1242, 1243, 1245, 1246, 1247, 1248, 1250, 1251, 1253, 1256, 1257, 1258, 1259, 1260, 1263, 1285, 1289, 1292, 1298, 1307, 1308, 1309, 1314, 1332, 1347, 1374, 1376, 1377, 1378, 1380, 1384, 1385, 1390, 1392, 1405, 1413, 1455, 1483, 1502, 1505, 1513, 1514, A1, A2, A58, A76, A89, A97, A103, A106 A109, A110, A118, A141, A145, A168]

T25. *Coefficient Bounds for Meromorphic Univalent (p-valent) Functions.*

[5, 215, 216, 235, 264, 277, 363, 364, 372, 408, 601, 619, 623, 624, 625, 629, 682, 709, 741, 804, 859, 866, 924, 925, 979, 983, 985, 986, 987, 996, 1020, 1071, 1072, 1076, 1214, 1215, 1233, 1246, 1259, 1332, 1333, 1392, 1451, 1452, 1490, 1499, 1502, 1503, 1504, 1506, 1513, A4, A92, A106, A119, A146]

T26. *Unrelated Functions* (Functions whose domain of values have no points in common).

[40, 41, 42, 44, 418, 604, 696, 779, 829, 830, 832, 976, 979, 1285].

T27. *Area Principle* (and generalizations).

[5, 36, 45, 51, 83, 105, 107, 117, 139, 141, 155, 183, 205, 277, 280, 363, 435, 478, 509, 576, 601, 646, 682, 728, 804, 831, 832, 846, 881, 925, 1009, 1082, 1187, 1192, 1193, 1194, 1195, 1198, 1267, 1286, 1323, 1327, 1332, 1333, 1394, A62, A100]

T28. *Univalence of Special Functions* (such as Bessel functions).

[161, 171, 172, 200, 201, 213, 243, 321, 349, 535, 536, 540, 541, 542, 652, 686, 738, 748, 750, 751, 752, 753, 885, 916, 917, 1109, 1110, 1112, 1147, 1161, 1176, 1398, 1442, 1478, 1511, 1512, A48, A125]

T29. *Regions of Variability* (of coefficients; values of functionals).

[31, 32, 33, 34, 46, 60, 63, 78, 117, 133, 170, 178, 204, 287, 399, 404, 448, 472, 523, 551, 552, 590, 601, 620, 723, 763, 766, 782, 784, 826, 828, 830, 979, 1038, 1114, 1125, 1205, 1207, 1225, 1235, 1239, 1267, 1340, 1374, 1383, 1413, 1454, 1502, A3, A9, A32, A78, A108, A113, A124]

T30. *Coefficient Bounds for Bounded Functions.*

[194, 255, 261, 379, 412, 583, 586, 587, 866, 979, 1073, 1097, 1098, 1144, 1201, 1309, 1455, A25, A58, A117]

T31. *Schwarzian Derivative-The Differential Equation* $w'' + pw \doteq 0$.

[38, 89, 90, 91, 171, 172, 193, 335, 336, 492, 536, 728, 786, 804,

876, 971, 980, 981, 982, 1046, 1058, 1160, 1265, 1267, 1273, 1364, A29, A52, A53, A54, A55, A148, A149, A150, A165, A166]

T32. *Univalent Polynomials–Remak Series–Hurwitz Class.*

[52, 175, 177, 179, 180, 181, 202, 240, 242, 330, 533, 651, 678, 848, 849, 850, 946, 947, 1001, 1002, 1003, 1004, 1036, 1067, 1124, 1139, 1145, 1266, 1267, 1301, 1496, A36, A47, A66]

T33. *Mean-valent Functions* (and generalizations).

(36, 45, 51, 138, 141, 145, 146, 147, 360, 393, 505, 507, 512, 521, 522, 610, 614, 756, 786, 792, 925, 1009, 1069, 1322, 1325, 1326, 1327, 1328, 1330]

T34. *Bieberbach-Eilenberg Functions* ($f(z_1)f(z_2) \neq 1$).

[40, 41, 44, 117, 228, 275, 370, 435, 605, 611, 627, 737, 805, 806, 833, 834, 835, 870, 1186, 1288, A60a]

T35. *Curvature, Level Curves*

[29, 109, 247, 255, 261, 533, 661, 663, 731, 732, 734, 785, 786, 1037, 1104, 1138, 1227, 1231, 1417, 1444, 1462, 1517, 1521, A141]

T36. *Coefficient Bounds for Starlike Functions.*

[32, 55, 92, 217, 235, 255, 261, 400, 430, 431, 552, 677, 846, 888, 898, 985, 986, 990, 996, 1022, 1071, 1072, 1073, 1075, 1076, 1105, 1106, 1107, 1108, 1142, 1144, 1147, 1154, 1159, 1215, 1267, 1317, 1427, 1433, 1449, 1504, 1506, A91]

T37. *Schwarz's Lemma* (and generalizations)

[10, 11, 21, 43, 87, 98, 102, 191, 320, 369, 483, 528, 581, 694, 814, 877, 956, 964, 968, 972, 1088, 1118, 1179, 1182, 1183, 1278, 1341, 1366]

T38. *Coefficient Bounds for Convex Functions.*

[217, 235, 255, 261, 349, 355, 430, 431, 552, 660, 687, 859, 888, 898, 1071, 1072, 1073, 1075, 1105, 1106, 1107, 1108, 1129, 1131, 1132, 1142, 1144, 1222, 1317, 1340, 1435, A65, A92, A117, A140]

T39. *Special Conditions on the Coefficients* (restrictions on amplitudes, moduli, gaps, real coefficients).

[34, 49, 71, 103, 105, 208, 211, 241, 242, 280, 283, 284, 295, 296, 297, 329, 364, 375, 381, 383, 408, 416, 430, 431, 445, 562, 563, 566, 607, 621, 638, 645, 675, 728, 798, 804, 852, 853, 859, 861, 888, 901, 924, 986, 997, 1020, 1036, 1038, 1071, 1120, 1133, 1149, 1152, 1155, 1156, 1157, 1159, 1187, 1200, 1211, 1224, 1265, 1275, 1284, 1301, 1309, 1349, 1362, 1368, 1380, 1381, 1403, 1413, 1430, 1431, 1448, 1497, A21, A22, A47, A86, A94, A98, A116]

T40. *Starlike In One Direction.*

[660, 1022, 1028, 1108, 1142, 1147, 1166, 1212, 1217, 1427, 1430, 1433, A121]

T41. *Convex In One Direction.*

[235, 238, 257, 301, 660, 913, 1022, 1105, 1129, 1142, 1144, 1155, 1156, 1157, 1160, 1162, 1212, 1337, 1428, 1522, A117, A134]

T42. *Bieberbach Conjecture* ($|a_n| \leq n$).

[61, 72, 73, 74, 75, 77, 83, 117, 119, 122, 184, 205, 207, 242, 246, 247, 267, 268, 269, 363, 373, 381, 407, 428, 431, 440, 445, 496, 511, 517, 521, 629, 639, 643, 651, 735, 736, 739, 780, 781, 788, 793, 794, 808, 809, 833, 834, 835, 844, 846, 852, 853, 862, 863, 869, 882, 888, 903, 906, 907, 934, 984, 990, 1042, 1106, 1129, 1142, 1143, 1146, 1187, 1234, 1245, 1247, 1248, 1250, 1255, 1280, 1317, 1329, 1349, 1368, 1369, 1374, 1379, 1380, 1425, 1450, 1479, 1490, A7, A17, A33, A64, A76, A85, A91, A101, A105]

T43. *Radius of Schlichtness* (p-valency).

[6, 7, 25, 47, 122, 175, 176, 179, 180, 182, 240, 242, 256, 259, 263, 330, 348, 351, 354, 500, 502, 540, 557, 558, 560, 562, 563, 564, 566, 568, 570, 572, 595, 637, 638, 641, 653, 672, 682, 733, 752, 791, 799, 800, 804, 877, 887, 889, 901, 916, 918, 942, 970, 975, 997, 1000, 1009, 1024, 1035, 1113, 1139, 1153, 1221, 1280, 1312, 1367, 1369, 1495, 1499, A48, A68, A77, A92, A122, 567]

T44. *Survey Articles.*

[104, 125, 126, 239, 403, 411, 514, 515, 537, 618, 864, 886, 939, 978, 1064, 1240, 1241, 1305, 1331, 1400, A24, A51, A115, A163 (Chapter 6)]

128 BIBLIOGRAPHY OF SCHLICHT FUNCTIONS

T45. *Odd Univalent Functions.*

[37, 49, 70, 75, 184, 207, 208, 241, 242, 268, 295, 296, 304, 363, 365, 374, 399, 413, 519, 557, 561, 565, 595, 625, 637, 638, 641, 675, 729, 733, 734, 804, 812, 844, 846, 863, 865, 888, 953, 967, 1022, 1081, 1140, 1141, 1147, 1149, 1190, 1192, 1193, 1194, 1195, 1198, 1200, 1233, 1325, 1340, 1368, 1370, 1426, 1450]

T46. *Angular Derivative, Spherical Derivative, Invariant Derivative.*

[193, 250, 251, 252, 308, 311, 312, 313, 528, 620, 647, 648, 650, 725, 839, 949, 1008, 1044, 1278, 1330, 1409, 1410, 1436, 1438, 1445, 1446, 1447, 1467, 1469, A20, A72, A73, A162, A167]

T47. *Convolution* (Faltung) *of Functions*

[143, 145, 229, 230, 278, 498, 518, 719, 721, 869, 985, 1031, 1063, 1071, 1157, 1164, 1270, 1477, A11, A143, A144, A160, A164]

T48. *Schlicht In An Annulus.*

[2, 4, 5, 54, 231, 255, 265, 266, 346, 549, 714, 718, 784, 921, 1053, 1054, 1432, 1518, A2, A34, A40]

T49. *Special Geometry of the Mapping* (such as k-fold symmetry).

[33, 34, 40, 56, 67, 78, 106, 123, 133, 138, 221, 234, 255, 260, 261, 301, 305, 306, 322, 352, 367, 383, 434, 458, 474, 479, 480, 487, 546, 562, 566, 570, 572, 573, 577, 578, 579, 580, 587, 588, 589, 607, 625, 654, 660, 732, 734, 770, 778, 779, 789, 793, 795, 796, 816, 844, 932, 941, 950, 976, 999, 1002, 1042, 1043, 1049, 1062, 1071, 1072, 1093, 1130, 1151, 1158, 1216, 1219, 1245, 1345, 1381, 1382, 1406, 1430, 1431, 1449, 1482, 1484, 1486, 1494, 1498, 1517, A74, A147, A158]

T50. *Coefficient Bounds for p-valent Functions.*

[48, 92, 135, 255, 261, 305, 376, 388, 389, 396, 428, 430, 431, 445, 496, 505, 576, 679, 682, 684, 1019, 1020, 1024, 1150, 1151, 1154, 1158, 1159, 1215, 1427]

T51. *Coefficient Bounds for Typically-Real Functions.*

[55, 170, 372, 374, 375, 413, 432, 436, 445, 580, 715, 901, 995, 1022, 1140, 1144, 1147, 1149, 1152, 1158, 1159, 1431, A7]

TOPIC REFERENCES 129

T52. *Coefficient Bounds for Odd Univalent Functions.*

[37, 49, 70, 184, 207, 208, 241, 242, 304, 363, 365, 374, 413, 519, 812, 844, 846, 863, 865, 1022, 1081, 1140, 1141, 1147, 1152, 1190, 1192, 1193, 1194, 1195, 1198, 1200, 1233, 1325, 1368, 1426, 1450]

T53. *Faber Polynomials, Grunsky Coefficients*

[141, 205, 234, 264, 267, 363, 393, 416, 485, 554, 569, 571, 957, 958, 1029, 1247, 1248, 1250, 1255, 1259, 1272, 1286, 1332, 1342, 1415, 1491, A27, A28, A56, A62, A69, A100, A114, A142]

T54. *Coefficient Bounds for the Class* (S) (the class of functions $f(z)$ regular and univalent in $|z| < 1$, with $f(0) = 0, f'(0) = 1$.

[72, 75, 107, 117, 184, 205, 206, 241, 242, 269, 271, 329, 363, 364, 365, 394, 400, 407, 408, 421, 511, 521, 567, 568, 599, 619, 651, 654, 727, 728, 740, 755, 763, 788, 789, 790, 793, 804, 833, 834, 846, 906, 907, 958, 1038, 1075, 1144, 1146, 1150, 1154, 1155, 1189, 1190, 1192, 1193, 1194, 1195, 1196, 1197, 1198, 1199, 1200, 1203, 1233, 1234, 1250, 1284, 1329, 1338, 1374, 1375, 1377, 1381, 1396, 1425, 1426, 1450, 1453, 1454, 1479, A46, A111, A119, 1143]

T55. *Continued Fractions.*

[913, 914, 915, 916, 1035, A49]

T56. *Distortion Theorems for Functions of Positive Real Part.*

[887, 1144, 1147, 1154, 1320, A31, A94]

T57. *Univalency Over Regiona Other than the Unit Circle*

[50, 471, 1059, 1150, 1212, 1213, 1302, 1303, 1304, 1363, 1398, 1487, 1488, 1489, A18, A40, A42, A131]

T58. *Distortion Theorems for Meromorphic Univalent* (p-valent) *Functions.*

[5, 73, 74, 75, 76, 155, 209, 387, 390, 395, 397, 400, 404, 407, 408, 416, 418, 422, 426, 434, 455, 481, 544, 595, 697, 709, 712, 804, 881, 921, 1071, 1076, 1083, 1125, 1189, 1255, 1313, 1316, 1332, 1333, 1390, 1397, 1481, 1499, 1513, A35, A117, A146, A159]

T59. *Related Results from Analytic Function Theory.*

[8, 9, 12, 13, 14, 15, 17, 18, 19, 27, 59, 62, 66, 84, 85, 93, 94, 95, 96, 112, 113, 114, 115, 127, 153, 154, 159, 160, 162, 163, 164, 165, 166, 185, 187, 188, 190, 192, 195, 210, 214, 219, 220, 222, 223, 224, 225, 232, 244, 245, 248, 249, 273, 276, 286, 287, 291, 292, 298, 299, 302, 303, 314, 316, 319, 323, 324, 326, 331, 332, 334, 340, 343, 344, 345, 347, 357, 358, 359, 361, 362, 398, 402, 425, 447, 449, 450, 451, 457, 459, 462, 464, 466, 486, 491, 494, 497, 499, 501, 504, 506, 510, 513, 520, 525, 527, 529, 531, 534, 535, 539, 547, 548, 555, 556, 582, 593, 594, 597, 608, 609, 616, 622, 631, 632, 633, 634, 635, 649, 657, 658, 666, 667, 669, 670, 671, 688, 689, 690, 695, 699, 701, 702, 704, 706, 726, 730, 744, 745, 771, 775, 776, 783, 811, 817, 818, 819, 823, 837, 840, 841, 845, 847, 860, 871, 878, 895, 900, 904, 905, 920, 930, 938, 944, 945, 948, 962, 966, 969, 973, 991, 992, 994, 1012, 1013, 1014, 1015, 1017, 1026, 1034, 1039, 1040, 1045, 1047, 1052, 1056, 1060, 1061, 1065, 1066, 1074, 1077, 1084, 1086, 1087, 1089, 1096, 1100, 1121, 1126, 1134, 1135, 1136, 1137, 1188, 1204, 1206, 1223, 1226, 1252, 1261, 1262, 1264, 1268, 1269, 1271, 1274, 1277, 1281, 1282, 1287, 1294, 1297, 1311, 1318, 1321, 1334, 1336, 1344, 1350, 1351, 1353, 1354, 1355, 1357, 1359, 1373, 1388, 1393, 1411, 1422, 1441, 1443, 1457, 1458, 1460, 1461, 1465, 1466, 1468, 1470, 1472, 1473, 1475, 1492, 1493, 1525, 1526, A13, A16, A19, A43, A44, A50, A57, A61, A63, A82, A84, A112, A113, A129, A132, A153, A154, A155, A156, A157, A161]

T60. *Univalent Functions in Multiply-Connected Domains.*

[11, 39, 45, 47, 277, 320, 377, 387, 467, 468, 469, 470, 471, 472, 477, 478, 482, 502, 596, 615, 617, 691, 692, 693, 707, 708, 710, 754, 769, 815, 821, 857, 899, 919, 926, 927, 968, 977, 1050, 1078, 1080, 1088, 1127, 1128, 1254, 1298, 1316, 1343, 1386, 1387, 1434, 1464, A6, A40, A61, A151, A152]

T61. *Distortion Theorems for Bounded Functions.*

[242, 255, 369, 396, 412, 482, 505, 532, 573, 585, 590, 591, 641, 758, 759, 760, 826, 867, 965, 1051, 1097, 1098, 1102, 1103, 1144, 1172, 1191, 1313, 1348, 1390, 1456, A25, A130]

T62. *Boundary Behavior of Univalent Functions.*

[23, 178, 221, 236, 237, 253, 279, 308, 317, 318, 341, 342, 346, 348,

511, 522, 613, 655, 700, 757, 872, 873, 874, 875, 1044, 1079, 1115, 1116, 1124, 1143, 1146, 1232, 1279, 1280, 1343, 1408, 1410, 1486, A8]

T63. *Distortion Theorems for Starlike Functions* (including generalizations of starlikeness).

[92, 142, 157, 255, 258, 260, 261, 282, 353, 356, 431, 434, 661, 677, 716, 853, 896, 897, 953, 1071, 1073, 1076, 1108, 1144, 1147, 1154, 1267, 1340, 1427, 1440, 1494, A37, A38, A71, A91, A121]

T64. *Distortion Theorems for Convex Functions* (including generalizations of convexity).

[118, 142, 247, 255, 261, 282, 355, 356, 383, 431, 453, 570, 661, 687, 716, 723, 761, 762, 766, 881, 884, 896, 897, 909, 953, 963, 1071, 1095, 1108, 1131, 1144, 1340, 1435, 1481, 1494, 1510, A65, A67, A78, A117]

T65. *Distortion Theorems for p-valent Functions.*

[48, 92, 157, 197, 246, 255, 258, 261, 282, 325, 353, 356, 367, 376, 388, 389, 431, 505, 507, 644, 679, 680, 682, 683, 684, 698, 737, 760, 1150, 1151, 1154, 1427, 1435, A37, A38]

T66. *Distortion Theorems for Typically-Real Functions.*

[4, 170, 375, 413, 416, 427, 436, 626, 858, 953, 1053, 1125, 1144, 1147, 1149, 1427, A41]

T67. *Distortion Theorems for Odd Univalent Functions.*

[49, 70, 75, 399, 519, 557, 733, 804, 953, 1147, 1340, 1426]

T68. *Distortion Theorems for the Class* (S).

[64, 65, 68, 71, 75, 76, 118, 142, 156, 234, 246, 247, 281, 306, 383, 385, 386, 390, 394, 397, 399, 400, 408, 415, 453, 455, 484, 500, 507, 557, 558, 559, 599, 711, 724, 742, 764, 788, 793, 804, 826, 828, 833, 834, 835, 846, 852, 892, 906, 909, 932, 940, 951, 959, 989, 1006, 1008, 1030, 1051, 1144, 1146, 1150, 1154, 1172, 1175, 1189, 1225, 1280, 1339, 1374, 1375, 1396, 1425, 1426, 1143]

Corrections

Following is a list of references in the bibliography indicating duplications or translations of the same papers.

Reference	same as	*Reference*
389	......	388
426	......	422
476	......	475
559	......	558
565	......	561
566	......	562
568	......	567
702	......	701
708	......	704
806	......	805
850	......	848
1146	......	1143
1260	......	1258
1292	......	1289
1497	......	211

Notes (a) Reference 282: Erzohi should be Ezrohi
(b) Reference 869: Loewner, C. is same author as Lowner, K.

BIBLIOGRAPHY OF SCHLICHT FUNCTIONS PART II (1966–1975)

CONTENTS

Preface 134
Bibliography.. 137
Topic References 254
Table 1. References Prior to Year 1966 270
Table 2. References in Year 1976 271
Table 3. Number of References Published Each Year 271
Table 4. Some Additional Math. Review (MR) Numbers 271
Corrections 273

PREFACE

The first extensive bibliography in the field of univalent (Schlicht) functions was published by this writer in May 1966. It is titled Bibliography of Schlicht Functions [Courant Institute of Mathematical Sciences, New York University, Technical Report No. NRO 41-019, IMM 351 (1966) ix + 157 pp.; MR 34 # 2849]. We shall refer to it as Bibliography I. Bibliography I contains 1694 references covering the years 1907 (when the first substantial results were published) through the year 1965, and includes the work of 570 authors whose papers appeared in 220 mathematical journals. The various results in the theory of Schlicht functions are, in Bibliography I, classified into 68 subtopics and cross-index listings are given between references and subtopics. Thus, each subtopic is followed by a list of those references pertaining to that subtopic, and conversely, following each reference is a list of those subtopics dealt with in that reference.

PREFACE 135

The favorable reception given to Bibliography I by individual mathematicians, graduate sutdents, university libraries, and industrial research laboratories encouraged this writer to continue the (laborious) collection, reading, topic classification and cataloguing the periodic flow of new research papers during the past ten years. These efforts have culminated with the publication of the present Bibliography of Schlicht Functions, Part II, to which we refer subsequently as Bibliography II.

Bibliography II contains 1563 references to the publications of 523 authors which have apeared in 200 mathematical journals and publishing companies in the U.S.A., the Soviet Union, England, Germany, France, Japan, China, and in many other countries. It includes many symposiums, colloquiums, congresses, dissertations, abstracts, technical reports, lecture notes, and books dealing with the theory of univalent (Schlicht) and multivalent mappings of simple and multiply-connected domains. This survey covers the years 1966 through 1975. Some papers published prior to the year 1966 and which were not included in Bibliography I are now listed in Table 1. Some papers published in the year 1976 are listed in Table 2.

The various results in the theory of Schlicht functions have been classified, in Bibliography II, into ninety subtopics. The first sixty-eight of these subtopics are the same as in Bibliography I. The additional subtopics reflect new developments or emphases that have taken place during the past ten years. As in Bibliography I, cross-index listings are given between references and subtopics. Thus, the following two examples illustrate the principal use of the Bibliography.

Example 1. Reference [1250] lists "Hadamard products of Schlicht functions and the Pólya-Schoenberg conjecture," by Ruscheweyh, S. and Sheil-Small, T. This paper appeared in the journal Comment. Math. Helv., Volume 48, year 1973, pages 119-135. A review of this paper is to be found in Mathematical Reviews (MR) of the American Mathematical Society, Volume 48, review # 6393. The numbers in brackets [1, 2, 4, ..., 64] indicate that this paper contains information regarding subtopics T1, T2, T4, ..., T64.

Example 2. The reader who is interested in Bieberbach-Eilenberg functions (or related classes) will find this subtopic listed as T34. The bracketed references [9, ..., 1493] indicate that information regarding this subtopic may be found in reference [9] which is a paper by Aharonov, Dov. A., ..., reference [1493] which is a paper by Volgnec, I. A.

The classification of the various references into subtopics was based on a detailed reading of 785 reprints (of full-length papers), and for the remaining references on a reading of their (brief) abstracts or reviews, or (in a few cases) from the titles only. Therefore, in fairness to the authors, we emphasize that the classifications do not always indicate the complete scope of the papers.

Forty-three of the references have no subtopic classification, due either to lack of available information or the contents simply did not fall within the fixed framework of the ninety subtopics.

The references include 152 abstracts. Some of these were never published in full, others may have been published perhaps under a different title or co-authorship, while others are too recent to have appeared in full. Abstracts, of course, have no Math. Review (MR) numbers.

One hundred and thirty-two references which are not abstracts have no Math. Review numbers. Many of them are lecture notes, dissertations, technical and special reports, papers not yet published, or too recently published for inclusion in the Reviews.

It is difficult in a work of this nature to cover every paper published on the subject of Schlicht functions. Reasons for omissions are numerous and varied. However, it is felt that this bibliography does include a major portion of the publications during the ten years following Bibiliography I.

I gratefully acknowledge the aid given me by two graduate sutdents, Mr. Mohammed Elmasri and Mr. Anthony DiNardo, who assisted me in locating journals, photocopying hundreds of papers, looking up MR references and numerous other clerical tasks.

I give many thanks also to the most cooperative personnel of the admirably stocked mathematics library of the Courant Institute of Mathematical Sciences at New York University, and to the Director of the Institute, Professor Peter Lax. Finally, I wish to express my sincere appreciation to Professor A. W. Goodman of the University of South Florida who encouraged me during the past ten years to complete this bibliography.

S. D. Bernardi
March 1977

BIBLIOGRAPHY OF SCHLICHT FUNCTIONS (PART II)

1. Abe, H. *On meromorphic and circumferentially mean univalent functions*. J. Math. Soc. Japan 16(1964). 342-351; MR 31 #2387. [9, 14, 19, 33]
2. Avhadiev, F. G.; Aksent'ev, L. A. *Functions of the Bazilevic class in the disc and in an annulus*. (Russian) Dokl. Akad. Nauk SSSr 214(1974), 241-244; MR 50 #2469. {Soviet Math. Dokl. 15(1974), 78-82}. [48, 70]
3. Abian, Alexander. *A contrast between complex and real-valued Taylor series*. J. Austral. Math. Soc. 18(1974), 458-460; MR 51 #5901. [20]
4. Acilov, H. *Certain theorems on convolutions of power series* (Russian. Uzbek Summary). Izv. Akad. Nauk UzSSR Ser. Fiz.-Mat. Nauk 9(1965), no. 5, 5-11; MR 33 #5902. [1, 10, 22, 47]
5. Acilov, H. *Certain questions of the theory of univalent conformal mappings*. (Russian. Uzbek Summary) Izv. Akad. Nauk UzSSR Ser. Fiz.-Mat. Nauk 10(1966), no. 5, 3-9; MR 34 #2848. [4]

6. Aharonov, Dov. *A remark on an areally mean p-valent function.* Israel J. Math. 6(1968), 119-120. [33]
7. Aharonov, Dov. *The theorem of Cartwright, Spencer and Hayman.* J. London Math. Soc. (2)1(1969), 119-126; MR 39 #4377. [33]
8. Aharonov, Dov. *A necessary and sufficient condition for univalence of a meromorphic function.* Duke Math. J. 36(1969), 599-604; MR 40 #2865. [4, 9, 31]
9. Aharonov, Dov. *A generalization of a theorem of Jenkins.* Math. Zeit. 110(1969), 218-222; MR 40 #325. [27, 34, 86]
10. Aharonov, Dov. *A note on slit mappings.* Bull. Amer. Math. Soc. 75(1969), 836-839; MR 41 #3725. [53]
11. Aharonov, Dov. *On Bieberbach Eilenberg Functions.* Bull., A.M.S., 76(1970), 101-104; MR 41 #1994. [34, 86, 87]
12. Aharonov, Dov. *Proof of the Bieberbach conjecture for a certain class of univalent functions.* Israel J. of Math. 8(1970), 102-104; MR 42 #486. [39, 42]
13. Aharonov, Dov. *Sequence of necessary conditions for univalence of meromorphic function.* Duke Math. J. 38(1971), 595-598; MR 43 #5015. [31, 53, 54]
14. Aharonov, Dov. *On the Bierberbach conjecture for functions with a small second coefficient.* Israel Journal of Math. IS, no. 2(1973), 137-139; MR 48 #520. [42, 54]
15. Aharonov, Dov. *On pairs of functions and related classes.* Duke Math. J. 40(1973), 669-676; MR 47 #8837. [2, 21, 26, 34, 86, 87]
16. Aharonov, Dov.; Friedland, S. *On an inequality connected with the coefficient conjecture for functions of bounded boundary rotation.* Ann. Acad. Sci. Fenn. Ser. A. I. Math., no. 524(1972), 14 pp. [1, 71, 77]; MR 48 #519
17. Aharonov, D,; Friedland, S. *On functions of bounded boundary rotation.* Ann. Acad. Sci. Fenn. Ser. AI Mat. no. 585(1974), 18 pp. [71]
18. Aharonov, Dov; Kirwan, W. E. *A method of symmetrization and applications.* I. Trans. Amer. Math. Soc. 163(1972), 369-377; MR 45 #516. [19, 24, 34, 35]
19. Aharonov, Dov; Kirwan, W. E. *A method of symmetrization and applications.* II. Trans. Amer. Math. Soc. 169(1972), 279-291; MR 47 #2042. [16, 19, 24]
20. Aharonov, Dov; Kirwan, W. E. *Covering theorems for classes of univalent functions.* Canad. J. Math. 25(1973), 412-419; MR 47 #2041. [6, 10, 19]
21. Ahlfors, Lars V. *An inequality between the coefficients a_2 and a_4 of*

a univalent function. (Russian). Certain Problems of Mathematics and Mechanics (Russian), pp. 71-74. Izdat. "Nauka", Leningrad, 1970; MR 45 #5338. [54]

22. Ahlfors, L. V. *Conformal invariants; topics in geometric function theory.* McGraw-Hill Series in Higher Mathematics. McGraw-Hill Book Co., New York-Dusseldorf-Johannesburg, 1973. IX + 157 pp.; MR 50 #10211. [44]
23. Ahlfors, L. V. *A remark on schlicht functions with quasiconformal extensions.* Proceedings of the Symposium on Complex Analysis (Univ. Kent, Canterburg 1973), pp. 7-10. London Math. Soc. Lecture Notes Ser., No. 12, Cambridge Univ. Press, London, 1974.
24. Aksent'ev, L. A. *An application of the principle of the argument to the study of univalence conditions.* I. (Russian) Izv. Vyss Ucebn. Zaved. Matematika 1968, no. 12 (79), 3-15; MR 39 #7078. [28, 57]
25. Aksent'ev, L. A. *An application of the principle of the argument to the study of univalence conditions.* II. (Russian) Izv. Vyss. Ucebn. Zaved. Matematika 1969, no. 3 (82), 3-15; MR 39 #7079. [28, 57]
26. Aksent'ev, L. A. *The univalent solvability of inverse boundary value problems.* (Russian) Trudy Sem. Kraev. Zadacam Vyp. 10(1973), 11-24; MR 50 #619. [4, 62]
27. Aksent'ev, L. A.; Avhadiev, F. G. *A certain class of univalent functions.* (Russian) Izv. Vyss. Ucebn. Zaved. Matematika 1970, no. 10 (101), 12-20; MR 43 #7607. [4, 9]
28. Aksent'ev, L. A.; Gaiduk, V. N.; Mikka, V. P. *Univalence criteria for n-symmetric functions.* (Russian) Izv. Vyss. Ucebn. Zaved. Matematika 1974, no. 4(143), 3-13; MR 50 #4922. [4]
29. Al-Amiri, H. S. *The a-points of Faber polynomials for a special function.* Notices, Amer. Math. Soc., vol. 15, no. 1, issue no. 103, January 1968, p. 145 (Abstract). [53]
30. Al-Amiri, Hassoon, S. *On p-close-to-star functions of order α.* Proc. Amer. Math. Soc. 29(1971), 103-108; MR 43 #7608. [6, 10, 11, 43, 63, 64]
31. Al-Amiri, H. *On the radius of univalence of bounded functions.* Colloq. Math., 25(1972), Fasc. 1, pp. 125-126; MR 46 #341. [22, 43]
32. Al-Amiri, H. S. *On the radius of univalence of certain analytic functions.* Colloq. Math. 28(1973), Fasc. 1, pp. 133-139; MR 48 #6383. [2, 11, 12, 22, 43, 61, 72, 82]
33. Al-Amiri, H. S. *On the radius of β-convexity of starlike functions of order α.* Proc. Amer. Math. Soc. 39(1973), 101-109; MR 47 #445. [4, 6, 10, 12, 56, 69]
34. Al-Amiri, Hasson S. *On the radius of starlikeness of certain analytic*

functions. Proc. Amer. Math. Soc. 42(1974), 466–474; MR 48 #8768. [2, 5, 6, 10, 11, 85]
35. Al-Amiri, H. S. *The radius of α-convexity for the class of starlike univalent functions in the circular region* $0 < |z| < 1$. Rev. Roumaine Math. Pures Appl. (20) (1975), no. 8, 863–868. [12, 69]; MR 52 #719.
36. Al-Amiri, H. S.; Reade, Maxwell O. *On a linear combination of some expressions in the theory of univalent functions.* Monatsh. Math. 80(1975), no. 4, 257–264. [82]
37. Aleksandrov, I. A. *The variation of nonunivalent analytic functions.* (Russian) Trudy Tomsk. Gos. Univ. 163(1963), 155–159. [24]
38. Aleksandrov, I. A. *Geometric properties of schlicht functions.* (Russian) Trudy Tomsk. Gos. Univ. Ser. Meh.-Mat. 175(1964), 29–38; MR 32 #2571. [44]
39. Aleksandrov, I. A. *The range of systems of Functionals* (Russian) Trudy Tomsk. Gos. Univ. Ser. Meh.-Mat. 182(1965), Vyp. 3, 59–70; MR 33 #7514.
40. Aleksandrov, I. O.; Baranova, V. V. *A family of holomorphic univalent functions with a shift symmetry.* (Ukrainian. English and Russian summaries). Dopovidi Akad. Nauk Ukrain. RSR SerA 1972, 387–390, 475; MR 46 #5597. [49]
41. Aleksandrov, I. A.; Chernikov, V. V.; Kufarev, P. O. *Russian Math.* Surveys, vol 24 (1970), no. 4; Amer. Math. Soc., MOS article code 6WWQ. [44]
42. Aleksandrov, I. A.; Gutljanskii, V. Ja. *Extremal problems for classes of analytic functions having a structural formula.* (Russian) Dokl. Akad. Nauk SSSR 165(1965), 983–986; MR 33 #5900 {English transl. Soviet Math. Dokl. 6(1965), 1531–1535}. [1, 2, 6, 13, 24]
43. Aleksandrov, I. A.; Gutljanskii, V. Ja. *Extremal problems for classes of analytic functions having a structural formula.* (Russian) Trudy Tomsk. Gos. Univ. Ser. Meh.-Mat. 189(1966), 111–122; MR 37 #1578. [1, 2, 6, 13]
44. Aleksandrov, I. A.; Gutljanskii, V.Ja. *Extremal properties of close-to-convex functions.* (Russian) Sibirsk. Mat. Z. 7(1966), 3–22; MR 33 #4270. [5, 6, 23, 24]
45. Aleksandrov, I. A.; Gutljanskii, V.Ja. *On the coefficient problem in the theory of univalent functions.* (Russian) Dokl. Akad. Nauk SSSR 188(1969), 266–268; MR 41 #455. [24, 42]
46. Aleksandrov, I. A.; Kocegurova, V. G. *The domain of values of polynomial expressions consisting of coefficients of functions of the classes S or* Σ. (Russian) Trudy. Tomsk. Gos. Univ. Ser. Meh.-Mat. 175(1964), Vyp. 2, 39–49; MR 30 #4925. [24, 29]

47. Aleksandrov, I. A.; Kopanev. S. A. *On mutual growth of the modulus of a univalent function and of the modulus of its derivative.* (Russian) Sibirsk. Mat. Z. 7(1966), 23–30; MR 32 #7723. [29, 49, 68]
48. Aleksandrov, I. A.; Kopanev. S. A. *The range of values of the derivative on the class of holomorphic univalent functions.* (Russian) Ukrain. Mat. Z. 22(1970), 660–664; MR 44 #419. [29]
49. Aleksandrov, I. O.; Krjuckov, B.Ja. *The first coefficients of bounded holomorphic univalent functions.* (Ukrainian. English and Russian summaries) Dopovidi Akad. Nauk Ukrain. RSR Ser. A 1973, 3–5, 91; MR 47 #7018. [29, 54]
50. Aleksandrov, I. O.; Nikul'sina, M. M. *On the theory of subordinate functions.* (Ukrainian. English and Russian summaries) Dopovidi Akad. Nauk Ukrain. RSR Ser A 1972, 195–198, 283; MR 46 #5600. [1, 57]
51. Aleksandrov, I. A.; Popov, V. I. *Solution of a problem of I. E. Bazilevic and G. V. Korickii on star-shaped arc of level curves.* (Russian). Sibirsk. Mat. Z. 6(1965), 16–37; MR 30 #3202. [6, 7]
52. Aleksandrov, I. A.; Popov, V. I. *Optimal controls and univalent functions.* (Russian. Polish and English summaries) Ann. Univ. Mariae Curie-Sklodowska Sect. A 22–24 (1968/70), 13–20 (1972); MR 50 #592. [24]
53. Aleksandrov, I. A.; Popova, G. A. *Extremal properties of univalent holomorphic functions with real coefficients.* (Russian) Sibirsk Mat. Z. 14(1973), 915–926, 1156; MR 49 #5337. [24, 39]
54. Aleksandrov, I. A.; Prohorova, A. E. *Estimates for the curvature of level curves on the class Sp* (Russian). Dokl. Akad. Nauk SSSR 203(1972), 267–269; erratum, ibid. 206(1972), vii; MR 46 #2027. {Soviet Math. Dokl. 13(1972), 359–362}. [6, 10, 29, 35, 43, 49]
55. Aleksandrov, I. A.; Sobolev, V. V. *Extremal problems for certain classes of functions that are univalent in the half-plane.* (Russian) Ukrain. Mat. Z. 22(1970), 291–307; MR 42 #1988. [24, 57]
56. Aleksandrov, I. A.; Sorokin, A. S. *On extending the variational method of G. M. Goluzin and P. P. Kufarev to a multiply connected region.* (Russian) Dokl. Akad. Nauk SSSR 175(1967), 1207–1210; MR 36 #1626 {English transl.: Soviet Mat. Dokl. 8(1967), 980–984}. [24, 60]
57. Aleksandrov, I. A. et al (62 participants). *Proceedings of the Fifth Conference on Analytic Functions.* (Univ. Mariae Curie-Sklodowska, Lublin, 24–29 August, 1970), 208 pp. Ann. Univ. Mariae Curie-Sklodowska Sect. A 22–24 (1968/70); MR 49 #10862. [44]
58. Alenicyn, Ju. E. *Univalent functions without common values in a*

multiply connected region. (Russian). Contemporary Problems in Theory Anal. Functions. Internat. Conf., Erevan, 1965 (Russian), pp. 9–11; MR 35 #5597 {Dokl. Nauk SSSR 167, 9–11; MR 33 #7515}. [26, 60]

59. Alenicyn, Ju. E. *On univalent functions without common values in a multiply connected domain.* (Russian) Dokl. Akad. Nauk SSSR 167(1966), 9–11; MR 33 #7515. {English translation: Soviet Math. Dokl. 7(1966), 305–307}. [22, 26]

60. Alenicyn, Ju. E. *Univalent functions without common values in a multiply connected domain.* (Russian) Trudy Mat. Inst. Steklov. 94(1968), 4–18; MR 37 #1579. {Dokl. Akad. Nauk SSSR 167(1966), 9–11; MR 33 #7515}, {Book: Problems of the Geometric Theory of Functions, Amer. Math. Soc. 1969}. [26]

61. Alenicyn, Ju. E. Proceedings of the Steklov Institute of Mathematics, No. 94(1968): Extremal problems of the geometric theory of functions. Edited by Ju. E. Alenicyn. Translated from the Russian by A. Yablonsky; Amer. Math. Sco., Providence, R. I. 1969. VI + 167 pp.; MR 39 #2963. [44]

62. Alenicyn, Ju. E. *Some theorems of areas for functions which are analytic in finitely connected domain.* Dokl. Akad. Nauk SSSR 209 (1973), S21–S24. {Soviet Math.-Doklady 14(1973), 415–419}; MR 48 #2372, #11462. [27]

63. Alenicyn, Ju. E. *Certain area theorems for analytic functions with quasiconformal continuation* (Russian). Mat. Sb. (N.S.) 94(136) (1974), 114–125, 160; Dokl. Akad. Nauk SSSR 215(1974), 1025–1028; MR 51 #3432. [18]

64. Alenicyn, Ju. E.; Kuzmina, G. V.; et al. *Extremal problems of the geometric theory of functions of a complex variable.* Consultants Bureau, New York, 1974, pp. 564–717 {Translated from the Russian. J. Soviet Math. 2(1974), no. 6}; MR 50 #581. [4]

65. Aleksandrov, I. A.; Sobolev, V. V. *The curvature of level curves and their orthogonal trajectories in certain conformal mappings of the half plane.* (Russian). Ukrain. Mat. Z. 26(1974), 510–516, 574. [35, 57]; MR 50 #4919

66. Anderson, J. M. *A note on starlike schlicht functions.* J. London Math. Soc. 40(1965), 713–718; MR 32 #203. [6, 62]

67. Antonjuk, G. K. *A certain corollary of Teichmuller's theorem.* (Russian). Proc. Sixth Interuniv. Sci. Conf. of the Far East on Physics and Mathematics, Vol. 3; Differential and Integral Equations. (Russian), pp. 14–19. Habarovsk. Gos. Ped. Inst. Khabarovsk, 1967; MR 41 #3744. [54]

68. Antonjuk, G. K. *On a certain result of Lewandowski and Zlotkie-*

wicz. (Russian). Trud. Tomsk. Gos. Univ. Ser. Meh.-Mat. 210(1969), 3-5; MR 44 #2920. [9, 24, 29]

69. Antonjuk, G. K. *A certain property of functions that are extremal in the coefficient problem.* (Russian) Dal'Nevostoc. Gos. Univ. Ucen. Zap. 69(1970), 150-151; MR 48 #11478.

70. Artemiadis, N. *On the coefficients of starlike functions of order α.* Notices A. M. S. Vol. 14, no. 1, issue no. 95, January 1967 (Abstract 642-171). [6, 36]

71. Avhadiev, F. G. *On sufficient conditions for the univalence of the solutions of inverse boundary value problems.* (Russian) Dokl. Akad. Nauk SSSR 190(1970), 495-498; MR 41 #1982; Soviet Math. Dokl. 11(1970), 109-112. [4, 9, 10, 58, 85]

72. Avhadiev, F. G. *Conditions for the univalence of analytic functions.* (Russian) Izv. Vyss. Ucebn. Zaved. Matematika 1970, no. 11(102), 3-13; MR 43 #6418. [4, 9, 10, 31, 85]

73. Avhadiev, F. G. *Radii of convexity and close-to-convexity of certain integral representations.* Math. Zametri 7(1970), 581-592; MR 44 #4197. {Math. Notes 7(1970), 350-357, English translation}. [12]

74. Avhadiev, F. G. *On the weak and the strong problem of univalence in inverse boundary value problems.* (Russian) Trudy Sem. Kraev. Zadcam. Vyp. 10(1973), 3-10; MR 50 #618. [62, 68]

75. Avhadiev, F. G. *Sufficient conditions for univalence in nonconvex domains.* (Russian) Sibirsk. Mat. Z. 15(1974), 963-971, 1180; MR 50 #4926. [4]

76. Avhadiev, F. G. *Sufficient conditions for the univalence of quasi-conformal mappings.* (Russian) Mat. Zametki 18 (1975), no. 6, 793-802. [4]

77. Avhadiev, F. G.; Aksent'ev, L. A. *Sufficient conditions for the univalence of analytic functions.* (Russian). Dokl. Akad. Nauk SSSR 198(1971), 743-746; MR 44 #2916. [4]

78. Avhadiev, F. G.; Aksent'ev, L. A. *A subordination principle in sufficient conditions for univalence.* (Russian) Dokl. Akad. Nauk SSSR 211(1973), 19-22; MR 48 #4288; Soviet Math. Dokl. 14(1973), 934-939. [1, 3, 4, 5, 9, 16, 31, 71]

79. Avhadiev, F. G.; Aksent'ev, L. A. *Fundamental results on sufficient conditions for the univalence of analytic functions.* (Russian) Uspehi Mat. Nauk 30(1975), no. 4 (184), 3-60. [4]

80. Avhadiev, F. G.; Gaiduk, V. N. *An application of close-to-convex functions to inverse boundary value problems.* (Russian) Izv. Vyss. Ucebn. Zaved. Matematika 1968, no. 6 (73), 3-10; MR 37 #4247. [4, 5]

81. Ayirtman, N. *On univalent functions.* (Turkish summary) Istanbul

Univ. Fen. Fak. Mec. Ser. A28 (1963), 9-17; MR 34 #7785. [8]
82. Babenko, K. I. *A contribution to the theory of the second variation of functionals on the class S of univalent functions.* Soviet Math. Dokl. 11(1970), no. 4, pp. 1037-1041; MR 45 #3690a. [24]
83. Babenko, K. I. *Some aspects of the theory of extremal problems for univalent functions of class S.* Soviet Math. Dokl., Vol. 11(1970), no. 5, pp. 1141-1144; MR 45 #3690b. [24]
84. Babenko, K. I. *On the structure of the coefficient-domain for univalent functions of class S.* Soviet Math. Dokl., vol. 11(1970), no. 5, pp. 1170-1173: MR 45 #3690c. [24]
85. Babenko, K. I. *The theory of extremal problems for univalent functions of class S.* Proceedings of the Steklov Institute of Mathematics, No. 101 (1972). Translated from the Russian by J. W. Noonan. Amer. Math. Soc., Providence, R. I., 1975. III + 327 pp.; MR 51 #8397. [44]
86. Baernstein, Albert, II. *Some extremal problems for univalent functions, harmonic measures, and subharmonic functions.* Proceedings of the Symposium on Complex Analysis (Univ. Kent, Canterbury 1973), pp. 11-15. London Math. Soc. Lecture Note Ser., No. 12, Cambridge Univ. Press, London, 1974. [44]; MR 52 #5944, #8430, #9207
87. Baernstein, Albert, II. *Integral means, univalent functions and circular symmetrization.* Acta Math. 133(1974), 139-169. [1, 3, 16, 24, 42]
88. Bahtin, O. K. *Certain extremal problems in conformal mapping.* (Russian) Ukrain. Mat. 26(1974), 517-522, 574; MR 50 #13490. [34, 60]
89. Bahtina, G. P. *A certain extremal problem on the conformal mapping of the unit disc onto nonoverlapping domains.* (Russian) Ukrain. Mat. Z. 26(1974), 646-648, 716; MR 51 #8398. [24, 26]
90. Bahtina, G. P. *The extremization of certain functionals in the problem of nonoverlapping domains.* (Russian) Ukrain. Mat. Z. 27(1975), 202-204, 285; MR 51 #8399. [24, 26]
91. Bajpai, P. L. *Some radii of starlikeness and convexity problems.* Indian Institute of Technology Kanpur, Kanpur, India. Doctoral dissertation, Dept. of Mathematics, August 1973. [2, 6, 9, 10, 11, 12, 13, 22, 56, 61, 72, 73, 82, 85]
92. Bajpai, P. L.; Singh, Prem. *The radius of starlikeness of certain analytic functions.* Proc. Amer. Math. Soc. 44(1974), 395-402; MR 49 #5330. [6, 11]
93. Bajpai, S. K. *A note on a class of starlike functions.* Indian J. Pure

Appl. Math. 3(1972), no. 5, 750-754; MR 48 #6384. [6, 12]
94. Bajpai, S. K. *Two convexity theorems for certain classes of analytic functions.* (Preliminary report) Notices A.M.S. 20(Feb. 1973), Abstract 73T-B62. [11, 85]
95. Bajpai, S. K. *The order of starlikeness of α-starlike functions.* Notices A.M.S. 20, August 1973, Abstract 73T-B233, P. A-491. [11, 69]
96. Bajpai, S. K. *A note on a class of meromorphic univalent functions.* Notices Amer. Math. Soc. 21, April 1974, Abstract 74T-B99. [9, 85]
97. Bajpai, S. K. *Influence of $1/2 f''(0)$ on the α-convexity of normalized starlike analytic functions $f(z)$ of order β.* Notices Amer. Math. Soc. 21, August. 1974, Abstract 74T-B141, p A-487. [69]
98. Bajpai, S. K. *On regions of α-convexity for starlike functions.* Proc. Amer. Math. Soc. 44(1974), 365-368; MR 49 #9178. [69]
99. Bajpai, S. K.; Dwivedi, S. P. *A subordination for certain classes of analytic functions.* Notices Amer. Math. Soc. 21(1974), Abstract 74T-B220, pA-544. [1, 6, 10, 85]
100. Bajpai, S. K.; Mehrok, S.J.T. *On regions of α-convexity for subclasses of starlike functions.* Notices A. M. S. 20 August 1973, Abstract 73T-B202, p. A-484. [12, 69]
101. Bajpai, S. K.; Mehrok, T.J.S. *On the coefficient structure and growth theorem for thee functions $f(z)$ for which $zf'(z)$ is spirallike.* Publ. Inst. Math. (Beograd)(N.S.) 16(30) (1973), 5-12; MR 50 #10222. [4, 6, 36]
102. Bajpai, S. K.; Mehrok, T.J.S. *On univalence of certain analytic functions associated with starlike, convex and close-to-convex functions.* Indian J. Pure Appl. Math. 4(1973), no. 1, 66-72; MR 48 #4289. [4, 5, 6, 10, 16, 47, 85]
103. Bajpai, S. K.; Mehrok, T.J.S. *On the radius of p-valent starlikeness and p-valent convexity.* Rev. Roumaine Math. Pure Appl. 19(1974), 973-976; MR 50 #4923. [4, 6, 11, 12, 14, 43]
104. Bajpai, S. K.; Mehrok, T.J.S. *A note on the class of meromorphic functions.* I. Ann. Polon. Math. 31(1975), no. 1, 43-46. [9]; MR 52 #720
105. Bajpai, S. K.; Silvia, Evelyn M. *On certain classes of analytic functions.* Notices A.M.S. 20(Jan. 1973), Abstract 701-30-17, pg. A-107. [11, 69, 85]
106. Bajpai, S. K.; Srivastava, R. S. L. *On the radius of convexity and starlikeness of univalent functions.* Proc. Amer. Math. Soc. 32(1972), 153-160; MR 45 #3687. [5, 6, 10, 11, 12]
107. Balasubrahmanyam, P.; Lakshminarasimhan, T. V. *On some*

classes of functions analytic in the unit disc. Notices Amer. Math. Soc. 22, no. 7, November 1975; Abstract 75T-B251, p. A-712. [2, 21]

108. Baranova, V. A. *An estimate for the coefficient c_4 of univalent functions depending on $|c_2|$.* (Russian) Mat. Zametki 12 (1972), 127-130. {Math Notes 12(1972), 510-512(1923)}; MR 47 #5245. [4, 54]

109. Baronova, V. A.; Lebedev, N. A. *On a lemma of G. M. Golozin.* (Russian) Zap. Naucn. Sem. Leningrad. Otdel. Mat. Inst. Steklov. (LOMI) 44(1974), 7-16, 186; MR 51 #10609. [68]

110. Baranova, V. V. *Parametric representation of functions with shift symmetry.* (Russian) Theory of optimal processes (Russian), pp. 54-63. Akad. Nauk Ukrain. SSR Inst. Kibernet., Kiev, 1972; MR 48 #6385. [49]

111. Barnard, Roger W. *On quasi-starlike functions.* Notices Amer. Math. Soc. 21, January 1974, Abstract 711-30-22, p. A-123. [22, 24]

112. Barnard, Roger W. *On bounded univalent functions whose ranges contain a fixed disk.* Notices. A.M.S., vol. 19, no. 1, issue no. 135, January 1972 (Abstract 691-30-7), p. A-112. [6, 16]

113. Barnard, R. W. *On a coefficient inequality for starlike functions.* Proc. Amer. Math. Soc. 47(1975), 429-430; MR 50 #4924. [5, 6, 41, 54, 75, 79]

114. Barnard, R. W. *On the radius of starlikeness of $(zf)'$ for f univalent.* Proc. Amer. Math. Soc. 53 (1975), 385-390. [6, 11]; MR 52 #3497.

115. Barnard, R. W. *A variational technique for bounded starlike functions.* Canad. J. Math. 27 (1975), 337-347; MR 51 #8393. [6, 22, 24]

116. Barnard, R. W. *On the radius of univalence of the images of linear operators on the class S.* Notices Amer. Math. Soc. 23, January 1976, Abstract 731-30-16, p. A-101. [43, 47]

117. Barnard, R. W.; Lewis, J. L. *A counterexample to the two thirds conjecture.* Proc. Amer. Math.Soc. 41(1973), 525-529; MR 48 #4290. [6, 12, 16, 19]

118. Barnard, Roger; Lewis, John L. *Coefficient bounds for some classes of starlike functions.* Pacific J. Math. 56(1975), 325-331. [6, 36]; MR 52 #730

119. Barr, Alvin F. *The radius of univalence of certain classes of analytic functions.* Notices, A.M.S., vol. 18, no. 7, issue no. 133, November 1971 (Abstract 689-B16), p. 1056. [6, 11, 43]

120. Basevic, B. V. *The univalence of certain functions.* (Russian) Trudy Tomsk. Gos. Univ. Ser. Meh.-Mat. 189(1966), 137–143; MR 37 #2964. [57]
121. Basevic, B. V. *The univalence of certain functions. I, II.* (Russian) Trudy Tomsk. Gos. Univ. Ser. Meh.-Mat. 200(1968), 3–13; ibid. 200(1968), 14–19; MR 42 #480. [28, 60]
122. Başgöze, T. *On the radius of univalence of a polynomial.* Math. Zeit. 105(1968), 299–300; MR 37 #2958. [32, 43]
123. Başgöze, Türkân. *On the univalence of certain classes of analytic functions.* J. London Math. Soc. (2) 1 (1969), 140–144; MR 39 #5781. [11, 12, 32]
124. Başgöze, T. *On the univalence of polynomials.* Compositio Math. 22(1970), 245–252; MR 42 #481. [4, 32]
125. Başgöze, T.; Frank, J. L.; Keogh, F. R. *On convex univalent functions.* Can. J. Math. 22(1970), 123–127; MR 41 #1983. [1, 10, 20, 22]
126. Başgöze, T.; Keogh, F. R. *The Hardy class of a spirallike function and its derivative.* Proc. Amer. Math. Soc. 26 (1970), 266–269; MR 41 #8680. [3, 4, 6, 36, 62, 63]
127. Bavrin, I. I. *Some estimates for the coefficients of bounded holomorphic functions.* (Russian) Dokl. Akad. Nauk SSSR 161 (1965), 503–506; MR 30 #4928. [22, 30]
128. Bazilevič, J. E. *On dispersion of coefficients of univalent functions.* (Russian) Mat. Sb. (N.S.) 68(110) (1965), 549–560; MR 33 #267. [9, 45]
129. Bazilevič, I. E. *A criterion for univalence of regular functions and for dispersion of their coefficients.* (Russian) Mat. Sb. (N.S.) 74(116) (1967), 133–146; MR 36 #2790. [4, 54, 62]
130. Bazilevič, I. E. *Supplement to the article "On a univalence criterion for regular functions and the dispersion of their coefficients".* Math. USSR-Sb., 13(1971), no. 4, pp. 626–630; MR 44 #5441. [4, 22, 23]
131. Bazilevich, I. E.; Dziubiński, I. *Löwner's general equations for quasi-α-starlike functions.* (Russian summary) Bull. Acad. Polon. Sci. Ser. Sci. Math. Astronom. Phys. 21(1973), 823–831; MR 49 #5331. [6, 24, 63]
132. Bazilevič, I. E.; Lebediv, N. A. *Dispersion of the coefficients of functions which are p-valent in the mean.* (Russian) Mat. Sb. (N.S.) 71(113) (1966), 227–235; MR 33 #7516. [33]
133. Becker, J. *Löwnersche Differentialgleichung und quasikonform fortsetzbare schlichte Funktionen.* J. Reine Angew. Math.

255(1972), 23–43; MR 45 #8828. [1, 4, 24]
134. Becker, J. *Löwnersche Differentialgleichung und Schlicht-heitskriterien.* Math. Ann. 202(1973), 321–335; MR 49 #9197. [1, 9, 24, 58 68]
135. Becker, J. *Über homöomorphe Fortsetzung Schlichter Funktionen.* Ann. Acad. Sci. Fenn. Ser A. I. Mathematica, no. 538(1973), 11 pp.; MR 49 #9196. [1, 31]
136. Becker, Jochen. *Über eine Golusinsche Ungleichung für quasikonform fortsetzbare schlichte Funktionen.* Math. Z. 131(1973), 177–182; MR 49 #7440. [1, 9, 53]
137. Behan, D. F. *Constant multiples of a holomorphic function which are subordinate to that function.* Abstract: Notices, A.M.S. 17(1970), pp. 186–187. [1]
138. Beresniewicz-Rajca, Olga; Sladkowska, Janina. *A certain area theorem.* (Polish. Russian and English summaries). Zeszyty Nauk. Politech. Slask. Mat.-Fiz. Zeszyt 24(1974), 127–141; MR 51 #10608. [34, 86]
139. Bernardi, S. D. *Special classes of subordinate functions.* Duke Math. J. 33(1966), 55–67; MR 32 #5858. [1, 10, 20, 47, 64, 85]
140. Bernardi, S. D. *Bibliography of Schlicht functions.* Courant Institute of Mathematical Sciences, New York University, Tech. Report No. NR041-019, IMM351(1966) IX + 157pp.; MR 34 #2849. [44]
141. Bernardi, S. D. *Convex and starlike univalent functions.* Trans. Amer. Math. Soc., vol. 135, January, 1969, pp. 429–446; MR 38 #1243. [2, 5, 6, 10, 85]
142. Bernardi, S. D. *The radius of univalence of certain analytic functions.* Proc. Amer. Math. Soc. 24(1970), 312–318; MR 40 #4433. [2, 5, 6, 10, 12, 43, 73, 82, 85]
143. Bernardi, S. D. *Univalent convex maps of the unit disk.* Notices A. M. S., vol. 17, no. 2, issue no. 120, Feb. 1970 (Abstract 70T-B54), p. 441. [4, 5,6, 10, 19, 38, 64]
144. Bernardi, S. D. *Univalent convex maps of the unit circle.* II. Notices, A. M. S., vol. 17, no. 3, issue no. 121, April, 1970 (Abstract 70T-B77) p. 566. [4, 5, 6, 10, 19, 38, 64]
145. Bernardi, S. D. *Univalent convex maps of the unit circle.* III. Notices, A. M. S., vol. 18, no. 4, issue no. 130, June 1971 (Abstract 71T-B113), p. 640. [4, 5, 6, 10, 19, 38, 64]
146. Bernardi, S. D. *The radius of univalence and starlikeness of certain classes of analytic functions.* Notices, Amer. Math. Soc., vol. 20, no. 3, April 1973 (Abstract 73T-B135, A-332). [6, 11, 43]

147. Bernardi, S. D. *New distortion theorems for functions of positive real part and applications to univalent convex functions.* Proc. Amer. Math. Soc. 45(1974), 113-118; MR 50 #10223. [2, 10, 11, 20, 56, 64]

148. Bielecki, A. *Quelques résultats récents sur les majorantes dans la théorie des fonctions holomorphes.* Colloq. Math., vol. 11, Fasc. 2(1964), 141-145. [1]

149. Bielecki, A.; Leandowski, Z. *Sur certaines majorantes des fonctions holomorphes dans le cercle unité.* Colloq. Math., vol. 9, Fasc. 2(1962), 299-303. [1]

150. Blerski, F. *L'évaluation des coefficients d'une classe de fonctions analytiques et univalentes dans l'anneau circulaire.* Zeszyty Nauk Univ. Jagiello. Prace Mat. No. 7 (1962), 13-16; MR 34 #2858. [39, 48]

151. Binmore, K. G.; Kirwan, W. E. *On the coefficients of typically real functions.* Duke Math. J. 36(1969), 455-464; MR 39 #7087. [13, 23, 51, 62]

152. Blevins, D. K. *Conformal mappings of domains bounded by quasiconformal circles.* Duke math. J. 40(1973), 877-883; MR 48 #4285. [19, 24, 54]

153. Blevins, D. K. *Covering theorems for univalent functions mapping onto domains bounded by quasi-circles.* Notices, Amer. Math. Soc. 23, January 1976, Abstract 731-30-20, p. A-102. [19, 49]

154. Bogowski, F. *Majoration en module et en domaine dans les classes de fonctions bornées pour les minorantes $f(z) \in H_o$.* (English and Russian summaries) Bull. Acad. Polon. Sci. Sér. Sci. Math. Astronom. Phys. 22(1974), 35-38; MR 50 #2478. [1, 22]

155. Bogowski, F.; Jablonski, F. F.; Stankiewicz, Jan. *Subordination en domaine et inégalités des modules pour certaines classes de fonctions holomorphes dans le cercle unité.* Ann. Univ. Mariae Curie-Sklodowska Sec. A 20 (1966), 23-28 (1971); MR 46 #3759. [1, 6, 10]

156. Bogowski, F.; Lewandowski, Z. *Sur la majoration en module et en domaine dan les classes de fonctions bornées.* (English and Russian summaries) Bull. Acad. Polon. Sci. Sér. Sci. Math. Astronom. 22(1974), 385-391; MR 49 #5340. [1, 22]

157. Bogowski, F.; Stankiewicz, Z. *Sur la majoration modulaire des fonctions et l'inclusion des domannes dans la class $S_{1/2}$.* Ann. Univ. Mariae Curie-Sklodowska Sect. A 25(1971), 5-14 (1973); MR 48 #11465. [1]

158. Bogowski, F.; Stankiewicz, Z. *Généralisation d'une problème relatif à la subordination en module et à la subordination en module et à la*

subordination en domaine dans le cas des minorantes de la class Ho. (Polish and Russian summaries) Ann. Univ. Mariae Curie-Sklodowska Sect. A 25 (1971), 15-25 (1973); MR 48 #11466. [1]
159. Bogucki, Zbigniew. *On a theorem of M. Biernacki concerning subordinate functions*. (Polish and Russian summaries) Ann. Univ. Mariae Curie-Sklodowska Sect. A 19(1965), 5-10 (1970); MR 41 #7081. [1]
160. Bogucki, Z.; Waniurski, J. *On the minorant sets for univalent functions*. Ann. Univ. Mariae Curie-Sklodowska, Sect. A, vols. 22/23/24 (1968/1969/1970), pp. 33-38. [6, 29]; MR 52 #722
161. Bogucki, Zbigniew; Waniurski, Jozef. *On a theorem of M. Biernacki concerning convex majorants*. (Polish and Russian summaries) Ann. Univ. Mariae Curie-Sklodowska Sect. A 19(1965), 11-15(1970); MR 41 #7082. [1, 10]
162. Bogucki, Z.; Waniurski, J. *The relative growth of subordinate functions*. Mich. Math. J. 18(1971), 357-364; MR 45 #8816. [1, 6, 10, 18, 62]
163. Bogucki, Z.; Waniurski, J. *On bounded spiral-like functions whose values cover a fixed disk*. Bull. Acad. Polon. Sci. Ser. Sci. Math. Astronom. Phys., 19(1971), no. 11, pp. 983-988; MR 46 #3760. [6, 19, 36, 63]
164. Bogucki, Z.; Waniurski, J. *The location of zeroes of subordinate functions and its influence on the radius of majorization*. Bull. Acad. Polon. Sci. Ser. Sci. Math. Astronom. Phys., 19(1971), no. 11, pp. 989-996; MR 46 #3776. [1]
165. Bogucki, Z.; Waniurski, J. *On the relative growth of subordinate functions*. (Polish and Russian summaries) Ann. Univ. Mariae Curie-Sklodowska Sect. A 22-24(1968/70), 29-32(1972); MR 49 #5332. [1, 6, 10]
166. Bogucki, Z.; Waniurski, J. *On univalent functions whose values cover a fixed disk*. Ann. Univ. Mariae Curie-Sklodowska Sect. A 22-24(1968/70), 39-44(1972). [6, 10, 22]
167. Bombieri, Enrico. *On the local maximum property of the Koebe function*. Invent. Math. 4(1967), 26-67; MR 36 #1635. [24, 42, 54]
168. Bombieri, Enroci. *Sulla seconda variazione della funzione di Koebe*. (English summary) Boll. Un. Mat. Ital. (3)22 (1967), 25-32; MR 35 #6813. [24, 42]
169. Bombieri, Enrico. *A geometric approach to coefficient inequalities for univalent functions*. Ann. Scuola Norm. Sup. Pisa (3)22(1968), 377-397; MR 39 #430. [24, 54]
170. Boyd, A. V. *Coefficient estimates for starlike functions of order* α.

Proc. Amer. Math. Soc. 17(1966), 1016–1018; MR 33 #7517. [6, 22, 30, 36, 49]
171. Brannan, D. A. *Coefficient regions for univalent polynomials of small degree.* Mathematika 14(1967), 165–169; MR 36 #3971. [4, 29, 32]
172. Brannan, D. A. *On functions of bounded boundary rotation.* I. Proc. of the Edinburgh Math. Soc. (2) 16 (1968/69), 339–347; MR 41 #8642. [3, 18, 71, 76, 77]
173. Brannan, D. A. *On functions of bounded boundary rotation.* II. Bull. London Math. Soc. 1(1969), 321–322; MR 41 #8643. [71, 77]
174. Brannan, D. A. *On univalent polynomials.* Glasgow Math. J. 11(1970), 102–107; MR 43 #2204. [4, 32]
175. Brannan, D. A. *On coefficient problems for certain power series.* Proceedings of the Symposium on Complex Analysis (Univ. Kent, Canterbury, 1973), pp. 17–27. London Math. Soc. Lecture Note Ser., No. 12, Cambridge Univ. Press, London, 1974. [44, 54]
176. Brannan, D. A.; Clunie, J.; Kirwan, W. E. *Coefficient estimates for a class of star-like functions.* Canad. J. Math. 22(1970), 476–485; MR 41 #5614. [1, 2, 6, 9, 10, 25, 36]
177. Brannan, D. A.; Clunie, J. G.; Kirwan, W. E. *On the coefficient problem for functions of bounded boundary rotation.* Ann. Acad. Sci. Fenn. Ser. A. I. Mathematics, no. 523 (1973), 1–18; MR 49 #3108. [1, 16, 71, 77]
178. Brannan, D. A.; Kirwan, W. E. *A covering theorem for typically real functions.* Glasgow Math. J. 10(1969), 153–155; MR 40 #7431. [11, 13, 19, 23, 43, 66]
179. Brannan, D. A.; Kirwan, W. E. *On some classes of bounded univalent functions.* J. London Math. Soc. (2)1(1969), 431–443; MR 40 #4439. [6, 7, 10, 22, 36, 71]
180. Brannan, D. A.; Kirwan, W. E. *The Maclaurin Coefficients of Bounded Convex Functions.* Bull. London Math. Soc., 2 (1970), 159–164; MR 42 #6211. [10, 22, 30]
181. Brannan, D. A.; Kirwan, W. E. *The growth of the maximum modulus of univalent functions.* Duke Math. J. 38(1971), 805–818; MR 45 #5334. [5, 6, 62]
182. Brickman, L. *Extreme points of the set of univalent functions.* Bull. Amer. Math. Soc. 76(1970), 372–374; MR 41 #448. [88]
183. Brickman, L. *Subordinate families of analytic functions.* Illinois J. Math. 15(1971), 241–248; MR 43 #2205. [1, 4, 10]
184. Brickman, Louis. *Extremal problems for certain classes of analytic functions.* Proc. Amer. Math. Soc. 35(1972), 67–73; MR 46 #7500.

[2, 6, 23, 24]
185. Brickman, Louis. ϕ-like analytic functions. I. Bull. Amer. Math. Soc. 79(1973), 555-558; MR 48 #8769. [4, 6]
186. Brickman, Louis. Certain algebraic functions and extreme points of S. Michigan Math. J. 22(1975), no. 3, 201-203. [88]
187. Brickman, L.; MacGregor, T. H.; Wilken, D. R. Convex hulls of some classical families of univalent functions. Trans. Amer. Math. Soc. 156(1971), 91-107; MR 43 #494. [16, 23, 42, 88]
188. Brickman, L.; Hallenbeck, D. J.; MacGregor, T. H.; Wilken, D. R. Convex hulls and extreme points of families of star-like and convex mappings. Trans. Amer. Math. Soc. 185 (1973), 413-428; MR 49 #3102. [1, 6, 10, 36, 88]
189. Brickman, Louis; Wilken, D. R. Support points of the set of univalent functions. Proc. Amer. Math. Soc. 42(1974), 523-528; MR 48 #6399. [62, 88]
190. Brodovic, M. T. A certain sufficient condition for the conformality of an arbitrary one-to-one mapping. (English and Russian summaries) Dopovidi Akad. Nauk Ukrain. RSR Ser. A 1974, 489-492, 572; MR 51 #8388. [4]
191. Bucka, Cz.; Ciozda, K. On a new subclass of the class S. Ann. Polon. Math. 28(1973), 153-161; MR 49 #3109. [6, 23, 29, 36, 63]
192. Bucka, Cz.; Ciozda, K. Sur une classe de fonctions univalentes. Ann. Polon. Math. 28(1973), 233-238; MR 48 #6386. [6, 23, 29, 36, 63]
193. Buckholtz, J. D. Zeroes of partial sums of power series. II. Notices, A.M.S. (Abstract), vol. 16, no. 3, issue no. 113, April, 1969, p. 540. [20]
194. Burdick, Gary R. On a ratio of a univalent function. Notices, A. M. S. 20(Jan. 1973), Abstract 701-30-24, pg. A-109 (Preliminary report). [5, 6, 10, 63, 64, 74]
195. Burstein, L. H. Extremal problems for certain classes of analytic functions. (Russian) Izv. Vyss. Ucebn. Zaved. Matematika 1969, no. 6(85), 9-16; MR 40 #4451. [2, 5, 6, 10, 24]
196. Burstein, L. H. On the question of conformal transformations of the disc onto nonoverlapping regions. (Russian) Mat. Zametki 6(1969), 417-424. MR 40 #5845. [26]
197. Burstein, L. H. The roots of the equation $f(z) = \alpha f(a)$ in a class of typically real functions. (Russian) Mat. Zametki 10(1971), 41-52; MR 44 #4198 {English translation: Math. Notes 10(1971), 449-455}. [13, 29]
198. Burstein, L. H. The solution of the equation $f(z) = \alpha f(\eta z)$ in the class of star-shaped functions. (Russian) Iav. Vyss. Ucebn. Zaved.

Mat. 1974, no. 12(151), 47–50. [6]; MR 52 #8398
199. Busklein, F.; Waadeland, H. *A geometric property for certain slit mappings*. (Norwegian. English summary) Noridsk Mat. Tidskr. 21(1973), 33–39, 56; MR 49 #549. [49]
200. Bustoz, J. *Jacobi polynomial sums and univalent Cesaro means*. Proc. Amer. Math. Soc. 50(1975), 259–264; MR 51 #5917. [16, 20]
201. Byers, R. B. *On a strenthened form of the 1/4 theorem for univalent functions*. Notices, Amer. Math. Soc., vol. 20, no. 3, April 1973 (Abstract 73T–B115, A–327). [19, 39]
202. Byers, R. B. *Some linear subordination results for classes of univalent functions*. Proc. Amer. Math. Soc. 47(1975), 143–146; MR 51 #873. [1, 4, 6, 10]
203. Calys, E. G. *A class of regular functions*. (Abstract), Notices, A. M. S., vol. 15, no. 6, issue no. 108, October 1968, p. 919. [5, 85]
204. Calys, E. G. *On a class of regular functions*. Notices, A. M. S., vol. 17, no. 2, issue no. 120, Feb. 1970 (Abstract). [2, 7, 18, 20, 21]
205. Calys, E. G. *The radius of univalence and starlikeness of some classes of regular functions*. Composito Math., 23(1971), Fasc. 4, pp. 467–470. [11, 43, 82, 85]
206. Campbell, D. M. *Locally univalent functions with locally univalent derivatives*. Trans. Amer. Math. Soc. 162(1971), 395–410; MR 44 #4199. [43]
207. Campbell, D. M. *Majorization-subordination theorems for locally univalent functions*. Bull. Amer. Math. Soc., 78 (1972), no. 4, pp. 535–538; MR 45 #8817. [1]
208. Campbell, Douglas Michael. *Majorization-subordination theorems for locally univalent functions*. II. Canad. J. Math. 25(1973), 420–425; MR 47 #3669. [1]
209. Campbell, D. *Uniform convergence on compacta and locally univalent analytic functions of finite order*. Monatsh. Math. 77(1973), 21–23; MR 47 #3664.
210. Campbell, D. M. *Eventually areally mean p-valent functions*. Israel J. Math. 16(1973), 216–236; MR 48 #11480. [33, 62]
211. Campbell, D. M. *The radius of convexity of a linear combination of functions. . . .* Can. J. Math. 25(1973), 982–985; MR 48 #6387. [12, 82]
212. Campbell, D. M. *Applications and proof of a uniqueness theorem for linear invariant families of finite order*. Rocky Mountain J. Math. 4(1974), 621–634; MR 50 #10235. [1, 10, 12, 19, 33, 43, 54, 62, 68]
213. Campbell, D. M. *Majorization-subordination theorems for locally*

univalent functions. III. Trans. Amer. Math. Soc. 198(1974), 297-306; MR 50 #2480. [1, 22, 61]
214. Campbell, D. M. *A survey of properties of the convex combination of univalent functions*. Rocky Mountain J. Math. 5(1975), no. 4, 475-492. [4, 5, 6, 10, 11, 12, 16, 43, 44, 71, 73, 82, 85]; MR 52 #3498
215. Campbell, D. M. *The limiting behavior of $zf''(z)/f'(z)$ and two conjectures on univalent functions*. Notices, Amer. Math. Soc., January 1975, Abstract 720-30-11, p. A-120. [16]
216. Campbell, D. M. *Generalized Bazilevič functions*. Notices Amer. Math. Soc. 23, January 1976, Abstract 731-30-30, p. A-104. [70]
217. Campbell, D. M.; Cima, Joseph A.; Pfaltzgraff, John A. *Linear spaces and linear-invariant families of locally univalent analytic functions*. Manuscripta Math. 4(1971), 1-30; MR 44 #5492. [4]
218. Campbell, D.; Eenigenburg, P.; Nelson, D. *The set of Univalent Functions as a subset of a certain Banach space*. Notices Amer. Math. Soc. 23, January 1976, Abstract 730-30-12, p. A-100.
219. Campbell, D. M.; Pearce, K. *An extension of the Bazilevič Functions*. (Preliminary report) Notices Amer. Math. Soc. 22, April 1975, Abstract 724-B2, p. A-428. [70]
220. Campbell, D. M.; Pfaltzgraff, J. A. *Properties of the generalized Koebe function*. Notices, A. M. S., vol. 19, no. 5, issue 139, August, 1972 (Abstract 696-30-6), p. A-637. [4, 5, 6, 12]
221. Campbell, D. M.; Pfaltzgraff, J. A. *Mapping properties of log $g'(z)$*. Colloq. Math. 32(1975), fasc. 2, 267-276, 310. [5, 6, 10, 49, 62, 73]; MR 51 #13208
222. Campbell, D. M.; Ziegler, M. R. *Argument of the derivative of linear-invariant families of finite order and the radius of close-to-convexity*. Notices, A. M. S. 20(Jan. 1973), Abstract 701-30-12, pg. A-106. [68, 73]
223. Cantrell, Grady Leon. *A distortion theorem for certain classes of analytic functions*. Notices, A. M. S., vol 16, October 1969, p. 906 (Abstract).
224. Caplinger, T. R. *On certain classes of analytic functions*. Notices, A. M. S. 19(1972), Abstract 72T-B202. pg. A-583. [2, 11, 43, 56, 85]
225. Caplinger, T. R.; Causey, W. M. *A class of univalent functions*. Proc. Amer. Math. Soc. 39(1973), 357-361; MR 47 #8833. [4, 6, 7, 12, 18]
226. Carrier, G. F.; Krook, M. N. B. *Functions of a complex variable: Theory and technique*. McGraw-Hill Book Co., New York-

Toronto, Ont.-London, 1966. IX + 438 pp.; MR 36 #5308. [44]
227. Causey, W. M. *The close-to-convexity and Univalence of an Integral.* Math. Z99(1967), 207–212; MR 35 #6807. [4, 5, 10, 31]
228. Causey, W. M. *The univalence of an integral.* Proc. Amer. Math. Soc. 27(1971), 500–502; MR 43 #6419. [31, 43, 85]
229. Causey, W. M.; Merkes, E. P. *Radii of starlikeness of certain classes of analytic functions.* J. Math. Anal. Appl. 31(1970), 579–586; MR 41 #8644. [6, 11]
230. Celikkanat, L. *Variational method and α-starlike functions.* (Turkish summary) Comm. Fac. Sci. Univ. Ankara Ser A 20(1971), 53–69; MR 48 #4291. [6, 9, 24, 63]
231. Cernei, N. I. *Tests for the stable convexity of a domain in the case of univalent conformal mappings.* I. (Russian) Ukrain. Mat. Z18(1966), no. 1, 86–91; MR 34 #2850. [10]
232. Cernei, N. I. *Criteria for stable convexity of regions under conformal mappings.* II. (Russian) Ukrain. Mat. Z. 18(1966), no. 5, 84–93; MR 34 #6058. [10]
233. Cernei, N. I. *Certain theorems on the stable convexity of closed curves under univalent mappings.* (Russian) Ukrain. Mat. Z. 23(1971), 276–280; MR 44 #2917 {English transl: Ukrainian Math. J. 23(1971), 241–244}.
234. Černikov, V. V. *Limitations on the comparative growth of the modulus of a function and the modulus of its derivative in a class of functions with real coefficients.* (Russian) Trudy Tomsk. Gos. Univ. 163(1963), 26–27. [39, 62]
235. Černikov, V. V. *On the domain of values of the functional $J = f(w)$ in the class of bounded functions with real coefficients.* (Russian). Trudy Tomsk. Gos. Univ. Ser. Meh.-Mat. 175(1964), Vyp. 2, 78–84; MR 31 #324. [24, 35]
236. Černikov, V. V. *An extremal problem in the class of bounded functions with real coefficients.* (Russian) Trudy Tomsk. Gos. Univ. Ser. Meh.-Mat. 182(1965), Vyp. 3, 96–105; MR 33 #5862. [22, 39, 61]
237. Černikov, V. V. *Ranges of values of certain functionals on the class of bounded functions with real coefficients.* (Russian). Trudy Tomsk. Gos. Univ. Ser. Meh.-Mat. 189 (1966), 206–210; MR 37 #1580. [22, 29, 39]
238. Černikov, V. V. *Domain of values of a certain functional in the class of bounded univalent functions with real coefficients.* (Russian) Sibirsk. Mat. Z 7(1966), 200–205; MR 33 #717. [22, 24, 29, 39]
239. Černikov, V. V. *The α-convexity of univalent functions.* (Russian)

Mat. Zametki 11(1972), 227–232; MR 45 #7035. {English transl: Math Notes 11(1972), 141–144}. [12, 69]

240. Černikov, V. V.; Kosevarov, V. F. *The range of values of a certain functional in the class of schlicht functions with real coefficients.* (Russian) Trud. Tomsk. Gos. Univ. Ser. Meh.-Mat. 210(1969), 122–129; MR 43 #5019. [29, 39]

241. Černikov, V. V.; Sizuk, P. I. *Distortion theorems for univalent functions with real coefficients.* (Russian) Trudy Tomsk. Gos. Univ. Ser. Meh.-Mat. 189(1966), 211–220; MR 37 #1581. [39, 68]

242. Černikov, V. V.; Sizuk, P. I. *A certain property of univalent functions with real coefficients.* (Russian) Trudy Tomsk. Gos. Univ. Ser. Meh.-Mat. 200(1968), 189–204; MR 41 #1984. [29, 39]

243. Černikov, V. V.; Sizuk, P. I. *Certain estimates of the coefficients of univalent functions.* (Russian) Trud. Tomsk. Gos. Univ. Ser. Meh.-Mat. 200(1968), 205–216; MR 41 #3745. [9, 25, 54]

244. Černikov, V. V.; Sizuk, P. I. *Certain geometric properties of schlicht functions with real coefficients.* (Russian) Trud. Tomsk. Gos. Univ. Ser. Meh.-Mat. 210(1969), 130–142; MR 43 #7609. [29, 39]

245. Chamberlain, E. W. *The Univalence of Functions Asymptotic to Nonconstant Logarithmic Monomials.* Proc. Amer. Math. Soc. 17(1966), 302–309; MR 32 #7724. [4, 49]

246. Chandra, Susheel. *On certain subclasses of regular and p-valent in the unit disc.* Notices, A. M. S. 20(Feb. 1973), Abstract 73T–B92. [14]

247. Chandra, S.; Singh, P. *On certain classes of the analytic functions.* Indian J. Pure Appl. Math. 4(1973), 745–748; MR 50 #2470. [2, 4, 10, 11, 12, 64]

248. Chandra, S.; Singh, P. *Certain subclasses of the class of functions regular and univalent in the unit disc.* Arch. Math. (Basel) 26(1975), 60–63; MR 51 #3413. [4, 5, 6, 10, 39, 82, 85]

249. Charzyński, Z.; Ławrynowicz, J. *On the coefficients of univalent polynomials.* Colloq. Math. 16(1967), 27–33; MR 35 #1768. [32]

250. Charzyński, Z.; Sladkowska, J. *Algebraic functions and analytic variations of univalent functions.* (Loose Russian summary) Bull. Acad. Polon. Sci. Ser. Sci. Math. Astronom. Phys. 16(1968), 793–794; MR 38 #4664. [24]

251. Charzyński, Z.; Sladkowska, J. *Fonctions algebriques et variations analytiques des fonctions univalentes.* Dissertationes Math. Rozprawy Mat. 70(1970), 77 pp.; MR 43 #495. [24]

252. Chase, W. E. *p-close-to-convex functions.* I. Notices, A. M. S.,

vol. 15, no. 5, issue no. 107, August 1968 (Abstract 658-105), p. 749. [4, 5, 16, 45, 67, 74]
253. Chase, W. E. *p-close-to-star functions*. Abstract: Notices, A. M. S. 17(1970), p. 172. [6, 11, 36, 43, 63]
254. Chen, Ming Po. *The radius of starlikeness of certain analytic functions*. Bull. Inst. Math. Acad. Sinical (1973), no. 2, 181-190; MR 49 #10866; Tamkang J. Math. 4 (1973), no. 2, 57-67. MR 50 #587. [2, 6, 10, 11]
255. Chen, Ming Po. *On the convexity of some analytic functions*. Tamkang J. Math. 4(1973), 61-71; MR 50 #13478. [2, 4, 12]
256. Chen, Ming Po. *On functions satisfying $Re\{f(z)/z\} > \alpha$*. Tamkang J. Math. 5(1974), 231-234; MR 51 #874. [2, 43, 56]
257. Chen, Ming Po. *The radius of univalence and starlikeness of certain classes of analytic functions*. Comment. Math. Univ. St. Paul 23(1974/75), fasc. 2, 139-145; MR 51 #10602. [2, 6, 11, 43]
258. Chen, Ming Po. *On the regular functions satisfying $Re\{f(z)/z\} > \alpha$*. Bull. Inst. Math. Acad. Sinica 3(1975), no. 1, 65-70. [2]
259. Chernei, N. I. *Some theorems of stable convexity of closed curves under one-sheeted mappings*. Ukrain. Mat. Z. 23(1971), 276-280. [10]; MR 44 #2917 (See reference 233)
260. Chiang, Pou-shun; Macintyre, A. J. *Upper bounds for a Bloch constant*. Proc. Amer. Math. Soc. 17(1966), 26-31; MR 32 #5883. [19]
261. Chernoff, P. R. *An area-width inequality for convex curves*. Amer. Math. Monthly, Vol. 76, no. 1, January, 1969, pp. 34-35. [10, 18]
262. Chiang, P. S.; Macintyre, A. J. *Some theorems of Bloch type*. Proc. A.M.S. 18(1967), 423-424; MR 35 #5592. [19, 34]
263. Chiang, P. S. *Computer investigation of Landau's theorem*. Math. comp. 23(1969), 185-188; MR 39 #2950. [19]
264. Chiba, T. *On convex univalent functions*. Kenkyu Kiyo-Gakushuin Kotoka No. 7, 1-8(1975). [10]
265. Chichra, Pran Nath. *An area theorem for bounded univalent functions*. Proc. Cambridge Philos. Soc. 66(1969), 317-321; MR 39 #5782. [2, 9, 22, 25, 27, 31, 56, 61, 68]
266. Chichra, Pran Nath. *A theorem on convex schlicht functions*. Ganita 20(1969), no. 2, 77-78; MR 42 #4717. [10, 19, 64, 68]
267. Chichra, P. N. *On the radii of starlikeness and convexity of certain classes of regular functions*. J. Austral. Math. Soc., 13(1972), part 2, pp. 208-218; MR 45 #8818. [2, 11, 12, 13, 56]
268. Chichra, Pran Nath. *Regular functions $f(z)$ for which $zf'(z)$ is α-spiral-like*. Proc. Amer. Math. Soc. 49(1975), 151-160; MR 50

#13479. [2, 4, 6, 10, 12, 16, 31, 56]
269. Chichra, Pran Nath; Singh, Ram. *Convex sum of univalent functions.* J. Austral. Math. Soc. 14(1972), 503-507; MR 47 #7014. [6, 10, 45, 63, 82, 85]
270. Cima, J. A. *On the dual of a space of locally schlicht functions.* Notices, A. M. S., vol. 18, no. 5, issue no. 131, (Abstract 687-30-3), August 1971, p. 771.
271. Cima, J. *On the log of a schlicht function.* Notices, Amer. Math. Soc., January 1975, Abstract 720-30-10, p. A-120. [68]
272. Cima, J. A. *Hadamard products of convex functions.* Notices, Amer. Math. Soc. 23, February 1976, Abstract 732-B1, p. A-304. [1, 10, 47]
273. Cima, J. A.; Pfaltzgraff, J. A. *A banach space of locally univalent functions.* Mich. Math. J. 17(1970), no. 4, pp. 321-334. [4, 5, 6, 10, 23, 88]; MR 43 #3784
274. Cima, J. A.; Pfaltzgraff, J. A. *Oscillatory behavior of $u'' + hu = 0$ for schlicht h.* Notices, A. M. S. (Abstract), vol. 17, no. 5, issue no. 123, August, 1970, p. 772. [31]
275. Cima, J. A.; Pfaltzgraff, J. A. *A normed linear space containing the schlicht functions.* Monatsh. Math., 75(1971), no. 4, pp. 296-302; MR 46 #7531.
276. Clunie, J. *On the derivative of a bounded function.* Proc. London Math. Soc. (3) 14A(1965), 58-68. [22]
277. Clunie, J. *On the coefficients of a class of univalent functions.* Entire Functions and Related Parts of Analysis (Proc. Sympos. Pure Math., La Jolla, Calif., 1966), pp. 171-178. Amer. Math. Soc., Providence, R. I. 1968. [54]
278. Clunie, J.; Duren, P. L. *Addendum: An arclength problem for close-to-convex functions.* J. London Math. Soc. 41 (1966), 181-182; MR 32 #7725. [2, 5, 7, 16, 23]
279. Clunie, J.; Hayman, W. K. *The spherical derivative of integral and meromorphic functions.* Comment. Math. Helv. 40(1966), 117-148; MR 33 #282. [9, 46, 62]
280. Clunie, J.; Hayman, W. K.; et al. *Proceedings of the Symposium on Complex Analysis.* Held at the University of Kent, Canterbury, 1973. Edited by J. Clunie and W. K. Hayman. London Math. Soc. Lecture Note Series, No. 12. Cambridge University Press, London-New York, 1974. VI + 180 pp. [44]; MR 52 #5944
281. Clunie, J.; Pommerenke, Ch. *On the coefficients of close-to-convex univalent functions.* J. London Math. Soc. 41 (1966), 161-165; MR 32 #7734. [4, 5, 6, 22, 30, 36, 38, 54]

282. Clunie, J.; Pommerenke, Ch. *On the coefficients of univalent functions.* Michigan Math. J. 14(1967), 71–78; MR 34 #7786. [3, 9, 14, 25, 27, 33, 50]
283. Coban, M. M. *Multivalued mappings and some associated problems.* Soviet Math. Dokl. 11, no. 1(1970), 105–108. [14]
284. Cochrane, P. C.; MacGregor, T. H. *Frechet differentiable functionals and support points for families of analytic functions.* Notices, A. M. S., vol. 22, no. 4, issue no. 162, June 1975 (Abstract 75T–B117), p. A–460. [88]
285. Cohn, Harvey. *Conformal mapping on Riemann Surfaces.* McGraw-Hill Book Co., New York-Toronto, Ont.-London, 1967, XIV + 325 pp.; MR 36 #3974. [44]
286. Coonce, H. B. *On the fourth coefficient for functions of bounded boundary rotation.* Abstract: Notices, A. M. S. 17(1970), p. 129. [71, 77]
287. Coonce, H. B. *Functions of bounded radius rotation.* Notices, A. M. S., vol. 18, no. 1, issue no. 127, January 1971 (Abstract 682-30-18), p. 149. [7, 11, 24, 62]
288. Coonce, H. B. *A variational method for functions of bounded boundary rotation.* Trans. Amer. Math. Soc. 157 (1971), 39–51; MR 43 #497. [4, 12, 16, 23, 24, 43, 71, 76, 77]
289. Coonce, Harry B. *Kaplan-Mocanu functions.* Notices, Amer. Math. Soc. 21, April 1974, Abstract 74T–B90, p. A–374. [4, 69, 70]
290. Coonce, H. B. *Some conditions for univalence.* Notices, Amer. Math. Soc. 23, January 1976, Abstract 731-30-31, p. A–104. [4, 69, 70]
291. Coonce, H. B.; Miller, S. S. *Distortion properties of p-fold symmetric alpha-starlike functions.* Proc. Amer. Math. Soc. 44(1974), 336–340; MR 49 #9179. [6, 10, 19, 45, 49, 69, 83, 84]
292. Coonce, H. B.; Ziegler, M. R. *The radius of close-to-convexity of functions of bounded boundary rotation.* Proc. Amer. Math. Soc. 35(1972), 207–210; MR 45 #5335. [5, 12, 71, 73]
293. Coonce, Harry B.; Ziegler, Michael R. *Functions with bounded Mocanu variation.* Rev. Roumaine Math. Pures Appl. 19(1974), 1093–1104; MR 50 #10224. [4, 6, 10, 23, 43, 69, 70, 71, 76, 83]
294. Cowling, V. F.; Royster, W. C. *Domains of Variability for Univalent Polynomials.* Proc. Amer. Math. Soc. 19, no. 4, August 1968, pp. 767–772; MR 37 #2976. [4, 32]
295. Craig, C.; Macintyre, A. J. *Inequalities for functions regular and bounded in a circle.* Pacific J. Math. 20(1967), 449–454; MR 34 #7806. [22, 49, 61]

296. Curtiss, J. H. *Faber polynomials and the Faber series.* The Amer. Math. Monthly 78(1971), 577–596. [4, 9, 25, 27, 53]; MR 45 #2183, #8893
297. Czubak, Jerzy. *Extremal limit functions for certain functionals in a family of univalent functions that are near the identity.* (Polish. French summary) Zeszyty Nauk. Univ. Łódźk. Nauki Mat. Przyrod. Ser. II Zeszyt 39 (Mat.) (1971), 3–30; MR 48 #11467. [22, 61]
298. Czubak, J. *On the radius of star-shapedness of some families of holomorphic functions k-symmetric in the unit disc.* (Polish and Russian summaries) Zeszyty Nauk Politech. Łódź. No. 208 Mat. No. 6(1975), 31–45. [11]; MR 52 #3499
299. Czubak, J. *Sharp estimation of (a certain functional) in the class univalent k-symmetric functions.* (Polish and Russian summaries) Zeszyty Nauk Politech. Łódź. No. 208 Mat. No. 6(1975), 77–105. [49]; MR 52 #3506
300. Dajovic, V. *On boundary values of typically real functions and the resultant of some classes of functions.* (Serbo-Croatian, French summary) Bull. Soc. Math. Phys. Serbie 15(1963), 51–55; MR 31 #5978. [13, 41, 47]
301. Dajovic, V. *Some properties of typically real functions.* (Serbo-Croatian. French summary) Mat. Vesnik 4(19) (1967), 169–172; MR 37 #1619. [13]
302. Dajovic, V. *Quelques theoremes d'existence des valerus limites du produit d'Hadamard.* Math. Balkanica 4(1974), 111–114; MR 51 #8384. [47]
303. Davis, Philip J. *The schwarz function and its applications.* The Carus Mathematical Monographs, No. 17. The Mathematical Association of America, Buffalo, N.Y., 1974. XI + 228 pp. [4]
304. De Temple, D. W. *Generalizations of the Grunsky-Nehari Inequalities.* Stanford Univ. (Dissertation) May 18, 1970, 99 pp. [53]
305. De Temple, Duane W. *On coefficient inequalities for bounded univalent functions.* Ann. Acad. Sci. Fenn. (Ser. A) I, vol. 469 (1970), 1–20; MR 43 #5020. [19, 22, 30]
306. De Temple, Duane W. *Generalizations of the Grunsky-Nehari Inequalities.* Arch. Rational Mech. Anal., vol. 44(1971/72), 93–120; MR 49 #556. [19, 22, 24, 30, 53, 54]
307. De Temple, D. W. *Grunsky-Nehari inequalities for a subclass of bounded univalent functions.* Trans. Amer. Math. Soc. 159 (1971), 317–328; MR 43 #5022. [1, 22, 34, 53]
308. De Temple, D. W. *An area method for systems of univalent functions whose ranges do not overlap.* Math. Z. 128(1972), 23–33; MR

48 #2366. [22, 26, 34, 49, 53, 54, 61]
309. De Temple, Duane W. *Further generalizations of the Nehari inequalities.* Trans. Amer. Math. Soc. 205(1975), 333–340; MR 51 #876. [22, 30, 45, 53]
310. Demahovskaja, R. I. *An extremal problem of the theory of special classes of schlicht functions.* (Russian) Problems of Math. Phys. and Theory of Functions, II (Russian), pp. 22–31. Naukova Dumka, Kiev, 1964; MR 32 #7726. [7]
311. Demahovskaja, R. I. *On the estimation of mean moduli in the theory of special classes of analytical functions.* (Ukrainian. Russian and English summaries) Dopovidi Akad. Nauk Ukrain. RSR 1965, 1266–1270; MR 32 #7727. [7, 18]
312. Dennler, G. *Eine Verallgemeinerung der Golusinschen Variationsmethode.* Dissertation, Jena, 1963. IV + 69 pp.; MR 32 #2573. [24]
313. Dobrowolska, Krystyna. *On some extremal problems in the class of quasi-starlike functions.* Ann. Univ. Mariae Curie-Sklodowska, Sect. A, vols. 22/23/24/(1968/1969/1970), pp. 45–52. [6, 19, 24]; MR 52 #723
314. Dobrowolska, Krystyna. *On meromorphic quasi-starlike functions.* Ann. Univ. Mariae Curie-Sklodowska, Sect. A, vols. 22/23/24 (1968/1969/1970), pp. 53–61. [6, 9, 25, 29]; MR 52 #724
315. Dobrowolska, Krystyna. *On k-symmetric quasi-starlike meromorphic functions.* Demonstratio Math. 4(1972), 251–266; MR 49 #10867. [9, 25]
316. Dobrowolska, K. *Estimation of the coefficients a_1, a_3, a_5 in the class of odd quasi-starlike meromorphic functions.* Demonstratio Math. 5(1973), 29–44; MR 48 #2367. [9, 25]
317. Dobrowolska, Krystyna. *The generalized Lowner equation in the class of quasi-α-starlike meromorphic k-symmetric functions.* Demonstratio Math. 1(1974), 93–107; MR 50 #4925. [9, 49]
318. Dobrowolska, Krystyna; Dziubiński, I. *On quasi α-convex functions.* (Polish and Russian summaries) Zeszyty Nauk. Politech. Lódź. No. 208 Mat. No. 6(1975), 5–21. [69]; MR 52 #8399
319. Domzal, Jan. *The ranges of certain functionals.* (Polish. English summary) Zeszyty Nauk. Univ. Lódźk. Nauki Mat. Przyrod. Ser. II Zeszyt 39(mat.)(1971), 31–42; MR 48 #11468. [2, 6, 13, 23, 29]
320. Domzal, Jan. *Extremal problems in the class $S(\alpha, \beta, \gamma)$.* (Polish and Russian summaries) Zeszyty Nauk. Politech. Lodz. No. 208 Mat. No. 6(1975), 23–29.
321. Doppel, K. *Eine Abschätzung für schlichte Funktionen.* Monatsh. Math., vol. 77, no. 4(1973), 299–301; MR 50 #13489. [68]

322. Doppel, Karl. *Über eine spezielle Klasse von schlichten Funktionen.* Osterreich. Akad. Wiss. Math.-Natur. K1. S.-B. II 181 (1973), 291–299; MR 48 #11481. [28]
323. Doppel, K.; Zinterhof, P. *Über die ϵ-Entropie und ϵ-Kapazität der Familie der schlichten Funktionen.* Monatsh. Math. 76(1972), 222–225; MR 46 #5604. [22]
324. Dundučenko, L. E. *The application of electro-modeling for estimating the curvature of level lines in convex mapping of a circular ring.* (Russian) Sibirsk. Mat. Z 7(1966), 270–284; MR 33 #1436. [7, 10]
325. Dundučenko, L. E. *On the inversion formula for a certain subclass of typically real functions.* (Russian) Ukrain. Mat. Z 18 (1966), no. 3, 107–112; MR 34 #2851. [13, 23]
326. Dundučenko, L. E. *On certain classes of regular functions univalent in an ellipse.* (Russian) Izv. Vyss. Ucebn. Zaved Matematika 1966, no. 2(51), 58–61; MR 33 #5863. [6, 10, 57]
327. Dundučenko, L. E. *Regular functions doubly spiral in a ring.* (Russian) Sibirsk. Mat. Z. 8(1967), 1272–1283; MR 36 #3969. [6, 48]
328. Dundučenko, L. E. *On some classes of analytic functions in an n-connected region.* (Ukrainian. Russian and English summaries) Dopovidi Akad. Nauk Ukrain. RSR Ser. A 1967, 109–112; MR 35 #357.
329. Dundučenko, L. O. *On a certain class of functions that are analytic in an n-connected circular domain* (Ukrainian. English and Russian summaries) Dopovidi Akad. Nauk Ukrain. RSR Ser A 1972, 398–400, 475; MR 46 #5598. [60]
330. Duren, P. L. *On the Marx conjecture for starlike functions.* Trans. Amer. Math. Soc. 118(1965), 331–337; MR 31. [1, 6, 12, 16]
331. Duren, Peter L. *Coefficients of meromorphic schlicht functions.* Proc. Amer. Math. Soc. 28(1971), 169–172; MR 42 #6212. [9, 25, 53]
332. Duren, Peter L. *Estimation of coefficients of univalent functions by a Tauberian remainder theorem.* J. London Math. Soc. (2) 8 (1974), 279–282; MR 51 #877. [54, 62]
333. Duren, P. L. *Coefficients of univalent functions.* Notices, Amer. Math. Soc. 23, January 1976, Abstract 731-30-4, p. A-98. [9, 25, 42, 44, 54]
334. Duren, P. L.; Lehto, O. *Schwarzian derivatives and homeomorphic extensions.* Ann. Acad. Sci. Fenn. (Series A) I, vol. 477(1970), 1–11; MR 43 #7610. [4, 31]
335. Duren, Peter L.; McLaughlin, Renate. *Two-slit mappings and the Marx conjecture.* Michigan Math. J., 19(1972), no. 3, pp. 267–273; MR 46 #3762. [1, 6, 16, 19]

336. Duren, P. L.; Schober, G. E. *On a class of schlicht functions.* Mich. Math. J. 18(1971), 353–356; MR 45 #532c. [16, 42, 45, 68]
337. Duren, P. L.; Shapiro, H. S.; Shields, A. L. *Singular measures and domains not of smirnov type.* Duke Math. J. 33(1966), 247–254. [4, 16, 31]
338. Dvořák, Oldrich. *Über schlichte Funktionen.* I. (Czech and Russian summaries) Casopis Pest. Mat. 92(1967), 162–192; MR 36 #6605. [25, 27, 42, 54, 68]
339. Dvořák, Oldrich. *Über schlichte Funktionen.* II. (Czech summary) Casopis Pest. Mat. 94(1969), 146–167, 222; MR 40 #5846. [4, 42, 45, 68]
340. Dwivedi, S. P. *Certain classes of univalent functions and generalizations of functions with bounded boundary rotation.* Indian Institute of Technology Kanpur, Dept. of Mathematics; Doctoral dissertation, September 1975. [1, 5, 6, 9, 10, 12, 23, 42, 43, 49, 71, 73, 76, 77]
341. Dziubiński, I. *L'equation des fonction extrêmales dans la famille des fonctions univalentes symétriques et bornées.* Panstwowe Wydawnictwo Naukowe, Lodz, 1960. 62 pp.; MR 33 #5885. [22, 24]
342. Dziubiński, I. *Variational formulas for quasi-starlike and quasi-convex functions.* Ann. Univ. Mariae Curie-Sklodowska, Sect. A, vols. 22/23/24(1968/1969/1970), pp. 63–67; MR 51 #5908. [6, 10, 24]
343. Dziubiński, I. *Quasi-starlike functions.* (Loose Russian summary) Bull. Acad. Polon. Sci. Ser. Sci. Math. Astronom. Phys. 16 (1968), 477–479; MR 38 #1244. [6, 36]
344. Dziubiński, I. *Quasi-starlike functions.* Ann. Polon. Math., 26 (1972), Fasc. 2, pp. 175–197; MR 46 #335. [6, 16]
345. Dziubiński, I. *Estimation of the coefficients of quasi-starlike functions.* Ann. Polon. Math. 30(1975), fasc. 3, 297–321. [6, 36]
346. Dziubiński, I.; Siewierski, L. *Sharp estimation of the functional $|a_4|$ in the class of quasi-convex functions.* Bull. Acad. Polon. Sci. Ser. Sci. Math. Astronom. Phys., vol. 20(1972), no. 5 pp. 365–366; MR 46 #3769. [10, 38]
347. Dzrbasjan, M. M. *The stratification of classes of univalent functions.* (Russian. Armenian and English summaries) Izv. Akad. Nauk Armjan. SSR Ser. Mat. 4(1969), no. 4, 225–243; MR 42 #6206. [6, 23, 36, 63]
348. Eenigenburg, P. *On a class of analytic functions whose images are star-shaped Jordan domains.* (Abstract), Notices, A. M. S., vol 16, no. 1, issue no. 111, Jan. 1969, p. 193. [6, 12, 36]

349. Eenigenburg, P. J. *On α-spiralike functions.* (Abstract) Notices, A. M. S. 17(1970), pp. 156-157. [6, 7, 62]
350. Eenigenburg, P. *On the radius of curvature for convex analytic functions.* Can. J. Math. 22(1970), 486-491; MR 41 #5608. [7, 10, 35]
351. Eenigenburg, P. J. *A class of starlike mappings of the unit disk.* Composito Math., 24(1972), Fasc. 2, pp. 235-238, MR 46 #336. [6, 12, 22, 61, 62, 82]
352. Eenigenburg, P. J. *Boundary behavior of starlike functions.* Proc. Amer. Math. Soc., 33(1972), no. 2, pp. 428-432; MR 45 #2177. [2, 4, 6, 10, 23, 62]
353. Eenigenburg, Paul J. *On α-convex functions.* Rev. Roumaine Math. Pures Appl. 19(1974), 305-310; MR 50 #2471. [22, 69]
354. Eenigenburg, P. J.; Keogh, F. R. *The Hardy class of some univalent functions and their derivatives.* Mich. Math. J. 17(1970), 335-346; MR 46 #5631. [1, 2, 3, 5, 6, 10, 62]
355. Eenigenburg, Paul J.; Miller, Sanford S. *The H^p classes of α-convex functions.* Proc. Amer. Math. Soc. 38(1973), 558-562; MR 46 #9318. [4, 16, 18, 23, 69]
356. Eenigenburg, P. J.; Miller, S. S.; Mocanu, P. T.; Reade, M. O. *On a subclass of Bazilevic functions.* Proc. Amer. Math. Soc. 45(1974), 88-92; MR 49 #9180. [4, 6, 70]
357. Eenigenburg, P. J.; Nelson, D. *On the order of starlikeness of α-convex functions.* Notices Amer. Math. Soc. 23, January 1976, Abstract 731-30-28, p. A-104. [6, 85]
358. Eenigenburg, P. J.; Silvia, E. M. *A coefficient inequality for Bazilevic functions.* (Polish and Russian summaries) Ann. Univ. Mariae Curie-Sklodowska Sect. A 27(1973), 5-12 (1975). [22, 30, 69, 70, 81, 84]; MR 52 #8408.
359. Ehrig, G. *The Bieberbach conjecture for univalent functions with restricted second coefficient.* J. London Math. Soc. (2) 8(1974), 355-360. [8, 42, 54]
360. Ehrig, G. *Coefficient estimates concerning the Bieberbach conjecture.* Math. Z. 140(1974), 111-126. [42, 54]
361. Eke, B. G. *Remarks on Ahlfors' distortion theorem.* J. Analyse Math. 19(1967), 97-134; MR 35 #6806. [62]
362. Eke, B. G. *The asymptotic behavior of areally mean valent functions.* Journal D'analyse Mathematique, vol. 20(1967), 147-212; MR 36 #5331. [8, 33]
363. Eke, B. G. *A note on L-strips.* Proc. Roy. Irish Acad. Sect. A 70, 33-34(1970); MR 43 #7604. [49]

364. Eke, B. G. *On multivalent functions with gap series.* Math. Nachr. 43(1970), 377-381; MR 41 #8652. [14, 33, 39]
365. Eke, B. G. *On the differentiability of conformal maps at the boundary.* Nagoya Math. J. 41(1971), 43-54. [62]
366. Eke, B. G. *Typically real mean univalent functions of large growth.* Israel J. Math. 22(1975), no. 1, 1-6. [13, 33]
367. Eke, B. G. *On multivalent functions of large growth in two directions.* Math. Scand. 37(1975), 105-110. [14, 62]
368. Eke, B. G.; Warschawski, S. E. *On the distortion of conformal maps at the boundary.* J. London Math. Soc. 44(1969), 625-630. [62]
369. Eke, V. R.; Eke, B. G. *Notes on the regularity of growth of multivalent functions.* J. Analyse Math. 28 (1975), 1-19. [62]
370. Essén, M. R. : Keogh. F. R. *The Schwarzian derivative and estimates of functions analytic on the unit disc.* Math. Proc. Cambridge Philos. Soc. 78(1975), part 3, 501-511. [4, 31]
371. Evgrafov, M. A. *Analytic Functions.* W. B. Saunders Company, 1966, 336 pp. [44]
372. Èzrohi, T. G. *On a theorem of Schild-Lewandowski.* (Russian. French and Romanian summaries) Bul. Inst. Politehn. Iasi(N.S.) 8(12) (1962), 15-18; MR 32 # [32]
373. Èzrohi, T. G. *A class of univalent functions.* (Russian) Problems of Math. Phys. and Theory of Functions, II (Russian), pp. 171-185. Naukova Dumka, Kiev, 1964; MR 32 #7728. [29, 39]
374. Èzrohi, T. G. *Certain estimates in special classes of univalent functions regular in the circle $|z| < 1$.* (Ukrainian. Russian and English summaries) Dopovidi Akad. Nauk Ukrain. RSR 1965, 984-988; MR 33 #7518. [12, 35]
375. Èzrohi, T. G. *On the curvature of level lines and their orthogonal trajectories in the class of functions of bounded rotation.* (Russian) Ukrain. Mat. Z. 17(1965), no. 6, 91-99; MR 33 #4245. [35]
376. Èzrohi, T. G. *On a class of univalent functions.* (Russian) Teor. Funkcii Funkcional Anal. i Prilozen Vyp. 2(1966), 198-204; MR 34 #325. [2, 7, 8, 10, 11, 12, 18, 19, 39, 56]
377. Èzrohi, T. G. *Certain special classes of p-valent functions in the disc $|z| < 1$ and in the circular region $1 < |z| < +\infty$.* (Russian) Izv. Vyss. Ucebn. Zaven Matematika 1966, no. 1(50), 174-181; MR 34 #2852. [6, 12, 14, 63]
378. Èzrohi, T. G. *Some properties of convex mappings of a disc.* Bull. Inst. Politehn. Iasi (N.S.) 12(16) (1966), fasc. 3-4, 33-37; MR 36 #5321. [10, 35]

379. Èzrohi, T. G. *A class of functions regular in multiply connected circular domains.* (Russian. Romanian summary) An. Sti. Univ. "Al. I. Cuza" Iasi Sect. I a Mat. (N.S.) 13(1967), 273-276; MR 40 #5847. [6, 60, 68]
380. Èzrohi, T. G. *On the radius of convexity and bounds for the curvature of level curves of certain classes of functions regular in the disc $|z| < 1$.* (Ukrainian. Russian and English summaries) Dopovidi Akad. Nauk Ukrain. RSR Ser A 1967, 113-117; MR 35 #380. [12, 35]
381. Èzrohi, T. G. *The boundaries of convexity of certain classes of analytic functions.* (Russian) Teor. Funkcii Funkcional. Anal. i Prilozen. Vyp. 8(1969), 117-125; MR 41 #8645. [12]
382. Èzrohi, T. G.; Acilov, H. *A transformation of regular functions which are typically real in the half-plane Re $\zeta > 0$.* (Russian. Uzbek summary) Izv. Akad. Nauk UzSSR Ser. Fiz.-Mat. Nauk 11 (1967), no. 4, 29-33; MR 35 #6840. [13, 23, 57]
383. Èzrohi, T. G.; Acilov, H. *A class of functions which are regular in a multiply connected circle region.* (Russian. Uzbek summary) Izv. Akad. Nauk UzSSR Ser. Fiz.-Mat. Nauk 12(1968), no. 3, 31-35; MR 38 #319. [60, 68]
384. Feldman, Ja. S. *Certain extremal regions for systems of functionals of bounded schlicht funcitons.* (Russian) Trudy Tomsk. Gos. Uiv. Ser. Meh.-Mat. 189(1966), 194-205. [22]
385. Feldman, Ja. S. *Certain extremal regions for systems of functionals of bounded schlicht functions.* (Russian) Trudy Tomsk. Gos. Univ. Ser. Meh.-Mat. 189(1966), 194-205; MR 37 #2965. [22]
386. Feng, J.; MacGregor, T. H. *Estimates on integral means of the derivatives of univalent functions.* Notices Amer. Math. Soc. 21, August 1974 Abstract 74T-B163, p. A-492. [3, 5, 10]
387. Feng, Jinfu; MacGregor, T. H. *Integral mean estiamtes for derivatives of functions with a positive real part.* University of Maryland, Dept. of Mathematics Technical report TR-74-11, February 1974. {Notices Amer. Math. Soc. 21, June 1974, Abstract 74T-B124, p. A-441}. [1, 2, 3, 56, 88]
388. Finkelstein, M. *Growth estimates of convex functions.* Proc. Amer. Math. Soc. 18(1967), 412-418; MR 35 #5598. [6, 10, 22, 37, 45, 61, 63, 64, 67, 72]
389. Fisher, S. *The convex hull of the finite Blaschke Products.* Bull. Amer. Math. Soc., 74(1968), 1128-1129.
390. FitzGerald, C. H. *On analytic continuation to a Schlict function.* Proc. Amer. Math. Soc. 18(1967), 788-792; MR 36 #2791. [57, 62]

391. FitzGerald, C. H. *Topics in geometric function theory*. Dept. of Math., Stanford Univ., Stanford, California. [5, 6, 19, 24, 43, 49, 54, 62]
392. FitzGerald, Carl H. *On analytic continuation to a starlike function*. Arch. Rational Mech. Analy. 35(1969), 397–401; MR 40 #4434 [6, 24]
393. FitzGerald, Carl H. *Analytic continuation of a function defined on a sequence of points*. Bull. London Math. Soc., 3(1971), no. 9, pp. 286–290. [4, 53]; MR 45 #3756
394. FitzGerald, C. H. *Conformal mappings onto w-swirly domains*. Pacific J. Math. 37(1971), 657–670; MR 47 #446. [6, 49]
395. FitzGerald, C. H. *Quadratic inequalities and coefficient estimates for schlicht functions*. Arch. Rational Mech. Anal. 46(1972), 356–368; MR 49 #557. [4, 16, 24, 39, 42, 53, 54]
396. FitzGerald, Carl H. *Exponentiation of certain quadratic inequalities for Schlicht functions*. Full. Amer. Math. Soc. 78(1972), 209–210; MR 46#5601. [53, 54, 68]
397. FitzGerald, Carl H. *Sums of slit mappings*. Notices Amer. Math. Soc. 23, January 1976, Abstract 731-30-36, p. A–105. [10, 49]
398. Flatto, L.; Newman, D. J.; Shapiro, H. S. *The level curves of harmonic functions*. Trans. Amer. Math. Soc. 123(1966), 425–436. [16, 35]
399. Frank, James L. *On the convex univalent functions*. II. J. Reine Angew. Math. 277(1975), 5–7. [1, 10]; MR 52 #8400
400. Frank, J. L.; Shaw, J. K. *Univalence of odd derivatives of even entire functions*. J. Reine Angew. Math. 277(1975), 1–4. [90]
401. Fricke, G. H. *A characterization of functions of bounded index*. Indian J. Math. 14(1972), 207–212; MR 50 #7525. [89]
402. Fricke, G. H. *A note on multivalence of a function of bounded index*. Proc. Amer. Math. Soc. 40(1973), 140–142; MR 47 #8854. [89]
403. Fricke, Gerd H. *Functions of bounded index and their logarithmic derivatives*. Math. Ann. 206(1973), 215–223; MR 48 #4308. [89]
404. Fricke, G. H. *Entire functions having positive zeroes*. Indian J. Pure Appl. Math. 5(1974), 478–485. [90]
405. Friedland, S. *On a conjecture of Robertson*. Archive for Rational Mechanics and Analysis 37(1970), 255–261; MR 41 #456. [16, 42, 45, 52, 54]
406. Friedland, S. *Generalized Hadamard inequality and its applications*. Linear and Multilinear Algebra 2(1975), 327–334; MR 51 #3420. [53, 90]
407. Friedland, S.; Nehari, Z. *Univalence conditions and Sturm-*

Liouville eigenvalues. Proc. Amer. Math. Soc. 24(1970), 595–603: MR 40 #4435. [4, 31]
408. Friedman, Neal. *Remarks on a quadratic form of Duren and Schiffer.* Proc. Amer. Math. Soc. 18(1967), 212–214; MR 34 #7787. [16, 24, 42]
409. Fuchs, Ilse. *Power series with multiply monotonic coefficients.* Math. Ann. 190(1971), 289–292; MR 43 #6424. [4, 23, 39]
410. Fuchs, Ilse. *Potenzreihn mit mehrfach monotonen Koeffizienten.* Arch. Math. (Basel) 22(1971), 275–278; MR 45 #5327. [4, 39]
411. Fuchs, W. H. J. *Topics in the theory of functions of one complex variable.* D. Van Nostrand Co., Inc., Princeton, N.J.-Toronto, Ont.-London, 1967. VI + 193 pp.; MR 36 #3954. [44]
412. Gackstatter, Fritz. *Gesamtdefekt und Krummungsmittelwerte bei meromorphen Funktionen.* Bayer. Akad. Wiss. Math.-Natur. Kl. S.-B. 1970, Abt. II, 79–102(1971); MR 48 #4314. [35]
413. Gaiduk, V. N.; Melkonjan, E. L. *Certain conditions for the univalence of n-symmetric functions.* (Russian) Izv. Vyss. Ucebn. Zaved. Matematika 1974, no. 2(141), 45–50; MR 49 #10874. [4]
414. Gaier, Dieter. *Konstruktive Methoden der konformen Abbildung.* Springer-Verlag, Berlin, 1964. XIII + 294 pp.; MR 33 #7507. [44]
415. Gaier, Dieter. *Estimates of conformal mappings near the boundary.* Indiana Univ. Math. J. 21(1971/1972), 581–595; MR 45 #2151. [62]
416. Gaier, D.; Pommerenke, Ch. *On the boundary behavior of conformal maps.* Michigan Math. J. 14(1967), 79–82; MR 34 #4470. [62]
417. Gale, D.; Nikaido, H. *The Jacobian matrix and global univalence of mappings.* Mathematische Annalen 159(1965), 81–93.
418. Gal'perin, I. M. *A distortion theorem for functions bounded in the unit disc.* (Russian) Ukrain. Mat. Z. 18 (1966), no. 1, 107–109; MR 33 #2807. [22, 61]
419. Gal'perin, I. M. *On estimates of several of the first coefficients of the Taylor series of functions univalent in the unit disc.* Teor. Funkcii Funkcional. Anal. i Prolozen. Vyp. 2(1966), 75–78; MR 33 #5876. [54]
420. Gal'perin, I. M. *Überdeckungssatz für die klasse θ-spiralischer Funktionen.* (Romanian summary) Bull. Inst. Politehn Iasi (N.S.) 18(22) (1972), fasc. 1–2, sect. 1, 59–60; MR 48 #2360. [6]
421. Garabedian, P. R. *Inequalities for the fifth coefficient.* Comm. Pure Appl. Math. 19(1966), 199–214; MR 36 #5326. [24, 42, 53]
422. Garabedian, P. R. *An extension of Grunsky's inequalities bearing on the Bieberbach conjecture.* J. D'analyse Math. 18(1967), 81–97; MR 37 #1583. [24, 42, 53, 54]

423. Garabedian, P. R. *Computer experiments with the Bieberbach conjecture.* Proc. Internat. Congr. Math. (Moscow, 1966), pp. 627-628. Izdat. "Mir", Moscow, 1968; MR 38 #3421. [42]
424. Garabedian, P. R.; Ross, G. G.; Schiffer, M. M. *On the Bieberbach conjecture for even n.* J. Math. Mech. 14(1965), 975-989; MR 32 #207. [4, 24, 29, 42, 53]
425. Garabedian, P. R.; Schiffer, M. *The local maximum theorem for the coefficients of univalent functions.* Arch. Rational Anal. 26 (1967), 1-32; MR 37 #1584. [24, 42, 53, 54]
426. Garabedian, P. R.; Swenson, E. V. *Remarks about local maxima of the coefficients of univalent functions.* J. Analyse Math. 23 (1970), 133-138; MR 42 #4724. [24, 42, 45, 52]
427. Garoutte, D. E.; Nickel, P. A. *A note on extremal properties characterizing weakly λ-valent principal functions.* Pacific J. Math. 25(1968), 109-115; MR 37 #404. [14]
428. Gavrilov, V. I. *Remarks on the radius of univalence of holomorphic functions.* (Russian) Mat. Zametki 7(1970), 295-298; MR 41 #5609 {English transl.: Math. Notes 7(1970), 179-181}. [43]
429. Gavrilov, V. I. *Behavior along chords of meromorphic functions in the unit disk.* (Russian). Dokl. Akad. Nauk SSSR 216(1974), 21-23. [9]
430. Giec, T. *The limit extremal functions for certain functionals in a family of univalent functions that are nearly the identity.* Zeszyty Nauk. Univ. Łódźk. Nauki Mat. Przyrod. Ser. II Zeszyt 20 Matematyka (1966), 161-183; MR 50 #13491. [22, 29]
431. Giec, Tadeusz. *The range of the functional $F(g) = F[g_k(\delta), \bar{g}_k(\delta)g'_k(0)]$ in the class $S_1^{[k]}$.* (Polish. French summary) Zeszyty Nauk Univ. Łódźk. Nauki Mat. Pryzrod. Ser. II Zeszyt 29(1968), 49-62; MR 40 #332. [22, 30, 49, 61]
432. Giec, Tadeusz. *Certain extremal problems in the family of bounded symmetric schlicht functions.* (Polish. French summary) Zeszyty Nauk. Univ. Łódźk. Nauki at. Przyrod. Ser. II Zeszyt 34(1969), 23-71; MR 41 #1985. [22, 24, 49, 61]
433. Giec, Tadeusz. *Estimation of the functional $|z^2 g'_k(z)/g_k^2(z)|$ in the class S_1^k.* (Polish. French summary) Zeszyty Nauk. Univ Łódźk. Nauki Mat. Przyrod. Ser. II Zeszyt 39 Mat. (1971), 43-51; MR 48 #8770. [22, 49, 61]
434. Giec, T. *On some extremal problems in the family of univalent bounded and symmetrical functions.* Bull. Acad. Polon. Sci. Ser. Sci. Math. Astronom. Phys., 20(1972), no. 2, pp. 137-144; MR 46 #7501. [22, 24, 61]

435. Gocal, Ryszard. *Certain properties of extremal polynomials.* (Polish. French summary) Zeszyty Nauk. Univ. Lodzk. Nauki Mat. Przyrod. Ser. II Zeszyt 52 Mat. (1973), 157–171; MR 48 #8780. [32]
436. Goel, Ram Murti. *On the partial sums of a class of univalent functions.* (Polish and Russian summaries) Ann. Univ. Mariae Curie-Sklodowska Sect. A 19(1965), 17–23(1970); MR 41 #3735. [4, 5, 20, 43]
437. Goel, R. M. *The radius of univalence of certain analytic functions.* Tohoku Math. J. (2) 18(1966), 398–403; MR #358. [2, 11, 43, 68]
438. Goel, R. M. *Radius of univalence and starlikeness for certain analytic functions.* Nieuw Arch. Wisk. (3) 14(1966), 255–260; MR 34 #6059. [6, 11, 22, 43]
439. Goel, R. M. *On the radius of univalence and starlikeness for certain analytic funcitons.* J. Math. Sci. 1(1966), 98–102; MR 34 #4474. [4, 6, 11]
440. Goel, R. M. *A class of univalent functions whose derivatives have positive real part in the unit disc.* Nieuw Arch. Wisk. (3) 15(1967), 55–63; MR 36 #5322. [2, 5, 12, 18, 19, 74, 75, 82]
441. Goel, R. M. *Radius of convexity of a certain class of meromorphically starlike functions.* Publ. Math. Debrecen 14(1967), 281–284; MR 36 #5323. [9, 12]
442. Goel, R. M. *On K-fold symmetric close-to-convex functions.* Ganita 18(1967), no. 2, 77–87; MR 39 #4370. [5, 12, 20, 49, 74]
443. Goel, R. M. *A class of close-to-convex functions.* (Loose Russian summary) Czechoslovak Math. 18(93)(1968), 104–116; MR 37 #2966. [5, 12, 74, 75]
444. Goel, R. M. *A class of univalent functions with fixed second coefficients.* J. Math. Sci. 4(1969), 85–92; MR 42 #482. [12, 47, 72]
445. Goel, R. M. *The coefficients of schlicht functions.* Indian J. Math. 11(1969), 25–27; MR 41 #3746. [23, 54]
446. Goel, R. M. *Corrigendum to: "A class of univalent functions whose derivatives have positive real part in the unit disc."* Nieuw Arch. Wisk. (3) 18(1970), 80–81; MR 42 #483. [2, 5, 12, 18, 19, 74, 75, 82]
447. Goel, R. M. *On a class of functions with starshaped images.* Czeckoslovak Math. J. 20(95) (1970), 34–38; MR 41 #449. [6, 11, 20]
448. Goel, R. M. *On functions satisfying $Re[f(z)/z] > \alpha$.* Publ. Math. Debrecen 18(1971), 111–117(1972); MR 46 #7494. [2, 20, 43, 72]
449. Goel, R. M. *The radius of convexity and starlikeness for certain classes of analytic functions with fixed second coefficients.* Ann.

Univ. Mariae Curie-Sklodowska Sect. A 25(1971), 33-39(1973); MR 48 #4292. [2, 11, 12, 43, 45, 72]
450. Goel, R. M. *A class of analytic functions whose derivatives have positive real part in the unit disc.* Indian J. Math 13(1971), 141-145; MR 48 #6425. [2, 21, 22, 30, 47, 56]
451. Goel, R. M. *On a class of functions schlicht in the unit circle.* Rev. Mat. Hisp.-Amer. (4) 31(1971), 20-33; MR 45 #523. [4, 11, 12, 20, 82]
452. Goel, R. M. *On partial sums of certain analytic functions.* Ganita 22(1971), no. 1, 11-20; MR 45 #7036. [2, 20]
453. Goel, R. M. *On radii of starlikeness, convexity, close-to-convexity for p-valent functions.* Arch. Rational Mech. Analy., 44(1972), no. 4, pp. 320-328; MR 48 #11469. [5, 6, 10, 11, 12, 14, 22, 61, 73]
454. Goel, R. M. *Radius of univalence and starlikeness for certain analytic functions.* Indian J. Math. 14(1972), 15-19; MR 48 #6388. [2, 11, 43]
455. Goel, R. M. *The radius of convexity for a certain class of analytic functions.* Indian J. Pure Appl. Math. 4(1973), 318-324; MR 48 #518. [2, 6, 10, 12, 36]
456. Goel, R. M. *On the coefficients of a class of close-to-convex functions.* Indian J. Pure Appl. Math. 5(1974), 128-131. [5, 75]; MR 52 #731
457. Goel, R. M. *Functions starlike and convex of order α.* J. London Math. Soc. (2) 9(1974/75), 128-130; MR 51 #3414. [6, 10, 11]
458. Goel, R. M. *On a class of analytic functions.* J. Austral. Math. Soc. 20(1975), part 1, 46-53; MR 51 #8391. [6, 11, 36, 45, 63, 67, 68]
459. Goel, R. M.; Singh, V. *Coefficient estimates for a class of starlike functions.* Indian J. Pure Appl. Math. 3(1972), no. 6, 1118-1130; MR 50 #10231. [2, 6, 16, 21, 36, 63]
460. Goel, R. M.; Singh, V. *On radii of univalence of certain analytic functions.* Indian J. Pure Appl. Math. 4(1973), 402-421; MR 51 #875. [2, 5, 6, 10, 11, 12, 43, 73, 82]
461. Göktürk, Z. *Estimates for univalent functions with quasiconformal extensions.* Ann. Acad. Sci. Fenn. Ser A I No. 589(1974), 21 pp.
462. Goldberg, J. L. *Bounds on the derivatives of positive functions.* SIAM Rev. 8(1966), 343-345; MR 35 #381. [2, 21]
463. Goldberg, J. L.; Ullman, J. L. *A note on the derivatives of functions positive in a half-plane.* (Abstract), Notices, A. M. S., vol. 14, no. 2, issue no. 96, February, 1967, p. 279. [2]
464. Goluzin, G. M. *Geometrical theory of functions of a complex variable* (Second edition). Izdat. "Nauka", Moscow, 1966. 628 pp.

{A. M. S.; Transl. of Math. Monographs vol. 26, 1969, 678 pp.}; MR 36 #2793; MR 40 #308. [44]

465. Goluzina, E. G. *On the ranges of certain systems of functionals in the class of typically real functions.* (Russian. English summary) Vestnik Leningrad. Univ. 20(1965), no. 7, 45–62; MR 32 #4277. [13, 23, 29]

466. Goluzina, E. G. *The mutual growth of the coefficients of a class of p-valent functions.* Dokl. Akad. Nauk SSSR 169(1966), 759–760; MR 34 #1508 {Soviet Math. Dokl. 7(1966), 992–993}. [6, 8, 14]

467. Goluzina, E. G. *On the coefficient regions of a certain class of functions which are meromorphic in a disc.* (Russian) Trudy Mat. Inst. Steklov. 94(1968), 33–46; MR 37 #1585. {Translated from the Russian: Extremal Problems of the Geometric Theory of Functions (A. M. S. 1969)}. [9, 29]

468. Goluzina, E. G. *Mutual growth of coefficients of a class of p-valent functions.* (Russian) Trudy Mat. Inst. Steklov. 94(1968), 27–32; MR 36 #6606. [6, 8, 14, 82]

469. Goluzina, E. G. *The coefficients of a certain class of functions that are regular in the disc and have an integral representation there.* (Russian) Zap. Naucn. Sem. Leningrad. Otdel. Mat. Inst. Steklov. (LOMI) 23(1972), 63–77; MR 46 #9328. [23, 28]

470. Goluzina, E. G. *The ranges of values of coefficient systems in the class of functions with positive real part in an annulus.* (Russian) Zap. Naucn. Sem. Leningrad. Otdel. Mat. Inst. Steklov. (LOMI) 44(1974), 17–25, 186; MR 51 #8394. [2, 29, 48]

471. Goluzina, E. G. *The range of values of certain coefficient systems in the class of typically real functions in an annulus.* (Russian) Zap. Naucn. Sem. Leningrad. Otdel. Mat. Inst. Steklv. (LOMI) 44(1974), 26–40, 186; MR 51 #8395. [13, 29, 48]

472. Goncar, A. A. *The rate of rational approximation and the property of univalence of an analytic function in the neighborhood of an isolated singular point.* (Russian) Mat. Sb. (N.S.) 94(136) (1974), 265–282, 336.

473. Goodman, A. W. *On the convexity of the level curves of a polynomial.* Proc. Amer. Math. Soc. 17(1966), 358–361; MR 32 #5847. [10, 16, 32, 35]

474. Goodman, A. W. *Curvature under an analytic transformation.* J. London Math.Soc. 43(1968), 527–533; MR 38 #313. [16, 35]

475. Goodman, A. W. *The valence of sums and products.* Canad. J. Math. 20(1968), 1173–1177; MR 38 #314. [4, 6, 10, 14, 16, 43, 82]

476. Goodman, A. W. *Open problems on univalent and multivalent*

functions. Bull. Amer. Math. Soc., vol. 74, no. 6, November 1968, pp. 1035-1050; MR 38 #315. [16, 44]
477. Goodman, A. W. *The valence of certain means.* J. Analyse Math. 22(1969), 355-361. MR 40 #326. [16, 57, 82]
478. Goodman, A. W. *Letter to the editors.* Ukrain. Mat. Z. 24(1972), 427; MR 45 #7037. [4, 9, 43]
479. Goodman, A. W. *A note on the Noshiro-Warschawski theorem.* J. Analyse Math. 25(1972), 401-408; MR 45 #8819. [2, 4, 10, 64]
480. Goodman, A. W. *On close-to-convex functions of higher order.* Ann. Univ. Sci. Budapest. Eötvös Sect. Math. 15(1972), 17-30 (1973); MR 48 #11470. [3, 4, 5, 7, 18]
481. Goodman, A. W. *Coefficients for the area theorem.* Proc. Amer. Math. Soc., 33(1972), no. 2, pp. 438-444; MR 45 #530. [23, 27]
482. Goodman, A. W. *The critical points of a typically-real function.* Proc. Amer. Math. Soc. 38(1973), 95-102; MR 47 #2043. [1, 14, 16, 23, 37, 43]
483. Goodman, A. W. *A note on the zeroes of Faber polynomials.* Proc. Amer. Math. Soc. 49(1975), 407-410; MR 51 #3409. [53]
484. Goodman, A. W.; Rahman, Q. I.; Ratti, J. S. *On the zeroes of a polynomial and its derivative.* Proc. Amer. Math. Soc. 21(1969), 273-274. [16, 32]; MR 39 #421
485. Goodman, G. S. *On the Determination of Univalent Functions with Prescribed Initial Coefficients.* Arch. Rational Mech. Anal. 24 (1967), 78-81; MR 34 #4478. [24, 29, 39, 42, 54]
486. Goodman, G. S. *A method for comparing univalent functions.* Bull. Amer. Math. Soc. 75, May 1969, pp. 517-521; MR 40 #7433. [24, 54]
487. Gopal, M. R. *Radius of starlikeness of certain classes of functions.* Notices A. M. S., vol. 17, no. 2, issue no. 120, Feb. 1970 (Abstract), pg. 409. [2, 11]
488. Gopalakrishna, H. S.; Shetiya, V. S. *Coefficient estimates for spirallike mappings.* J. Karnatak Univ. Sci. 18(1973), 297-307; MR 50 #588. [6, 36]
489. Gopengauz, B. E. *Certain classes of analytic functions.* (Russian). Trudy Tomsk. Gos. Univ. Ser. Meh.-Math. 189 (1966), 144-160; MR 37 #2967. [4, 6, 9, 39]
490. Gopengauz, B. E. *Certain classes of analytic functions.* II. (Russian). Trudy Tomsk. Gos. Univ. Ser. Meh.-Math. 200(1968), 20-30; MR 41 #3736. [9, 39, 58, 68]
491. Gopengauz, B. E. *Estimates of the coefficient of functions from $Kn(E)$ with real coefficients.* (Russian) Trudy Tomsk. Gos. Univ.

Ser. Meh.-Mat. 200(1968), 62–70; MR 41 #2024. [39, 54]
492. Gopengauz, B. E. *Certain theorems on functions whose divided difference is different from zero.* (Russian) Trudy Tomsk. Gos. Univ. Ser. Meh.-Mat. 200(1968), 31–42; MR 41 #2023. [39]
493. Gopengauz, B. E. *Functions whose nth derivative has positive real part.* (Russian) Trudy Tomsk. Gos. Univ. Ser. Meh.-Mat. 200 (1968), 43–61; MR 41 #2018. [2, 19, 54, 68]
494. Gopengauz, B. E. *Some remarks on functions that have a positive real part of the n-th derivative.* (Russian) Trudy Tomsk. Gos. Univ. Ser. Meh.-Mat. 210(1969), 9–17; MR 43 #5037. [2, 71]
495. Gopengauz, B. E. *A certain generalization of the Schwarzian derivative and its application.* (Russian) Mat. Zametki 10(1971), 229–238; MR 44 #5464 (English transl. Math. Notes 10(1971)), 559–564. [4, 31]
496. Gorjainov, V. V. *A rotation theorem in the class of bounded univalent functions.* (Russian) Mat. Zametki 18(1975), no. 5, 633–640. [22, 61]
497. Gorjainov, V. V.; Gutljans'kii, V. Ja. *The radius of starlikeness in conformal mapping.* (Ukrainian. English and Russian summaries). Dopovidi Akad. Nauk Ukrain. RSR Ser A 1974, 100–102, 187; MR 49 #9181. [11, 22]
498. Górski, J. *Some applications of the method of extremal points in the theory of analytic functions of one complex variable.* Colloq. Math. 11(1963/64), 151–156; MR 30 #4917. [24, 54]
499. Górski, J. *Some sharp estimations of coefficients of univalent functions.* J. Analyse Math. 14(1965), 199–207; MR 31 #3591. [54]
500. Górski, J. *Application of the extremal points method needed to some variational problems in the theory of schlicht functions.* Ann. Polon. Math. 17(1965), 141–145; MR 32 #5859. [24, 54, 88]
501. Górski, Jerzy. *A sharp estimation of a_5 in a subclass of the class S.* Zeszyty Nauk Univ. Jagiello. Prace Mat. Zeszyt 11(1966), 27–29; MR 44 #2921. [54]
502. Górski, J. *A certain minimum problem in the class S.* Ann. Univ. Mariae Curie-Sklodowska, Sect. A, Vols. 22/23/24(1968/1969/1970), pp. 73–77(1972); MR 49 #10872. [54]
503. Górski, J. *Local inequalities for some functionals in the class S.* Ann. Polon. Math. 24(1971), Fasc. 2, 219–224 MR 44 #6949. [54]
504. Górski, Jerzy. *Some local properties of Re a_3, RE a_4 and Re a_5 in the class S.* (Polish summary) Univ. Slaskiw Katowicach-Prace Mat. 2(1972), 19–24; MR 46 #7496. [54]
505. Górski, J. *Local inequalities of coefficients in the class S.* (Polish

summary) Univ. Slaskiw Katowicach-Prace Mat. 4(1973), 23-26; MR 48 #2368. [54]
506. Górski, J.; Poole, J. T. *Some sharp estimations of coefficients of univalent functions.* J. Math. Mech. 16(1966), 577-582; MR 35 #361. [9, 25, 54]
507. Greene, R. E.; Wu, Hung-Hsi. Bloch's theorem for meromorphic functions. Math. Z. 116(1970), 247-257. [19, 22, 46]; MR 42 #4777
508. Greene, R. E.; Wu, H. *Curvature and complex analysis.* Bull. Amer. Math. Soc. 77(1971), 1045-1049. [35]; MR 44 #473
509. Grinšpan, A. Z. *The coefficients of univalent functions that do not assume any pair of values w and $-w$.* (Russian) Mat. Zametki 11 (1972), 3-14; MR 45 #3691 (English transl.: Math Notes 11(1972), 3-11). [34, 54, 86, 87]
510. Grinšpan, A. Z. *Logarithmic coefficients of functions of class S.* (Russian) Sibirsk. Mat. Z. 13(1972), 1145-1157, 1199; MR 48 #6403 {English transl.: Siberian Math. J. 13(1972), 793-801 (1973)}. [42]
511. Grinšpan, A. Z. *The application of the area principle to the Bieberbach-Eilenberg functions.* Mat. Zametri 11(1972), 609-618 [Math. Notes 11(1972), 371-377]; MR 46 #5602. [27, 34, 86, 87]
512. Grinšpan, A. Z.; Kolomoiceva, Z. D. *Certain estimates in the class of functions that do not take any pair of values w and $-w$.* (Russian. English summary) Vestnik Leningrad. Univ. No. 19 Mat. Meh. Astronom. Vyp. 4(1973), 28-34, 151; MR 49 #3110. [26, 31, 34, 87]
513. Gromova, L. L. *On a certain class of analytic functions.* (Russian) Volz. Mat. Sb. Vyp. 3(1965), 110-113; MR 33 #4271. [19, 23, 41, 49]
514. Gromova, L. L.; Lebedev, N. A. *Non-overlapping domains that lie in a disk.* (Russian. English summary) Vestnik Leningrad. Univ. 24 (1969), no. 19, 7-12; MR 41 #8646. [26]
515. Gromova, L. L.; Lebedev, N. A. *Area theorems for nonoverlapping finitely connected regions.* II. (Russian. English summary) Vestnik Leningrad. Univ.25(1970), no. 1, 18-29; MR 41 #8647. [26, 27, 34]
516. Gromova, L. L.; Zjuzin, V. E. *On certain classes of analytic functions.* Volz. Mat. Sb. Vyp. 3(1965), 114-116; MR 33 #4255. [11, 23, 49]
517. Gromova, L. L.; Zybina, T. T. *Certain averages for starshaped functions.* (Russian) Saratov. Gos. Ped. Inst. Ucen. Zap. Vyp. 46 (1968), 6-12; MR 42 #484. [3, 6]
518. Grunsky, Helmut. *Zur konformen Abbildung von Gebieten, die in einer Richtung konvex sind.* J. Math. Anal. Appl. 34(1971), 685-701; MR 43 #7605. [10]

519. Grunsky, H.; Jenkins, J. *A Hilbert space method in the theory of schlicht functions*. Proceedings of the Symposium on Complex Analysis (Univ. Kent, Canterbury 1973), pp. 75–79. London Math. Soc. Lecture Note Ser., No. 12, Cambridge Univ. Press, London, 1974. [44]

520. Gudz', L. A. *Sharp bounds for the curvature of level curves and their orthogonal trajectories in classes of almost convex functions*. (Ukrainian. English and Russian summaries). Dopovidi Akad. Nauk Ukrain. RSR A 1974, 494–499, 573: MR 51 #3415. [5, 35]

521. Gupta, R. S. *On meromorphic circularly-symmetric functions*. Ganita 18(1967), no. 2, 31–38; MR 39 #4371. [9, 13, 25, 49]

522. Gupta, R. S. *Radius of univalence and starlikeness of a class of analytic functions*. J. Austral. Math. Soc. 14 (1972), Part 1, 1–8; MR 47 #5239. [2, 11, 22, 43, 56]

523. Gupta, R. S. *An extremal problem for functions with positive real part*. Proc. Amer. Math. Soc. 33(1972), 455–462; MR 45 #2167; MR 48 #4323. [2, 56]

524. Gupta, R. S. *On certain classes of analytic functions*. J. Indian Math. Soc. 36(1972), 79–88; MR 47 #7015. [5, 9, 16, 40, 41, 45, 56, 63, 67, 75, 78, 79]

525. Gupta, R. S. *Radius of convexity of convex sum of univalent functions*. Publ. Math. Debrecen, vol. 19, nos. 1–4, 1972, 39–42; MR 48 #2361. [6, 29, 35, 68]

526. Gupta, R. S. *Radius of convexity of a class of univalent functions*. J. Indian Math. Soc. 36(1972), 291–296; MR 48 #6389. [5, 6, 10, 12, 82]

527. Gupta, R. S. *On the integrals of univalent funcitons*. Rev. Mat. Hisp.-Amer. (4)34(1974), 269–275; MR 50 #13480. [6, 11, 12, 85]

528. Gutljanskii, V. Ja. *The range of values of certain functionals and the properties of level lines on classes of schlicht functions*. (Russian) Trudy Tomsk. Gos. Univ. Ser. Meh.-Mat. 200(1968), 71–87; MR 41 #3737. [6, 29, 35, 68]

529. Gutljanskii, V. Ja. *Parametric representation of univalent functions*. (Russian) Dokl. Akad. Nauk SSSR 194(1970), 750–753; MR 42 #6207 {Soviet Math. Dokl. 11(1970), 1273–1276}. [68]

530. Gutljanskii, V. Ja. *The stratification of the class of univalent analytic functions*. (Russian) Dokl. Akad. Nauk SSSR 196(1971), 498–501; MR 42 #7877. {Soviet Math. Dokl. 12(1971), 155–159}. [24]

531. Gutljanskii, V. Ja. *The rotation theorem in a class of schlicht p-symmetric functions*. (Russian). Mat. Zametki 10(1971), 239–

242; MR 45 #524 [English transl.: Math. Notes 10(1971), 565–566]. [49, 68]
532. Gutljanskii, V. Ja. *Integro-differential parametric representations and variational formulae.* (Ukrainian. English and Russian summaries). Dopovidi Akad. Nauk Ukrain. RSR Ser. A 1973, 781–783, 859; MR 48 #8778. [24]
533. Gutljanskii, V. Ja.; Scepetev, V. A. *On theorems of distortion and rotation on the class holomorphic univalent functions.* (Russian) Sibirsk. Mat. Z. 14(1973), 867–872, 911; MR 49 #560. [68]
534. Gutljanskii, V. Ja.; Scepetev, V. A. *A generalized area theorem for a certain class of q-quasiconformal mappings.* (Russian). Dokl. Akad. Nauk SSSR 218(1974), 509–512; MR 51 #3434. [27]
535. Guz, Eugeniusz. *Certain estimates in the family of schlicht functions satisfying auxiliary conditions.* (Polish. French summary) Zeszyty Nauk. Univ. Lodzk. Nauki Mat. Przyrod. Ser. II Zeszyt 29(1968), 31–47; MR 39 #5783. [9, 58, 68]
536. Guzek. R. F. *The coefficients of k-symmetric univalent functions.* (Polish. English summary). Zeszyty Nauk. Univ. Lodzk. Nauki Mat. Przyrod. Ser. II Zeszyt 20 Matematyka (1966), 147–160; MR 50 #2477. [22, 30, 49]
537. Haddad, David C. *Asymptotic values of finitely valent functions.* Duke Math. J. 39(1972), 361–368. [14]
538. Haddad, David C. *Angular limits of locally finitely valent holomorphic functions.* Pacific J. Math. 48(1973), 107–112; MR 48 #8802. [14, 62]
539. Haifawi, M. M. *On the behavior of hyperbolic curvature under univalent bounded transformation.* Rend. Circ. Mat. Palermo (2) 16(1967), 57–63; MR 39 #2964. [35]
540. Halász, G. *Tauberian theorems for univalent functions.* Studia Sci. Math. Hungar. 4(1969), 421–440; MR 40 #5842. [6, 20]
541. Hallenbeck, David J. *Extreme points of some families of univalent functions.* Notices, A. M. S., vol. 19, no. 1, issue no. 135, January, 1972 (Abstract 691-30-32), p. A–119. [88]
542. Hallenbeck, D. J. *Convex hulls and extreme points of some families of univalent functions.* Trans. Amer. Math. Soc. 192(1974), 285–292; MR 49 #3103. [1, 2, 4, 5, 6, 10, 12, 16, 21, 23, 39, 41, 56, 88]
543. Hallenbeck, D. J. *Some inequalities for convex, starlike, and close-to-convex mappings.* Czechoslovak Math. J. 24 (99) (1974), 411–415; MR 50 #589. [5, 6, 10, 63, 64]
544. Hallenbeck, D. J. *On the Marx conjecture for the convex hulls of families of starlike and convex mappings.* Proc. Amer. Math. Soc.

42(1974), 135-139; MR 48 #4293. [1, 6, 10, 12, 43, 88]
545. Hallenbeck, D. J. *Extreme points of classes of functions defined by subordination.* Proc. Amer. Math. Soc. 46(1974), 59-64; MR 50 #10225. [1, 6, 36, 88]
546. Hallenbeck, David J. *Convex hulls and extreme points of families of starlike and close-to-convex mappings.* Pacific J. Math. 57 (1975), 167-176. [1, 2, 5, 6, 13, 16, 23, 36, 45, 75, 88]; MR 52 #725
547. Hallenbeck, D. J.; Livingston, A. E. *Applications of extreme point theory to classes of multivalent functions.* Notices Amer. Math. Soc. 21(1974), Abstract 74T-B221, p. A-544. [14, 88]
548. Hallenbeck, D.: Livingston, A. *Application of extreme point theory to classes of multivalent functions.* Notices Amer. Math. Soc. 23, January 1976, Abstract 731-30-15, p. A-101. [1, 5, 14, 75, 88]
549. Hallenbeck, D. J.; Livingston, A. E. *Subordination chains and p-valent functions.* Notices Amer. Math. Soc. 23, January 1976, Abstract 731-30-6, p. A-99. [1, 4, 14]
550. Hallenbeck, D. J.; Livingston, A. E. *A coefficient estimate for multivalent functions.* Proc. Amer. Math. Soc. 54(1976), 201-206. [1, 5, 6, 14, 16, 42, 50, 88]; MR 52 #8401
551. Hallenbeck, D. J.; MacGregor, T. H. *Subordination and extreme-point theory.* Pacific J. Math. 50(1974), 455-468; MR 50 #13481. [1, 2, 3, 5, 6, 10, 13, 23, 88]
552. Hallenbeck, D. J.; Ruscheweyh, S. *Subordination by convex functions.* Proc. Amer. Math. Soc. 52(1975), 191-195; MR 51 #10603. [1, 4, 6, 10, 82]
553. Hansen, Lowell J. *The Hardy class of a spiral-like function.* Mich. Math. J., 18(1971), 279-282; MR 44 #4211. [6, 36, 62]
554. Hartmann, F. W.; MacGregor, T. H. *Matrix transformations of univalent power series.* J. Australian Math. Soc. 18(1974), 419-435; MR 51 #10597. [4, 54, 68]
555. Haruki, H. *On an application of a theorem of Apollonius.* Portugal. Math. 33(1974), 167-170; MR 50 #13476. [31]
556. Hayman, W. K. *Corrigendum: Survey article—coefficient problems for univalent functions and related function classes.* J. London Math. Soc. 41(1966), 550; MR 33 #7519. [44]
557. Hayman, W. K. *Research problems in function theory.* The Athlone Press (Univ. of London), London, 1967, VII + 56 pp.; MR 36 #359. [16, 44]
558. Hayman, W. K. *Mean p-valent functions with gaps.* Colloq. Math. 16(1967), 1-21; MR 35 #4395. [14, 33, 39]

559. Hayman, Walter K. *Les fonctions multivalentes*. Les Presses de l'Universite de Montréal, Montréal, Que., 1968. 52 pp.; MR 40 #5848. [14, 44]
560. Hayman, W. K. *On the second Hankel determinant of mean univalent functions*. Proc. London Math. Soc. (3rd series), 18 (1968), 77-94; MR 36 #2794. [33]
561. Hayman, W. K. *Mean p-valent functions with mini-gaps*. Math. Nachr. 39(1969), 312-324; MR 40 #336. [33]
562. Hayman, W. K. *Tauberian theorems for multivalent functions*. Acta Math. 125(1970), 269-298; MR 42 #3270. {See also MR 48 #505, MR 50 #593}. [14, 33, 62]
563. Hayman, W. K. *Research problems in function theory; progress on the previous problems; new problems*. Proceedings of the Symposium on Complex Analysis (Univ. Kent, Canterbury, 1973), pp. 143-154, 155-180. London Math. Soc. Lecture Note Ser., No. 12, Cambridge Univ. Press, London, 1974. [44]; MR 52 #5944, #8385, #8386
564. Hayman, W. K. *Differential inequalities and local valency*. Pacific J. Math. 44(1973), 117-138; MR 47 #5240. [14]
565. Hayman, W. K.; Nicholls, P. J. *O the minimum modulus of functions with given coefficients*. Bull. London Math. Soc. 5(1973), 295-301; MR 48 #4301. [33]
566. Hayman, W. K.; Storvick, D. A. *A question of M. L. Cartwright*. J. London Math. Soc. (2)5(1972), 419-422; MR 47 #5238. [62]
567. Hayman, W. K.; Weitsman, A. *On the coefficients and means of functions omitting values*. Math. Proc. Cambridge Philos. Soc. 77 (1975), 119-137; MR 50 #13495. [3, 14, 62]
568. Heins, Maurice. *On a theorem of Study concerning conformal maps with convex images*. Mathematical Essays dedicated to A. J. Macintyre, pp. 171-176. Ohio Univ. Press, Athens, Ohio, 1970; MR 42 #7878. [10]
569. Heins, Maurice. *A note on the Lowner differential equations*. Pacific J. Math. 39(1971), 173-177; MR 46 #3771. [24]
570. Heins, M. *On a theorem of Study concerning conformal maps with convex images*. II. Math. Scand. 32(1973), 245-257; MR 49 #551. [10]
571. Hengartner, Walter; Schober, Glenn. *On schlicht mappings to domains convex in one direction*. Comment. Math. Helv. 45(1970), 303-314; MR 43 #3436. [2, 19, 21, 41, 78, 79]
572. Hengartner, Walter; Schober, Glenn. *Analytic functions close to mappings convex in one direction*. Proc. Amer. Math. Soc.

28(1971), 519–524; MR 43 #3437. [5, 12, 41, 43, 78, 79]
573. Hengartner, W.; Schober, G. E. *A remark on the Pólya-Schoenberg conjecture*. Notices, A. M. S., vol. 19, no. 4, issue no. 138, June 1972 (Abstract 72T–B150), p. A–518. [47]
574. Hengartner, W.; Schober, G. *Points extrémaux des familles de fonctions univalentes*. C. R. Acad. Sci. Paris Sér. A–B 274(1972), A837–A838; MR 45 #7038. [88]
575. Hengartner, W.; Schober, G. *Extreme points for some classes of univalent functions*. Trans. Amer. Math. Soc. 185(1973), 265–270; MR 49 #559. [4, 41, 88]
576. Hengartner, W.; Schober, G. *Compact families of univalent functions and their support points*. Michigan Math. J. 21(1974), 205–217 (1975); MR 51 #898. [24, 88]
577. Hengartner, W.; Schober, G. *Propriétés des points d'appui des familles compactes de fonctions univalentes*. C. R. Acad. Sci Paris Sér. A. 279(1974), 551–553; MR 51 #880. [88]
578. Hengartner, W.; Schober, G. *Some new properties of support points for compact families of univalent functions*. Notices Amer. Math. Soc. 23, January 1976, Abstract 731–30–18, p. A–102. [88]
579. Herzog, F.; Piranian, G. *The counting function for points of maximum modulus*. Proc. of Symposia In Pure Math., vol. 11(1968), 240–243. [68, 90]
580. Hindmarsh, L.; Volk, B. *Schlicht cubes on $|z| < 1$*. The Amer. Math. Monthly, March 1976, 207–209. [32]
581. Holland, F. *Some properties of a class of regular functions*. Proc. Roy. Irish Acad. Sect. A 69, 85–95(1970); MR 41 #7084. [6, 23, 62]
582. Holland, F. *Some asymptotic relations for starlike functions*. Proc. Roy. Irish Acad. Sect. A, 72(1972), no. 1, pp. 1–16; MR 47 #451. [6, 62]
583. Holland, Finbarr. *On the coefficients of starlike functions*. Proc. Amer. Math. Soc. 33(1972), 463–470; MR 45 #531. [2, 6, 9, 18, 23, 25, 27]
584. Holland, Finbarr. *Some extremum problems for polynomials with positive real part*. Bull. London Math. Soc. 5(1973), 54–58; MR 47 #8824. [2, 32]
585. Holland, F. *The extreme points of a class of functions with positive real part*. Math. Ann. 202(1973), 85–88; MR 49 #562. [2, 21, 23, 88]
586. Holland, F.; Thomas, D. K. *The area theorem for starlike functions*. J. London Math. Soc. (2)1(1969), 127–134; MR 39 #7080. [6, 18, 27, 36]
587. Holland, F.; Thomas, D. K. *On the order of a starlike functions*.

Trans. Amer. Math. Soc. 158(1971), 189-201; MR 43 #3438. [3, 6, 36, 62, 63]
588. Holland, F.; Twomey, J. B. *On coefficient means of certain subclasses of univalent functions.* Trans. Amer. Math. Soc. 185 (1973), 151-164; MR 48 #6396. [2, 5, 6, 21, 22, 30, 62, 75]
589. Horgan, C. O.; Knowles, J. K. *A note on conformal mappings of convex mappings.* Utilitas Math. 5(1974), 75-78; MR 49 #9176. [27, 53, 54]
590. Horn, R. A. *On boundary values of a schlicht mapping.* Proc. Amer. Math. Soc. 18(1967), 782-787; MR 36 #2792. [53, 62]
591. Hornich, Hans. *Über die Fixpunkte der schlichter Funktionen.* (English and Italian summaries) Rent. Ist. Mat. Univ. Trieste 2 (1970), 54-58; MR 42 #1989.
592. Hornich, Hans. *Eine partielle Differentialgleichung in Zusammenhang met den schlichten Funktionen.* Monatsh. Math. 76(1972), 121-123; MR 46 #2029.
593. Horowitz, D. *Coefficient estimates for univalent polynomials.* Notices Amer. Math. Soc., January 1975, Abstract 720-30-28, p. A-125. [32, 42]
594. Horowitz, David. *A refinement for coefficient estimates for univalent functions.* Proc. Amer. Math. Soc. 54(1976), 176-178. [16, 42, 54]
595. Hübner, Otto. *Die Faktorisierung konformer Abbildungen und Anwendungen.* Mat. Z. 92(1966), 95-109; MR 33 #2808. [58, 68]
596. Huckemann, F. *On Schiffer's variational lemma.* Arch. Rational Mech. Anal. 22(1966), 310-312; MR 33 #1446. [62]
597. Huckemann, F. *Extremal elements in certain classes of conformal mapping of an annulus.* Acta Math. 118(1967), 193-221; MR 35 #3048. [57]
598. Huckemann, F. *On Schiffer's variational lemma.* II. Arch. Rational Mech. Anal. 33(1969), 246-248; MR 39 #1652. [62]
599. Hummel, J. A. *Multivalent starlike functions.* J. Analyse Math. 18 (1967), 133-160; MR 35 #359. [6, 14]
600. Hummel, J. A. *Extremal properties of weakly starlike p-valent functions.* Amer. Math. Soc. Trans. 130(1968), 544-551; MR 36 #5332. [6, 14]
601. Hummel, J. A. *The Coefficients of starlike functions.* Proc. A. M. S. 22(1969), 311-315; MR 40 #4440. [4, 6, 36, 54]
602. Hummel, J. A. *Bounds for the coefficient body of univalent functions.* Archive for Rational Mechanics and Analysis 36(1970), 128-134; MR 42 #6213. [6, 36]

603. Hummel, James A. *A counterexample to the Marx conjecture for starlike functions.* Notices, A. M. S., vol. 19, no. 1, issue no. 135, January 1972 (Abstract 691-30-15), p. A-114. [1, 6, 16]
604. Hummel, James A. *Lectures on variational methods in the theory of univalent functions.* University of Maryland, Dept. of Mathematics, Lecture Note #8, 189 pp. [24]
605. Hummel, J. A. *Inequalities of Grunsky type for Aharonov pairs.* J. Analyse Math. 25(1972), 217-257; MR 47 #455. [4, 22, 34, 53]
606. Hummel, J. A. *The Marx conjecture for starlike functions.* Michigan Math. J., 19(1972), no. 3, pp. 257-266; MR 46 #3761. [6, 16]
607. Hummel, James A. *Lectures on variational method in the theory of univalent functions.* Univ. of Maryland (Student bookstore), College Park, Maryland 20742. [24]
608. Hummel, J. A.; Schiffer, M. *Coefficient inequalities for Bieberbach-Eilenberg functions.* Arch. Rational Mech. Anal. 32 (1969), 87-99; MR 39 #426. [4, 34, 53, 87]
609. Il'ina, L. P. *The relative growth of nearby coefficients of schlicht functions.* (Russian) Mat. Zametki 4(1968), 715-722; MR 39 #1644. [8]
610. Il'ina, L. P. *Estimates for the coefficients of univalent functions in dependence on the second coefficient.* (Russian) Mat. Zametki 13 (1973), 351-357; MR 47 #7019. [42]
611. Il'ina, L. P.; Kolomoiceva, Z. D. *The estimation of $|c_4|$ as a function of $|c_2|$ in the class S.* (Russian. English summary) Vestnik Leningrad. Univ. No. 1 Mat. Meh. Astronom. Vyp. 1(1974), 27-30, 164; MR 49 #9185. [27, 53, 54]
612. Inove, T. *An application of the variational method to hyperbolic capacity.* Math. Japan. 12(1967), 35-40; MR 37 #397. [24]
613. Izdebski, Lucjan. *A contribution to the theory of subordination.* Ann. Univ. Mariae Curie-Sklodowska Sect. A 18(1964), 9-12(1967); MR 38 #2298. [1, 16, 68]
614. Jabłoński, F. F. *Sur la subordination en module et en domaine des fonctions holomorphes.* Ann. Univ. Mariae Curie-Sklodowska, Sect. A, vols. 22/23/24(1968/1969/1970), p. 79-83; MR 49 #5339. [1, 6]
615. Jabłoński, F. F.; Lewandowski, Z. *Caractérisation de certaines classes de foncitons holomorphes par la subordination modulaire.* Ann. Univ. Mariae Curie-Sklodowska Sect. A 19(1965), 25-31. [1, 6]
616. Jabłoński, F. F.; Wesolowski, A. *Sur une famille de fonctions holomorphes dans le cercle unité.* (Polish and Russian summaries). Ann. Univ. Mariae Curie-Sklodowska Sect. A 21(1967), 91-99 (1972); MR 48 #8771. [5, 6, 12, 63]

617. Jabotinsky, E. *Universal relations between the elements of Grunsky's matrix.* J. Analyse Math. 17(1966), 411-417; MR 35 #6808. [53]
618. Jabotinsky, E. *Universal relations between the elements of Grunsky's matrix.* J. Analyse Math. 17(1966), 411-417; MR 35 #6808. [53]
619. Jack, I. S. *Functions starlike and convex of order α.* J. London Math. Soc. (2)3(1971), 469-474; MR 43 #7611. [6, 10, 11, 12, 16, 63, 64]
620. Jaenisch, S. *Length distortion of curves under conformal mappings.* Mich. Math. J. 15(1968), 121-128; MR 36 #5319. [7, 16, 35, 62]
621. Jakubowski, Z. J. *Théorèmes sur la déformation dans la famille de fonctions univalentes au pôle simple à l'infini et bornées inférieurement dans le cercle $K(\infty, 1)$.* Panstwowe Wydawnictwo Naukowe, Łódź, 1962. 43 pp.; MR 34 #326. [9, 58, 22]
622. Jakubowski, Z. J. *Sur les fonctions univalentes p-symétriques et bornées dans un cercle unitaire.* (Russian summary) Bull. Acad. Polon. Sci. Sér. Sci. Math. Astronom. Phys. 14(1966), 637-642; MR 34 #6071. [22, 49, 68]
623. Jakubowski, Z. J. *The maximum of the functional $A_3 + \alpha A_2$ in the family of univalent function with real coefficients.* (Polish. French summary) Zeszyty Nauk. Univ. Lodzk. Nauki Mat. Przyrod. Zeszyt 20 Matematyka (1966), 43-61; MR 50 #7502. [39, 54]
624. Jakubowski, Z. J. *Sur les coefficients des fonctions univalentes et symétriques dans un cercle unitaire.* (Russian summary) Bull. Acad. Polon. Sci. Sér. Sci. Math. Astronom. Phys. 14(1966), 643-646; MR 34 #6072. [22, 30, 39]
625. Jakubowski, Z. J. *Sur les coefficients des fonctions univalentes dans le cercle unité.* Ann. Polon. Math. 19(1967), 207-233; MR 35 #6814. [22, 30, 39, 54]
626. Jakubowski, Z. J. *Les fonctions univalentes, p-symmetriques et bornees dans le cercle unité.* Ann. Polon. Math. 20(1968), 119-148; MR 37 #2968. [22, 24]
627. Jakubowski, Z. J. *On the coefficients of Carathéodory functions.* (Russian summary) Bull. Acad. Polon. Sci. Sér. Sci. Mat. Astronom. Phys. 19(1971), 805-809; MR 46 #3770. [2, 21, 56]
628. Jakubowski, Z. J. *On some application of the Clunie method.* Ann. Polon. Math., 26(1972), pp. 211-217; MR 46 #2064. [2, 13, 21, 41, 51, 79]
629. Jakubowski, Z. J. *On the coefficients of starlike functions of some classes.* Ann. Polon. Math., 26(1972), No. 3, pp. 305-313 {Also: Bull. Acad. Polon. Sci. Sér. Sci. Math. Astronom. Phys., 19(1971),

no. 9, pp. 811-815}; MR 46 #2030, #7497. [1, 6, 21, 36]
630. Jakubowski, Z. J. *On the upper bound of the functional $|f^{(n)}(z)|$ ($n = 2, 3, \ldots$) in some classes of univalent functions.* Comment. Math. Prace Mat. 17(1973), 65-69; MR 48 #2369. [54, 68]
631. Jakubowski, Z. J. *On some properties of extremal functions of Carathéodory.* Comment. Math. Prace Mat. 17(1973), 71-80; MR 48 #523. [2, 56]
632. Jakubowski, Z. J. *On some special class of regular functions.* Math. Balkanica 4(1974), 299-300.
633. Janczar, B. *The structure of boundary functions in the classes quasi-starlike and quasi-convex functions.* Demonstratio Math. 5 (1973), 17-27; MR 48 #2362. [6, 10, 24]
634. Janczar, B. z_o-*quasi-convex functions.* (Polish and Russian summaries) Zeszyty Nauk. Politech. Łódź. No. 186 Mat. no. 5(1974), 27-42. [10]
635. Jankovics, Ronald. *Uber Funktionen mit der Eigenschaft $Re[e^{i\alpha}\{f(z)/z - \beta] > 0$.* Math. Z. 143(1975), Heft 3, 235-242; MR 51 #8392. [6, 11, 12, 20]
636. Janowski, W. *Sur une certaine famille de fonctions univalentes.* (Russian summary) Bull. Acad. Polon. Sci. Sér. Sci. Math. Astronom. Phys. 13(1965), 707-713; MR 35 #4386. [24, 29]
637. Janowski, W. *A certain family of univalent functions.* (Polish. French summary) Zeszyty Nauk. Univ. Lodzk. Nauki Mat. Przyrod. Ser. II Zeszyt 20 Matematyka (1966), 3-41; MR 50 #7498. [29]
638. Janowski, W. *Sur une certain famille de fonctions univalentes.* Ann. Polon. Math. 18(1966), 171-203; MR 34 #327. [24, 29, 62]
639. Janowski, W. *On the radius of starlikeness of some families of regular funcitons.* (Loose Russian summary) Bull. Acad. Polon. Sci. Sér. Sci. Math. Astronom. Phys. 17(1969), 503-508; MR 40 #7438. [6, 11]
640. Janowski, W. *Extremal problems for a family of functions with positive real part and for some related families.* (Loose Russian summary) Bull. Acad. Polon. Sci. Sér. Sci. Math. Astronom. Phys. 17(1969), 633-637; MR 41 #1986. [2, 12, 21, 56]
641. Janowski, Witold. On the radius of starlikeness of some families of regular functions. Comment. Math. Prace Mat. 14(1970), 137-149; MR 42 #6208. [6, 11]
642. Janowski, W. *Extremal problems for a family of functions with positive real part and for some related families.* Ann. Polon. Math. 23(1970/71), 159-177; MR 42 #2005. [2, 6, 11, 12, 13, 19, 21, 24, 40, 41]

643. Janowski, W. *Some extremal problems for certain families of analytic functions.* I. Bull. Acad. Polon. Sci. Sér. Math. Astronom. Phys. 21(1973), 17-26; MR 47 #3659. [2, 6, 12, 56, 63]
644. Janowski, W. *Some extremal problems for certain families of analytic functions.* I. Ann. Polon. Math. 28(1973), 297-326; MR 48 #6401. [2, 12, 56]
645. Jenkins, James A. *On normalization in the general coefficient theorem.* Proc. Internat. Congr. Mathematicians (Stockholm, 1962), pp. 347-350. Inst. Mittag-Leffler, Djursholm, 1963; MR 31 #328. [24]
646. Jenkins, J. A. *On Bieberbach-Eilenberg functions.* III. Trans. Amer. Math. Soc. 119(1965), 195-215; MR 31 #5969. [26, 34, 86]
647. Jenkins, J. A. *On a result of Nehari.* Proc. Amer. Math. Soc. 17 (1966), 62-66; MR 32 #5865. [9, 19, 25, 27, 29, 54]
648. Jenkins, James A. *On certain extremal problems for the coefficients of univalent functions.* J. D'analyse Math. 18(1967), 173-184; MR 35 #3051. [24, 54]
649. Jenkins, James A. *On an inequality considered by Robertson.* Proc. Amer. Math. Soc. 19, no. 3, June 1968, pp. 549-550; MR 37 #401. [8, 42, 54]
650. Jenkins, James A. *A uniqueness result in conformal mapping.* Proc. Amer. Math. Soc. 22(1969), 324-325; MR 39 #2958. [7, 9, 19]
651. Jenkins, James A. *A remark on "pairs" of regular functions.* Proc. Amer. Math. Soc. 31(1972), 119-121; MR 45 #519. [26, 34, 86]
652. Jenkins, James A. *The representation of certain slit mappings.* Notices Amer. Math. Soc. 23, January 1976, Abstract 731-30-21, p. A-102. [9]
653. Jenkins, James A.; Oikawa, Kôtaro. *A remark on p-valent functions.* J. Austral. Math. Soc. 12(1971), 397-404; MR 45 #3693. [14, 33]
654. Jenkins, James A.; Oikawa, Kôtaro. *On results of Ahlfors and Hayman.* Ill. J. Math. 15(1971), 664-671; MR 45 #5332. [24, 33]
655. Jenkins, J. A.; Ozawa, M. *On local maximality for the coefficient a_8.* Illinois J. Math. 11(1967), 596-602; MR 36 #1636a. [42, 53, 54]
656. Jenkins, J. A.; Ozawa, M. *On local maximality for the coefficient a_6.* Nagoya Math. J. 30(1967), 71-78; MR 36 #1636b. [27, 54]
657. Juneja, O. P.; Mogra, M. L. *A class of univalent functions.* Notices Amer. Math. Soc. 22, February 1975, Abstract 75T-B64, p. A-316. [2, 12, 21, 56]
658. Juneja, O. P.; Mogra, M. L. *On starlike functions of order α and type β.* Notices Amer. Math. Soc. 22, April 1975, Abstract

75T-B80, p. A-384. [4, 6, 12, 36, 63]
659. Kac, B. A. *Convexity and starlikeness of the level curves of polynomials.* (Russian) Mat. Zametki 15(1974), 701-710; MR 50 #2459. [13]
660. Kac, I. S. *Integral and exponential representations of analytic functions mapping the upper half-plane into itself.* (Russian) Problems of Math. Phys. and Theory of Functions, II (Russian), pp. 51-62. Naukova Dumka, Kiev, 1964; MR 32 #7721. [23]
661. Kaczmarski, J. *The least upper bound of a certian functional in the family of bounded univalent functions.* (Polish. French summary) Zeszyty Nauk. Univ. Lodzk. Nauki Mat. Przyrod. Ser. II Zeszyt 20 Matematyka (1966), 63-101; MR 50 #7505. [22, 24, 29]
662. Kaczmarski, J. *Sur l'equation $f(z) = pf(c)$ dans la classe des fonctions etoilees.* (Loose Russian Summary) Bull. Acad. Polon. Sci. Sér. Sci. Math. Astronom. Phys. 15(1967), 455-464; MR 36 #3970. [6, 24]
663. Kaczmarski, J. *Sur l'equation $f(z) = pf(c)$ dans la famille des fonctions univalentes a coefficients reels.* (Russian summary) Bull. Acad. Polon. Sci. Ser. Sci. Math. Astronom. Phys. 15 (1967), 245-251; MR 35 #4387. [24, 39]
664. Kaczmarski, J. *On the coefficients of some classes of starlike functions.* (Loose Russian summary) Bull. Acad. Polon. Sci. Sér. Sci. Math. Astronom. Phys. 17(1969), 495-501; MR 40 #7437. [2, 9, 25]
665. Kaczmarski, J. *On the radius of β-μ-N-spiral-starlikeness of the family $S^*(\alpha, \lambda, M)$ of spiral-starlike functions in the disc $|z| < 1$.* (Loose Russian summary) Bull. Acad. Polon. Sci. Sér. Sci. Math. Astronom. Phys. 18(1970), 467-473; MR 42 #4718. [6, 11]
666. Kaczmarski, J. *On some extremal problems for certain classes of close-to-starlike functions.* (Russian summary) Bull. Acad. Polon. Sci. Sér. Sci. Math. Astronom. Phys. 21(1973), 133-140; MR 47 #3660. [2, 6, 11]
667. Kaczmarski, J. *Some radius of convexity problems in certain family of functions with bounded distortion.* Bull. Acad. Polon. Sci. Sér. Math. Astronom. Phys. 21(1973), 27-34; MR 47 #5241. [2, 12]
668. Kaczmarski, J. *On the radius of convexity for certain regular functions.* Comment. Math. Prace Mat. 17(1973/74), 373-383; MR 50 #2472. [2, 12]
669. Kaczmarski, J. *On some extremal problem for certain classes of close-to-starlike functions.* Comment. Math. Prace Mat. 17 (1973/74), 385-398; MR 50 #2473. [6, 11]
670. Kahramaner, Suzan. *Sur les coefficients des fonctions univalentes.*

(Turkish summary) Istanbul Univ. Fen. Fak. Mec. Ser. A 28(1963), 1-7; MR 34 #4479. [50]
671. Kahramaner, Suzan. *Sur l'argument des fonctions univalentes.* (Turkish summary) Istanbul. Univ. Fen. Fak. Mecm. Ser. A 32 (1967), 1-22(1971); MR 46 #9319. [68]
672. Kakehashi, T. *A note on schlicht functions.* Math. Japan. 11(1967), 149-151; MR 35 #3049. [2, 24]
673. Kakehashi, T. *On a certain class of univalent functions.* Proc. Japan Acad. 43(1967), 469-471; MR 36 #6607. [49, 54]
674. Kakehashi, T.; Matsumura, Y. *Note on the coefficients of Taylor expansion.* Math. Japan 12(1968), 131-132; MR 37 #2960. [54]
675. Kalme, C. I. *Remarks on a paper by Lipman Bers.* Ann. of Math. (2)91(1970), 601-606; MR 42 #7879. [28]
676. Kan, V. I. *The variation of Faber polynomials.* (Russian) Trudy Tomsk. Gos. Univ. Ser. Meh.-Mat. 175(1964), Vyp. 2, 50-58; MR 34 #6119. [24, 53]
677. Kan, V. I. *Extremal properties of the Faber polynomials for starlike continua.* (Russian) Trudy Tomsk. Gos. Univ. Ser. Meh.-Mat. 182(1965), Vyp. 3, 71-85; MR 34 #2901. [53]
678. Karunakaran, V. *On the arithmetic mean of certain classes of analytic funcitons.* (Serbo-Croatian summary). Glasnik Mat. Ser. III 10(30)(1975), no. 2, 249-256. [82]
679. Karunakaran, V. *A subclass of functions of positive real part.* (Preliminary report) Notices Amer. Math. Soc. 22, February 1975, Abstract 75T-B55, p. A-314. [2, 56]
680. Karunakaran, V. *A certain generalization of functions with positive real part in the theory of univalent functions.* Notices Amer. Math. Soc. 23, February 1976, Abstract 76T-B28, p. A-276. [2, 56]
681. Kas'janjuk, S. A. *On a certain interpolation problem for positive functions harmonic in the unit circle.* (Russian) Uspehi Mat. Nauk 20(1965), no. 6(126), 98-101; MR 33 #2812. [2, 23]
682. Kas'janjuk, S. A. *Estimates for the curvature of level curves under a conformal mapping with bounded distortion.* (Russian) Izv. Vyss. Ucebn. Zaved. Matematika, 1967, no. 6(61), 54-58; MR 35 #3046. [35]
683. Kas'janjuk, S. A.; Finogenova, V. G. *The ranges of values of typically real functions with fixed coefficients and their regions.* (Russian). Izv. Vyss. Ucebn. Zaved. Matematika 1968, no. 4(71), 40-47; MR 37 #2977. [13, 29]
684. Kas'janjuk, S. A.; Tracuk, G. I. *A property of functions with positive real part.* (Russian) Teor. Funkcii Funkcional Anal. i

Prilozen. Vyp. I(1965), 224-227; MR 34 #4508. [2, 39]
685. Kayser, Hans-Jürgen. *Über Abschatzungsprobleme bei gewissen holomorphen Abbildungen*. Bonn. Math. Schr. No. 49(1967), VIII + 55 pp.; MR 47 #452. [19, 54]
686. Keogh, F. R. *A strengthened form of the $\frac{1}{4}$ theorem for starlike univalent functions*. Mathematical Essays Dedicated to A. J. Macintyre, pp. 201-211. Ohio Univ. Press, Athens, Ohio, 1970; MR 43 #496. [1, 6, 19]
687. Keogh, F. R. *A subordinate property of univalent functions*. Bull. London Math. Soc. 3(1971), 181-184; MR 45 #2155. [1, 4, 10, 19, 20]
688. Keogh, F. R. *On spiral-like univalent functions*. Notices, A. M. S., vol. 19, no. 2, issue 136, February, 1972 (Abstract 692-B24), p. A-362. [6, 9, 25, 36]
689. Keogh, F. R. *On spiral-like univalent functions*. Notices A. M. S. 20(Jan. 1973), Abstract 701-30-7, pg. A-105. [6, 7]
690. Keogh, F. R. *A characterisation of convex domains in the plane*. Notices Amer. Math. Soc., January 1975, Abstract 720-30-1, p. A-118. [20]
691. Keogh, F. R. *A characterization of convex domains in the plane*. Notices Amer. Math. Soc. 23, January 1976, Abstract 731-30-22, p. A-103. [20]
692. Keogh, F. R.; Başgöze, T. *The Hardy class of a spiral-like function and its derivative*. Proc. Amer. Math. Soc. 26(1970), no. 4, 266-269. [5, 6]; MR 41 #8680
693. Keogh, F. R.; Merkes, E. P. *A coefficient inequality for certain classes of analytic functions*. Proc. Amer. Math. Soc. 20, January, 1969, pp. 8-12; MR 38 #1249. [5, 6, 10, 22, 30, 36, 38, 54, 75]
694. Keogh, F. R.; Miller, S. S. *On the coefficients of Bazilevič functions*. Proc. Amer. Amth. Soc. 30(1971), 492-496; MR 44 #424. [1, 4, 5, 49, 70, 81]
695. Kikuchi, K. *Starlike and convex mappings in several complex variables*. Pacific J. Math. 44(1973), 569-580. [1, 6, 10]
696. Kim, W. J. *The Schwarzian derivative and multivalence*. Pacific J. Math. 31(1969), 717-724; MR 40 #5849. [4, 31]
697. Kim, Y. J.; Merkes, E. P. *On an integral of powers of a spirallike function*. Kyungpook Math. J., vol. 12, no. 2, December 1972, pp. 249-253; MR 47 #8834. [4, 6, 85]
698. Kim, Y. J.; Merkes, E. P. *On certain convex sets in the space of locally schlicht functions*. Trans. Amer. Math. Soc 196(1974), 217-224; MR 50 #2474. [5, 10, 85]

699. King, Amy C. *A class of entire functions of bounded index.* Abstract: Notices A. M. S. 17(1970), p. 110-111. [11, 43, 73, 89, 90]
700. Kir'jackii, E. G. (Kirjackis, E.) *Certain classes of univalent functions.* (Russian. Lithuanian and German summaries) Litovsk. Mat. Sb. 12(1972), no. 3, 75-84, 207; MR 47 #5243. [28]
701. Kir'jackii, E. G. *A certain class of rational univalent functions.* (Russian. Lithuanian and German summaries) Litovsk. Mat. Sb. 13(1973), no. 2, 79-89, 259; MR 48 #2363. [28]
702. Kirwan, W. E. *Extremal problems for the typically real functions.* Amer. J. Math. 88(1966), 942-954; MR 34 #2853. [12, 13, 43]
703. Kirwan, W. E. *A note on extremal problems for certain classes of analytic functions.* Proc. Amer. Math. Soc. 17(1966), 1028-1030; MR 34 #2854. [2, 5, 6, 10, 24]
704. Kirwan, W. E. *The Koebe constant for a class of bounded domains.* Ann. Univ. Mariae Curie-Sklodowska Sect. a, vol. 21(1967), Nos. 1-8, pp. 5-9(1972); MR 48 #4294. [1, 10, 16, 22, 71]
705. Kirwan, W. E. *On the rate of growth of typically real functions.* Duke Math J. 35(1968), 9-20; MR 36 #2805. [13, 62]
706. Kirwan, W. E. *On the coefficients of functions with bounded boundary rotation.* Mich. Math. J. 15(1968), 277-282. MR 38 #1250. [3, 71, 77]
707. Kirwan, W. E. *Extremal problems for functions with bounded boundary rotation.* Ann. Acad. Sci. Fenn. Ser. AI Mat. No. 595 (1975), 19 pp. [4, 23, 24, 43, 71, 76]
708. Kirwan, W. E. *Extremal problems for functions meromorphic and univalent in the unit disc.* Notices Amer. Math. Soc. 23, January 1976, Abstract 731-30-19, p. A-102. [9, 25, 58]
709. Kirwan, W. E.; Schober, G. *On extreme points and support points for some families of univalent functions.* Duke Math. J. 42(1975), 285-296; MR 51 #3416. [6, 9, 88]
710. Kishi, M. *Sur le principe de majoration de K. Yosida.* Nagoya Math. J. 37(1970), 33-36. [1]; MR 41 #3796
711. Klein, M. *Estimates for the transfinite diameter with applications to conformal mapping.* Pacific J. Math 22(1967), 267-279; MR 37 #1577. [19]
712. Klein, M. *Functions starlike of order α.* Trans. Amer. Math. Soc. 131(1968), 99-106; MR 36 #2795. [6, 9, 10, 19, 25, 36, 38]
713. Klein, Melvyn. *Meromorphic starlike multivalent functions.* Notices Amer. Math. Soc. (Abstract 68T-311), vol. 15, no. 2, issue no. 104, February, 1968, p. 391. [6, 9, 25, 14]

714. Klein, M. *Star radii for multivalent meromorphic functions*. Math. Japan. 20(1975), no. 1, 59–63. [11, 14]; MR 52 #8402
715. Kocak, Cevdet. *An extremum problem in conformal mapping*. (Turkish summary) Istanbul Tek. Univ. bul. 24(1971), no. 1, 122–127; MR 48 #8779. [82]
716. Kocetkov, A. P. *On certain extremal problems for analytic functions with positive real part*. (Russian) Studies Contemporary Problems Constructive Theory of Functions (Proc. Second All-Union Conf. Baku, 1962)(Russian), pp. 249–254. Izdat. Akad. Nauk Azerbaidzan. SSR, Baku, 1965; MR 33 #4272. [2]
717. Kocetkov, V. K. *The multiple differentiation of one parameter families of schlicht functions*. (Russian) Sibirsk. Mat. Z. 12(1971), 367–373; MR 45 #525 [English transl.: Siberian Math. J. 12(1971), 261–265.]
718. Kocur, M. F. *On a new Q-method for the solution of extremal problems in certain special classes of analytic functions that are connected with functions having positive real part in the disk*. (Russian. English summary) Vycisl. Prikl. Mat. (Kiev) Vyp. 17(1972), 152–162; MR 46 #5628. [2]
719. Koczan, L.; Szapiel, W. *Sur certaines sous-classes de fonctions typiquement réelles*. (Russian summary). Bull. Acad. Polon. Sci. Sér. Sci. Math. Astronom. Phys. 22(1974), 115–120; MR 49 #5356. [13]
720. Kohlhammer, Hans-Peter. *Schlichtheitsradien von Potenzreihenabschnitten schlichter Funktionen*. Gesellschaft fur Mathematik und Datenverarbeitung, Bonn. Ber. No. 99, Gesellschaft fur Mathematik und Datenverarbeitung. Bonn, 1975. 90 pp. [20, 43]
721. Kokilasvili, V. M. *On the precise order of the best approximations of analytic functions representable by generalized gap series with respect to Faber polynomials*. (Russian. Georgian summary). Soobsc. Akad. Nauk Gruzin. SSR 41(1966), 529–534; MR 34 #7811. [53]
722. Kolokol'nikov, A. S. *The logarithmic derivative of a meromorphic function*. (Russian) Mat. Zametki 15 (1974), 711–718; MR 50 #7529. [9]
723. Komatu, Y. *On mean distortion for analytic functions with positive real part in a circle*. Nagoya Math. J. 28(1966), 221–228; MR 34 #7807. [2, 56]
724. Komatu, Y.; Nishimiya, H. *A remark on distoriton for fourth derivative of functions regular and univalent in the unit circle*. Sci. Rep. Saitama Univ. Ser. A6, (1968), 3–4; MR 39 #427. [42, 68]

725. Kopanev. S. A. *A certain functional on the class of typically real functions.* (Russian) Trud. Tomsk. Gos. Univ. Ser. Meh.-Mat. 200 (1968), 100–111; MR 41 #2019. [13, 29]
726. Korobkova, I. K. *Extremal problems in certain classes of univalent functions with gaps of initial coefficients in their power expansions.* (Ukranian. English and Russian summaries) Dopovidi Akad. Nauk, Ukrain. RSR Ser. A 1970, 684–688, 764; MR 44 #420. [6, 9, 12]
727. Korobkova, I. K. *The boundaries of the convexity of certain classes of analytic functions with purely imaginary characteristics.* (Ukrainian. English and Russian summaries) Dopovidi Akad. Nauk Ukrain. RSR Ser. A 1970, 589–592, 667; MR 44 #2918. [6, 9, 12]
728. Korobkova, I. K. *A certain problem on a bound extremum.* (Ukrainian. English and Russian summaries) Dopovidi Akad. Nauk Ukrain. RSR Ser. A(1973), 796–800, 861; MR 49 #552. [2, 11]
729. Korobkova, I. K. *A certain property of analytic majorants and convolutions of schlicht functions.* (Ukrainian. English and Russian summaries) Dopovidi Akad. Nauk Ukrain. RSR Ser A 1973, 1076–1079, 1150; MR 49 #553. [1, 47]
730. Korobkova, I. K.; Zmorovic, V. A. *A certain class of univalent mappings of an annulus.* (Ukrainian. English and Russian summaries) Dopovidi Akad. Nauk Ukrain. RSR Ser. A (1972), 612–615, 670; MR 46 #9321. [48]
731. Korobockin, B. I. *On the theory of univalent functions.* (Russian. English summary) Latvijas Valsts Univ. Zinatn. Raksti 47(1963), laid. 1, 29–50; MR 34 #6060. [26]
732. Kortram, Ronald. *On an extended class of bounded univalent functions.* Ann. Acad. Sci. Fenn. Ser AI No. 578(1974), 16 pp. [22, 30]
733. Kortram, R.; Tammi, O. *On the first coefficient regions of bounded univalent functions.* Ann. Acad. Sci. Fenn. Ser. AI Math. No. 592(1974), 27 pp. (1975); MR 51 #5914. [22, 24, 29, 39]
734. Kortram, R.; Tammi, O. *On the second coefficient region for bounded univalent functions.* Ann. Acad. Sci. Fenn. Ser. AI Math. 1(1975), no. 1, 155–175. [22, 24, 29, 39]; MR 52 #8409
735. Koseki, Keniti. *Über die Koeffizienten der schlichten Functionen.* V. Math. J. Okayama Univ. 13(1967), 35–83; MR 37 #2978. [54]
736. Koval'cuk, R. N. *On a generalization of Kellog's theorem.* (Russian) Ukrain. Mat. Z. 17(1965), no. 4, 104–108; MR 33 #2809.
737. Kövari, T.; Pommerenke, Ch. *On Faber polynomials and Faber Expansions.* Math. Z. 99(1967), 193–206. [53]
738. Kövari, T.; Pommerenke, Ch. *On the distribution of Fekete points.*

Mathematika 15(1968), 70-75; MR 38 #316. [68]
739. Kronstadt, E. *Compact families of univalent functions.* Preliminary report. Notices Amer. Math. Soc. 23, January 1976, Abstract 731-30-25, p. A-103.
740. Kruskal', S. L. *Some extremal problems for univalent analytic functions.* (Russian) Dokl. Akad. Nauk SSSR 182(1968), 754-757; MR 38 #3429 {Soviet Math. Dokl. 9(1968), 1191-1194}.
741. Krzyz, Jan. *On the region of variability of the ratio $f(z_1)/f(z_2)$ within the class S of univalent functions.* (Polish and Russian summaries) Ann. Univ. Mariae Curie-Sklodowska Sect. A. 17(1963), 55-64(1965); MR 34 #328. [24, 29, 68]
742. Krzyz, J. *On a theorem of Kubo concerning functions regular in an annulus.* Colloq. Math. 16(1967), 43-47; MR 35 #1767. [24]
743. Krzyz, J. G. *On close-to-convex functions.* Symposia on Theoretical Physics and Mathematics, Vol. 10(Inst. Math. Sci., Madras, 1969), pp. 23-27. Plenum, New York, 1970; MR 41 #5610. [5]
744. Krzyz, J. *An extremal length problem.* Ann. Univ. Mariae Curie-Sklodowska, Sect. A., vols. 22/23/24(1968/1969/1970), 95-104. [49]
745. Krzyz, Jan G. *The Green function of domains containing a fixed ellipse.* Michigan Math. J. 20(1973), 13-19; MR 46 #9314. [1, 6, 10, 16, 19, 49]
746. Krzyz, Jan; Rahman, Q. I. *Univalent polynomials of small degree.* Ann. Univ. Mariae Curie-Sklodowska Sect. A 21 (1967), 79-90 (1972); MR 48 #11471. [16, 20, 32, 43, 73, 85]
747. Krzyz, Jan; Reade, M. O. *Koebe domains for certain classes of analytic functions.* J. D'analyse Math. 18(1967), 185-195; MR 35 #3050. [5, 6, 19, 39, 41, 45, 49]
748. Krzyz, J.; Zlotkiewicz, E. *Koebe sets for univalent functions with two preassigned values.* Ann. Acad. Sci. Fenn. Ser. A I No. 487 (1971), 12 pp.; MR 43 #7612. [5, 6, 10, 19]
749. Kubota, Y. *On extremal problems which correspond to algebraic univalent functions.* Kodai Math. Sem. Rep. 25(1973), 412-428; MR 49 #563. [9, 24, 25, 29, 53, 54]
750. Kubota, Y. *On the fourth coefficient of meromorphic univalent functions.* Kodai Math. Sem. Rep. 26(1974/75), 267-288. [9, 25]; MR 52 #732]
751. Kubota, Y. *A coefficient inequality for certain meromorphic-univalent functions.* Kodai Math. Sem. Rep. 26 (1974/75), 85- 94; MR 51 #5915. [9, 25, 39]
752. Kudelski, F. *A variational formula for starshaped functions with*

real coefficients. Bull. Acad. Polon. Sci. Ser. Sci. Math. Astronom. Phys. 12(1964), 617-619; MR 30 #3203. [6, 24, 39]

753. Kudelski, F. *On the univalence of Taylor sums for a class of univalent functions.* (Polish and Russian summaries) Ann.Univ. Mariae Curie-Sklodowska Sect. A 17(1963), 65-67(1965), MR 33 #5864. [20, 43]

754. Kudelski, F. *Sur quelques problèmes de la théorie des fonctions subordonnées.* (Polish and Russian summaries). Ann. Univ. Mariae Curie-Sklodowska Sect. A 27(1973), 43-48(1975). [1]; MR 52 #8411

755. Kudrjasov, S. N. *Certain sufficient conditions for the univalence of the solution of an exterior inverse boundary problem.* (Russian) Izv. Vyss. Ucebn. Zaved. Matematika 1965, no. 3(46), 105-110; MR 31 #5970. [4]

756. Kudrjasov, S. N. *certain sufficient conditions for the univalence of a solution of the inverse problem in the theory of filtration.* (Russian) Izv. Vyss. Ucebn. Zaved. Matematika 1966, no. 5(54), 88-99; MR 34 #6061. [4]

757. Kudrjasov, S. N. *Certain criteria for the univalence of analytic functions.* (Russian) Mat. Zametki 13(1973), 359-366 {Math. Notes 13(1973), 219-223}; MR 47 #7022. [4, 9]

758. Kufarev, P. O. *The variational formula of G. M. Goluzin.* (Russian) Trudy Tomsk. Gos. Univ. 163(1963), 58-62. [24]

759. Kufarev, P. P.; Soboleva, S. V. *A property of a family of univalent functions.* (Russian) Trudy Tomsk. Gos. Univ. Ser. Meh.-Mat. 175(1964), Vyp. 2, 5-7; MR 31 #5979. [9]

760. Kufarev, P. P.; Sobolev, V. V.; Sporyseva, L. V. *A certain method of investigation of extremal problems for functions which are univalent in the half-plane.* (Russian) Trudy Tomsk. Gos. Univ. Ser. Meh.-Mat. 200(1968), 142-164; MR 41 #1987. [24, 57]

761. Kühnau, Reiner. *Über die schlichte konforme Abbildung auf nichtüberlappende Gebiete.* Math. Nachr. 36(1968), 61-71; MR 37 #2969.

762. Kühnau, Reiner. *Koeffizientenbedingungen bei quasikonformen Abbildungen.* (Polish and Russian summaries) Ann. Univl Mariae Cruie-Sklodowska Sect. A 22-24(1968/70), 105-111(1972); MR 49 #7441. [4, 31, 53]

763. Kühnau, Reiner. *Geometrisch-funktionen theoretische Lösung eines Extremal problems der knoformen Abbildung.* J. Reine Angew. Math. 229(1968), 131-136; MR 40 #7434.

764. Kühnau, Reiner. *Schranken fur die Koeffizienten gewisser schlicht*

abbildender Laurentscher Reihen. Math. Nachr. 41(1969), 177–183; MR 40 #327. [9, 25]
765. Kühnau, Reiner. *Diskretisierungen zu den Rundungsund Sternschranken bei konformer Abbildung.* Rev. Roumaine Math. Pures Appl. 15(1970), 1229–1234; MR 42 #4719. [12]
766. Kühnau, Reiner. *Weitere elementare Bemerkungen zur Theorie der konformen und quasikonformen Abbildungen.* Math. Machr. 51(1971), 377–382; MR 47 #3657.
767. Kühnau, Reiner. *Eine Bemerkung zu zwei Arbeiten von O. Dvořák.* Math. Nachr. 48(1971), 225–226; MR 45 #532b. [42]
768. Kühnau, Reiner. *Über vier Klassen schlichter Funktionen.* Math. Nachr. 50(1971), 17–26; MR 46 #9315. [53]
769. Kühnau, Reiner. *Über zwei Klassen schlichter konformer Abbildungen.* Math. Nachr. 49(1971), 173–185; MR 48 #4300. [24]
770. Kühnau, R. *Koeffizientenbedingungen für schlicht abbildende Laurentsche Reihen.* (English and Russian summaries) Bull. Acad. Polon. Sci. Ser. Sci. Math. Astronom. Phys. 20(1972), 7–10; MR 48 #6397. [48, 53]
771. Kühnau, R. *Zum Koeffizientenproblem bei den quasikonform fortsetzbaren schlichten konformen Abbildungen.* Math. Nachr. 55(1973), 225–231; MR 48 #522. [9, 25, 54]
772. Kühnau, Reiner. *Eine Klasse nichtschlichter konformer Abbildungen mit einer schlichten quasikonformen Fortsetzung.* Math. Nachr. 59(1974), 261–263; MR 49 #7442.
773. Kuhnau, Reiner. *Zur Abschätzung der Schwarzschen Ableitung bei schlichten Funktionen.* Math. Nachr. 59(1974), 195–198; MR 50 #594. [4, 31]
774. Kulshrestha, P. K. *Coefficient bounds for k-fold symmetric univalent functions of bounded rotation.* Notices, A. M. s., vol. 19, no. 5, issue 139, August, 1972 (Abstract 696-30-7), p. A–637. [4, 9, 71, 77]
775. Kulshrestha, P. K. *Distoriton of spiral-like mappings.* Proc. Roy. Irish Acad. Sect. A 73, 1–5(1973), MR 47 #7016. [6, 63]
776. Kulshrestha, P. K. *Bounded Robertson functions.* Notices A. M. S. 20, August 1973, Abstract 706-30-5, p. A–521. [6, 12, 36, 63, 73]
777. Kulshrestha, P. K. *Generalized convexity in conformal mpapings.* J. Math. Anal. Appl. 43(1973), 441–449; MR 49 #9182. [6, 29, 69, 83, 84]
778. Kulshrestha, P. K. *Coefficients for alpha-convex univalent functions.* Bull. Amer. Math. Soc. 80(1974), 341–342; MR 49 #7434. [69, 84]

779. Kulshrestha, P. K. *Coefficient problem for α-convex univalent functions.* Arch. Rational Mech. Anal. 54(1974), 205-211; MR 51 #878. [23, 69, 70, 84]
780. Kulshrestha, P. K. *Coefficient problem for a class of Mocanu-Bazilevic functions.* Ann. Polon. Math. 31(1975/76), no. 3, 291-299. [69, 70, 84]
781. Kuroda, Inao. *On circles and hyperbolas.* (Japanese. English summary) Bull. Yamagata Univ. Natur. Sci. 5(1960), no. 1, 119-132; MR 41 #3730. [49]
782. Kuroda, Inao. *On discs and strips.* (Japanese. English summary). Bull. Yamagata Univ. Natur. Sci. 5(1961), no. 2, 241-258; MR 41 #3732. [49]
783. Kuroda, Inao. *On a theorem due to Viktors Linis.* (Japanese summary) Bull. Yamagata Univ. Natur. Sci. 5(1962), no. 3, 525-528; MR 41 #7085a. [39]
784. Kuroda, Inao. *Several sources of the family of functions due to Friedman or Linis.* (Japanese summary) Bull. Yamagata Univ. Natur. Sci. 6(1963), no. 1, 1-8; MR 41 #7085b. [39]
785. Kuroda, Inao. *On discs and on figures bounded by two circular arcs.* (Japanese. English summary) Bull. Yamagata Univ. Natur. Sci. 5(1963), no. 4, 807-821; MR 41 #3732. [49]
786. Kuroda, Inao. *A view of the family of functions regular, schlicht and normalized in the unit disc.* (Japanese summary) Bull. Yamagata Univ. Natur. Sci. 6(1964), 129-137; MR 31 #2388. [18, 44]
787. Kuroda, Inao. *A view of the fmaily of functions regular, schlicht and normalized in the unit disc.* (Japanese summary) Bull. Yamagata Univ. Natur. Sci. 6(1964), no. 2, 129-137; MR 41 #3738. [18, 44]
788. Kuroda, Inao. *On the functions $\tanh^{-1} z$ and $\tan^{-1} z$ regular, schlicht and normalized in the unit disc.* (Japanese summary) Bull. Yamagata Univ. Natur. Sci. 6 (1965), no. 3, 237-246; MR 41 #3733. [28]
789. Kuroda, Inao. *A view of the family of functions regular schlicht and normalized in the unit disc. II.* (Japanese summary) Bull. Yamagata Univ. Natur. Sci. 6(1967), no. 4, 383-393; MR 41 #3739. [18, 44]
790. Kuz'mina, G. V. *Covering theorems for functions holomorphic and univalent within a disk.* (Russian) Dokl. Akad. Nauk SSSR 160 (1965), 25-28; MR 30 #3204 {Soviet Math. Dokl. 6(1965), 21-25}. [6, 19, 39]
791. Kuz'mina, G. V. *Estimates of the transfinite diameter of a certain*

family of continua and covering theorems for schlicht functions. (Russian) Trudy Mat. Inst. Steklov. 94(1968), 47–65; MR 37 #402 {Translated from the Russian: Extremal Problems of the Geometric Theory of Functions (A. M. S., 1969)}. [6, 9, 19, 39]

792. Kuz'mina, G. V. *Extremal properties of the hyperbolic transfinite diameter and schlicht functions in a ring.* (Russian) Trudy Mat. Inst. Steklov. 94(1968), 66–78; MR 36 #5320. {Translated from the Russian: Extremal Problems of the Geometric Theory of Functions (A. M. S. 1969)}. [24, 48]

793. Kuz'mina, G. V. *Estimation of the conformal modulus of a certain family of domain and covering theorems for univalent functions.* (Russian) Zap. Naucn. Sem. Leningrad. Otdel. Mat. Inst. Steklov (LOMI) 24(1972), 148–172; MR 46 #9312. [19]

794. Kuz'mina, G. V. *Extremal problems of the geometric theory of funcns of a complex variable.* II. A collection of articles edited by G. V. Kuz'mina. Zap. Naucn. Sem. Leningrad. Otdel. Mat. Inst. Steklov. (LOMI)44(1974). Izdat. "Nauka" Leningrad. Otdel., Leningrad, 1974. 188 pp. [44]

795. Labelle, G.; Rahman, Q. I. *Remarque sur la moyenne arithmétique de fonctions univalentes convexes.* Canad. J. Math. 21(1969), 977–981; MR 39 #7081. [10, 12, 64, 82]

796. Lameier, S. H. *On domains of univalence for certain meromorphic functions.* Notices Amer. Math. Soc. 21, November 1974, Abstract 717-B1, p. A-609. [9]

797. Lappan, Peter. *A p-valent functions with a non-normal derivative.* Math. Z. 144(1975), Heft 2, 147–148. [9, 14, 62] MR 52 #734

798. Launonen, Eero. *On exponentiated Grunsky inequalities for bounded univalent functions.* Ann. Acad. Sci. Fenn. Ser. A I Math. Dissertation No. 1(1975), 34 pp. [22, 53]

799. Ławrynowicz, Julian. *On the coefficient problem for univalent polynomials.* Proc. Cambridge Philos. Soc. 64(1968), 87–98; MR 36 #5327. [32]

800. Ławrynowicz, Julian; Tammi, Olli. *On estimating of a fourth order functional for bounded univalent functions.* Ann. Acad. Sci. Fenn. Ser. A I, 490(1971), pp. 3–17; MR 44 #5447. [22, 30]

801. Ławrynowicz, J.; Tammi, O. *On estimating a fifth order functional for bounded univalent functions.* Colloq. Math. 25(1972), 307–313, 326; MR 48 #4297. [16, 22, 30]

802. Leach, R. J. *On odd functions of bounded boundary rotation.* Canad. J. Math. 26(1974), 551–564; MR 49 #3111. [4, 5, 6, 23, 45, 62, 63, 71, 73, 76, 77]

803. Leach, Ronald J. *Multivalent Bazilevič functions.* Notices Amer. Math. Soc. 21, January 1974, Abstract 711-30-13, p. A-121. (To appear in Revue Romaine des Maths. Pures et Appl.) [5, 14, 50, 65]
804. Leach, R. J. *Coefficients of symmetric functions of bounded boundary rotation.* Canad. J. Math. 26(1974), 1351-1355; MR 50 #7499. [14, 71, 77]
805. Leach, R. J. *Multivalent andd meromorphic functions of bounded boundary rotation.* Canad. J. Math. 27(1975), 186-199. [4, 12, 14, 39, 62, 71, 76, 77]
806. Leach, R. J. *The Marx conjecture for some alpha-convex functions.* Michigan Math. J. 22(1975), no. 2, 188-190. [69]; MR 52 #3500
807. Leach, Ronald J. *On some classes of multivalent starlike functions.* Trans. Amer. Math. Soc. 209(1975), 267-273; MR 51 #5909. [6, 14, 50, 65, 88]
808. Leach, Ronald J. *On alpha-convex functions.* Rev. Roumaine Math. Pures Appl. 20(1975), no. 5, 545-549. [23, 24, 69, 70]
809. Leach, R. J.; Noonan, J. W. *On functions with bounded boundary rotation.* Notices, A. M. S., vol. 18, no. 4, isue no. 130, June 1971 (Abstract 71T-B158), p. 654. [71]
810. Lebedev, N. A. *An application of the area principle to problems of nonoverlapping finitely connected domains.* Dokl. Akad. Nauk SSSR 167(1966), 26-29; MR 33 #5865 {Soviet Math. Dokl. 7(1966), 323-327}. [27]
811. Lebedev, N. A. *An addendum to the article "Application of the area principle to problems on non-overlapping regions."* (Russian. English Summary). Vestnik Leningrad. Univ. 22(1967), no. 7, 64-73; MR 35 #4389. [26, 27, 53, 54]
812. Lebedev, N. *Certain corollaries of an inequality of Grunsky.* (Russian. English summary) Vestnik Leningrad. Univ. 1972, no. 7, 45-55; MR 45 #8820. [53]
813. Lebedev. N. A.; Din'-van-F'eu. *The variational formula of G. M. Goluzin.* (Russian) Zap. Naucn. Sem. Leningrad. Otdel. Mat. Inst. Steklov. (LOMI) 24(1972), 173-181; MR 46 #7502. [24]
814. Lebedev, N. A.; Din'-van-F'eu. *Certain corollaries of Grunsky's inequality and of the fundamental area theorem.* (Russian) Zap. Naucn. Sem. Leningrad. Otdel. Mat. Inst. Steklov. (LOMI) 44(1974), 100-126, 187; MR 51 #8400. [9, 27, 53]
815. Lebedev, N. A.; Mamai, L. V. *Certain sufficient conditions for univalence of regular functions.* (Russian. English summary) Vestnik Leningrad. Univ. 22(1967), no. 1, 40-51; MR 35 #6809. [4]
816. Lebedev, N. A.; Mamai, L. V. *A generalization of a certain ine-*

quality of P. Garabedian andd M. Schiffer. (Russian. English summary) Vestnik Leningrad. Univ. 25(1970), no. 19, 41-45; MR 43 #6422. [26, 27, 53]
817. Lee, Boo Sang. *On univalent entire functions.* Kodai Math. Sem. Rep. 24(1972), 168-171; MR 46 #3772. [4, 10, 31, 90]
818. Lee, Suk Young. *Quasi-subordinate functions and coefficient conjectures.* J. Korean Math. Soc. 12(1975), no. 1, 43-50. [1, 16]; MR 52 #8412
819. Leeman, G. B. *The constrained coefficient problem for typically real functions.* Trans. Amer. Math. Soc. 186 (1973), 177-189; MR 49 #3112. [13, 16, 23, 51]
820. Leeman, G. B. Jr. *Some regularity theorems for typically real functions.* Proc. Amer. Math. Soc. 40(1973), 191-198; MR 48 #2370. [13, 51, 66]
821. Leeman, G. B. *A local estimate of typically real functions.* Pacific J. Math. 52(1974), 481-484; MR 50 #10232. [13, 51]
822. Leeman, George B., Jr. *A new proof for an inequality of Jenkins.* Proc. Amer. Math. Soc. 54(1976), 114-116. [54]
823. Lehto, Olli. *Schlicht functions with a quasiconformal extension.* Ann. Acad. Sci. Fenn. Series A I, No. 500 (1971), pp. 3-9. [9, 25, 27, 31]; MR 45 #3692
824. Leutwiler, H.; Schober, Glenn. *Toeplitz forms and the Grunsky-Nehari inequalities.* Mich. Math. J. 20(1973), 129-136; MR 48 #2371. [2, 21, 53]
825. Lewandowski, Z. *Some remarks on a paper by M. S. Robertson.* (Polish and Russian summaries) Ann. Univ. Mariae Curie-Sklodowska Sect. A 17(1963), 43-46(1965); MR 33 #5867. [1, 6]
826. Lewandowski, Z. *On circular symmetrization of starshaped domains.* (Polish and Russian summaries) Ann. Univ. Mariae Curie-Sklodowska Sect. A 17(1963), 35-38(1965); MR 33 #5866. [6, 19]
827. Lewandowski, Zdzislaw. *On some problems of M. Biernacki concerning subordinate functions and on some related topics.* (Polish and Russian summaries) Ann. Univ. Mariae Curie-Sklodowska Sect. A 19(1965), 33-46(1970); MR 41 #7086. [1]
828. Lewandowski, Z. *On a problem of M. Biernacki.* (Polish and Russian summaries) Ann. Univ. Mariae Curie-Sklodowska Sect. A 17 (1963), 39-41(1965); MR 33 #4249. [43, 85]
829. Lewandowski, Zdzislaw. *Some results concerning univalent majorants.* (Polish and Russian summaries) Ann. Univ. Mariae Curie-Sklodowska Sect. A 18(1964), 13-18(1967); MR 38 #2299. [1, 54, 68]
830. Lewandowski, Zdzislaw. *Modular and domain majorants of*

regular functions. (Polish and Russian summaries) Ann. Univ. Mariae Curie-Sklodowska Sect. A 18(1964), 19–22(1967); MR 38 #2300. [1, 6, 32]

831. Lewandowski, Z.; Miazga, J. *An extremal problem for a class of functions with positive real part.* Bull. Acad. Polon. Sci. Sér. Sci. Math. Astronom. Phys. 21(1973), 233–240; MR 47 #3668. [1, 2, 24, 56]

832. Lewandowski, Z.; Miazga, J. *Sur l'extension d'une method pour la determination des ensembles de Koebe.* (English and Russian summaries) Bull. Acad. Polon. Sci. Sér. Sci. Math. Astronom. Phys. 22(1974), 663–666; MR 50 #2479. [19]

833. Lewandowski, Z.; Miazga, J.; Szynal, J. *On an application of Sakaguchi's method to extremal problems for functions with positive real part with vanishing coefficients.* Bull. Acad. Polon. Sci. Sér. Sci. Math. Astronom. Phys. 21(1973), 241–244; MR 47 #5248. [2, 24, 56]

834. Lewandowski, Z.; Miazga, J.; Szynal, J. *Koebe domains for univalent functions with real coefficients under Montel's normalization.* Ann. Polon. Math. 30(1975), fasc. 3, 333–336. [6, 10, 19, 39, 41]; MR 51 #13210

835. Lewsandowski, Z.; Miller, S.; Zlotkiewicz, E. *Generating functions for some classes of univalent functions.* Proc. Amer. Math. Soc. (to appear). [2, 4, 5, 6, 69]

836. Lewandowski, Z.; Reade, M. O.; Zlotkiewicz, E. *On a certain condition for univalence.* (Romanian and Russian summaries) An. Sti. Univ. "Al. I. Cuza" Iasi Sect. Ia Mat. (N.S.)11B(1965), 119–123; MR 34 #329. [4, 5, 10, 20]

837. Lewandowski, Zdzislaw; Stankiewicz, Jan. *On mutually adjoint clost-to-convex functions.* (Polish and Russian summaries) Ann. Univ. Mariae Curie-Sklodowska Sect. A 19(1965), 47–51(1970); MR 41 #5611. [1, 5, 6, 82]

838. Lewandowski, Z.; Stankiewicz, Jan. *On the region of variability of* log $f'(z)$ *for some classes of close-to-convex functions.* Ann. Univ. Mariae Curie-Sklodowska Sect. A 20(1966), 45–52(1971); MR 46 #9322. [5, 29, 49, 74]

839. Lewandowski, Z.; Stankiewicz, J. *Les majorantes modulaires étoilées et l'inclusion.* (Russian summary) Bull. Acad. Polon. Sci. Sér. Sci. Math. Astronom. Phys. 19(1971), 923–929; MR 46 #2033. [1]

840. Lewandowski, Z.; Stankiewicz, J. *Majoration modulaire des fonctions et inclusion des domaines.* (Russian summary) Bull. Acad.

Polon. Sci. Sér. Sci. Math. Astronom. Phys. 19(1971), 917–922; MR 46 #2032. [1]

841. Lewandowski, Z.; Szynal, J.; Wajler, S. *On the covering sets and the majorization of functions.* (Russian summary) Bull. Acad. Polon. Sci. Sér. Sci. Math. Astronom. Phys. 22(1974), 29–34; MR 50 #2475. [1, 6, 10]

842. Lewandowski, Z.; Zlotkiewicz, E. *On the domain of variability of the second coefficient for a class of meromorphic, univalent functions.* Bull. Acad. Polon. Sci. Sér. Sci. Math. Astronom. Phys. 13(1965), 21–25; MR 31 #3593. [9, 24, 29]

843. Lewandowski, Z.; Zlotkiewicz, E. *Variational formulae for functions meromorphic and univalent in the unit disc.* (Polish and Russian summaries) Ann. Univ. Mariae Curie-Sklodowska Sect. A 17 (1963), 47–53(1965); MR 33 #4256. [9, 24, 25]

844. Lewandowski, Z.; Zlotkiewicz, E. *On some classes of starlike functions.* Notices Amer. Math. Soc. 21, April 1974, Abstract 74T–B89, p. A–374. [4, 6]

845. Lewin, M. *On a coefficient problem for bi-univalent functions.* Proc. Amer. Math. Soc. 18(1967), 63–68; MR 34 #6074. [16, 27, 53, 54]

846. Lewin, M. *Bounds for functionals of a certain class of analytic functions.* Bull. London Math. Soc. 3(1971), 329–330; MR 45 #7050. [22, 68, 72]

847. Lewis, John L. *On a problem of Gronwall for Bazilevič functions.* Trans. Amer. Math. Soc. 195(1974), 231–242; MR 49 #7432. [70, 80]

848. Lewis, John L. *An n-space analogue of a theorem of Suffridge.* Proc. London Math. Soc. (3)30(1975), part 4, 385–404. [6]

849. Lewis, J. L. *Note on an arc length problem.* J. London Math. Soc. (2)12(1975/76), no. 4, 469–474. [7]

850. Lewis, J. L.; Barnard, R. W. *Subordination theorems for some classes of univalent functions.* Notices A. M. S. 20, August 1973, Abstract 73T–B210, p. A–486. [1, 6]

851. Li Ti-min'. *Influence of the arguments of a function analytic in the unit circle on its univalence.* Sci. Sinica 14(1965), 666–678; MR 32 [4, 39]

852. Libera, R. J. *Univalent α-spiral functions.* Canadian J. Math. 19(1967), 449–456; MR 35 #5599. [6, 36]

853. Libera, Richard J. *Disk-like functions.* J. Austral. Math. Soc. 11(1970), 251–256; MR 42 #1990. [4, 6, 40, 41, 49]

854. Libera, R. *Some inequalities for bounded univalent functions.*

(Polish and Russian summaries) Ann. Univ. Mariae Curie-Sklodowska Sect. A 26(1972), 17–29(1974); MR 50 #10236. [22, 53, 61]
855. Libera, R. J.; Livingston, A. E. *On the univalence of some classes of regular functions.* Proc. Amer. Math. Soc. 30(1971), 327–336; MR 44 #5442. [5, 6, 10, 11, 2, 73, 82]
856. Libera, R. J.; Livingston, A. E. *Bounded functions with positive real part.* Czechoslovak Math. J. 22(97)(1972) 195–209; MR 45 #7039. [2, 22]
857. Libera, R. J.; Livingston, A. E. *Weakly starlike meromorphic univalent functions.* Trans. Amer. Math. Soc. 202(1975), 181–191; MR 50 #13482. [6, 9, 36, 62, 63]
858. Libera, R. J.; Ziegler, M. R. *Regular functions $F(z)$ for which $zF'(z)$ is alpha-spiral.* Trans. Amer.Math. Soc., 166(1972), pp. 361–370; MR 45 #526. [4, 6, 12, 31, 63, 73]
859. Libera, R. J.; Zlotkiewicz, E. J. *Loewner-type approximations for convex functions, Preliminary report.* Notices, A. M. S., vol. 22, no. 4, issue no. 162, June 1975 (Abstract 75T-B101), p. A-455. [1, 10, 82]
860. Libera, R. J.; Zlotkiewicz, E. J. *Loewner's differential equation for spirallike functions.* Notices, A. M. S. vol. 22, no. 5, issue no. 163, August 1975 (Abstract 75T-B154), p. A-509. [6, 24]
861. Lick, Don R. *Sets of non-uniform convergence of schlicht functions.* Math. Japon. 14(1968), 31–32; MR 41 #1989. [62]
862. Litvincuk, Ju. A.; Milin, I. M. *Estimation of exterior arcs under a univalent mapping* (Russian). Mat. Zametki 18(1975), no. 3, 367–378.
863. Liu, Ming-chit. *On functions of bounded boundary rotation.* Proc. Amer. Math. Soc. 29(1971), 345–348; MR 44 #4200. [7, 18, 62, 71]
864. Liu, Ming Chit. *On the derivative of some analytic functions.* Math. Z. 132(1973), 205–208; MR 48 #2351. [68]
865. Liu, Shu-ch'in. *Some inequalities for symmetric schlicht functions.* (Chinese) Shuxue Jinzhan 7(1964), 223–227; MR 39 #1645. [49, 54]
866. Livingston, A. E. *p-valent close-to-convex functions.* Trans. Amer. Math. Soc. 115(1965), 161–179; MR 33 #7520. [5, 6, 14, 50]
867. Livingston, A. E. *Meromorphic multivalent close-to-convex functions.* Trans. Amer. Math. Soc. 119(1965), 167–177; MR 32 #5860. [5, 9, 14, 38, 75]
868. Livingston, A. E. *On the radius of univalence of certain analytic functions.* Porc. Amer. Math. Soc. 17(1966), 352–357; MR 32 #5861. [2, 4, 5, 6, 10, 11, 12, 16]

869. Livingston, A. E. *The coefficients of multivalent close-to-convex functions.* Proc. A.M.S. 21(1969), 545-552; MR 39 #4378. [2, 5, 14, 21, 50]
870. Lohwater, A. J. *Boundary behavior of the derivative of a univalent function.* (Russian) Dokl. Akad. Nauk SSSR 195 (1970), 1033-1035; MR 42 #6241; Soviet Math.-Dokl. 11(1970), p. 1620. [62]
871. Lohwater, A. J. *The boundary behavior of derivatives of univalent functions.* Math. Z. 119(1971), 115-120; MR 43 #7606. [16, 62]
872. Lohwater, A. J.; Ryan, Frank. *A distortion theorem for a class of conformal mappings.* Mathematical Essays Dedicated to A. J. Macintyre, pp. 257-262. Ohio Univ. Press, Athens, Ohio, 1970; MR 45 #7033. [4, 23, 62]
873. London, R. R.; Thomas, D. K. *An area theorem for starlike functions.* Porc. London Math. Soc. (3)20(1970), 734-748; MR 41 #7087. [6, 11, 18, 23]
874. London, R. R.; Thomas, D. K. *An asymptotic formula for an integral in starlike function theory.* Trans. Amer. Math. Soc. 215 (1976), 393-406. [3, 6, 7, 16, 18, 23, 62]; MR 52 #8403
875. Lonka, H.; Tammi, O. *On the use of step-functions in extremum problems of the class with bounded boundary rotation.* Ann. Acad. Sci. Fenn. Ser. A I No. 418(1968), 18 pp.; MR 36 #6601. [24, 71, 77]
876. Lohwater, A. J.; Pommerenke, Ch *On normal meromorphic functions.* Ann. Acad. Sci. Fenn. Ser. A I., no. 550(1973), 12 pp. [46, 62]
877. Lozovik, V. G. *On functions of bounded rotation outside the unit circle.* (Ukrainian. Russian and English summaries) Dopovidi Akad. Nauk Ukrain. RSR 1962, 856-858; MR 34 #2855. [2, 9]
878. Lozovik, V. G. *On the coefficients of multivalent functions.* (Ukrainian. Russian and English summareis) Dopovidi Akad. Nauk Ukrain. RSR 1964, 1019-1021; MR 31 #325. [14, 24, 33, 50]
879. Lozovik, V. G. *On certain properties of typically real functions.* (Russian) Uspehi Mat. Nauk 20(1965), no. 3 (123), 189-195; MR 32 #204. [13, 23]
880. Lucas, K. W. *A two-point modulus bound for areally mean p-valent functions.* J. London Math. Soc., 43(1968), 487-494; MR 37 #1587. [33]
881. Lucas, K. W. *On successive coefficients of areally mean p-valent functions.* J. London Math. Soc. 44(1969), 631-642; MR 39 #4379. [8, 33]
882. Lustfield, Charles D. *Mass functions of bounded variation and*

starlikeness. Portugal. Math. 30(1971), 73-82; MR 46 #3763. [6, 11, 12, 43, 71]
883. McCarty, Carl P. *Functions with real part greater than* α. Proc. Amer. Math. Soc. 35(1972), 211-216; MR 45 #7066. [2, 19, 21, 22, 56, 61, 72]
884. McCarty, Carl P. *Functions with real part greater than* α. Notices A. M. S., vol. 20(Feb. 1973), Abstract 73T-B45. [2, 6, 12, 72]
885. McCarty, Carl P. *Some classes of analytic functions.* Notices Amer. Math. Soc. 21, August 1974, Abstract 74T-B142, p. A-487. [2, 12, 21, 56]
886. McCarty, Carl P. *Analytic functions with initial zero coefficients.* Notices Amer. Math. Soc. 21, February 1974, Abstract 74T-B69, p. A-309. [2, 39]
887. McCarty, C. P. *Two radius of convexity problems.* Proc. Amer. Math. Soc. 42(1974), 153-160; MR 48 #4295. [2, 6, 12, 21, 36, 56, 72]
888. McCarty, Carl P. *Starlike functions.* Proc. Amer. Math. Soc. 43(1974), 361-366; MR 48 #11472. [4, 6, 12, 36, 63]
889. McCarty, C. P.; Tepper, D. E. *A note on the $\frac{2}{3}$ conjecture for starlike functions.* Proc. Amer. Math. Soc. 34(1972), 417-421; MR 46 #3764. [6, 12, 16, 19]
890. McCoy, T. L. *A surface variation for extremal schlicht functions.* Notices, A. M. S., vol. 19, no. 1, issue no. 135, January, 1972 (Abstract 691-30-26), p. A-117, [24]
891. McGregor, M. T. *On the orders of convexity and starlikeness of convex domains.* J. London Math. Soc. (2)2(1970), 111-120; MR 40 #5843. [6, 10, 19, 22, 61, 64]
892. McLaughlin, Renate. *On the Marx conjecture for starlike functions of order* α. Trans. Amer. Math. Soc. 142(1969), 249-256; MR 40 #328. [1, 6, 23, 24]
893. McLaughlin, Renate. *Estremal probleme fur eine Familie schlichter Funktionen.* Math. Z. 118(1970), 320-330; MR 43 #2206. [6, 10, 24]
894. McLaughlin, Renate. *Some extremal problems for functions univalent in an annulus.* Math. Scand. 28(1971), 129-138; MR 46 #5606. [24, 46, 48]
895. McLaughlin, Renate. *Extremal problems for a class of symmetric functions.* Arch. Rational Mech. Analy., 24(1972), no. 4, pp. 310-319. [10, 22, 24, 35, 48, 61]
896. McLaughlin, R. *A variational method for a class of odd functions in an annulus.* Colloq. Math. 28(1973), 299-305. [24, 45]
897. McLeavey, J. O. *Extremal problems of analytic univalent functions*

with quasi-conformal extensions. Trans. Amer. Math. Soc. 195(1974), 327–343; MR 49 #10880. [24, 31, 53]
898. McMillan, J. E. *Boundary behavior of a conformal mapping.* Acta Math. 123(1969), 43–67; MR 41 #1981. [62]
899. McMillan, J. E. *On the boundary correspondence under conformal mapping.* Duke Math. J. 37(1970), 725–739; MR 45 #3684. [62]
900. McMillan, J. E.; Pommerenke, Ch. *On the boundary behavior of analytic functions without Koebe arcs.* Math. Ann. 189(1970), 275–279; MR 44 #4210. [4, 19, 62]
901. McMillan, J. E. *Distortion under conformal and quasiconformal mappings.* Acta Math. 126(1971), p. 121–141. [49]; MR 42 #3273
902. McMillan, J. E.; Pommerenke, Ch. *On the asymptotic values of locally univalent meromorphic functions.* J. Reine Angew. Math. 249(1971), 31–33; MR 44 #434. [62]
903. MacGregor, T. H. *Translations of the image domains of analytic functions.* Proc. Amer. Math. Soc. 16(1965), 1280–1286; MR 33 #2810. [1, 4, 19, 49, 54]
904. MacGregor, T. H. *Majorization by univalent functions.* Duke Math. J. 34(1967), 95–102; MR 34 #6062. [1, 6, 10, 13, 16, 22, 30, 36, 38, 42, 45, 51, 52, 54, 61]
905. MacGregor, T. H. *On the minimum difference quotient of univalent functions.* J. London Math. Soc. 42(1967), 267–268; MR 35 #360. [10, 12, 16, 64, 68]
906. MacGregor, T. H. *Certain Integrals of Univalent and Convex Functions.* Math. Z. 103(1968), 48–54; MR 36 #5324. [1, 4, 5, 6, 10, 16, 47, 64, 85]
907. MacGregor, T. H. *An inequality concerning analytic functions with a positive real part.* Canadian J. Math., 21(1969), 1172–1177; MR 40 #2878. [2, 21]
908. MacGregor, T. H. *Univalent power series whose coefficients have monotonic properties.* Math. Z. 112(1969), 222–228; MR 40 #2839. [4, 39]
909. MacGregor, T. H. *The univalence of a linear combination of convex mappings.* J. London Math. Soc., 44(1969), 210–212; MR 38 #4665. [4, 6, 9, 10, 43, 63, 68, 82]
910. MacGregor, T. H. *Approximation by polynomials subordinate to a univalent function.* Trans. Amer. Math. Soc. 148(1970), 199–209; MR 41 #2029. [1, 4, 5, 10, 20, 32, 43]
911. MacGregor, T. H. *Geometric problems in complex analysis.* Amer. Math. Monthly 79(1972), 447–467; MR 45 #7034. [44]
912. MacGregor, Thomas H. *Applications of extreme-point theory to*

univalent functions. Michigan Math. J. 19(1972), no. 4, pp. 361–376; MR 47 #447. [1, 2, 3, 5, 6, 10, 56, 63, 64, 74, 75, 88]
913. MacGregor, Thomas H. *Hull subordination and extremal problems for starlike and spirallike mappings.* Trans. Amer. Math. Soc. 183 (1973), 499–510; MR 49 #3104. [1, 6, 10, 20, 88]
914. MacGregor, Thomas H. *Rotations of the Range of an analytic function.* Math. Ann. 201(1973), 113–126; MR 48 #6390. [4, 16, 24, 49, 54, 68]
915. MacGregor, T. H. *A subordination for convex functions of order α.* J. London Math. Soc. (2), 9(1975), 530–536; MR 51 #3417. [1, 6, 10]
916. MacGregor, T. H. *Fréchet differentiable functionals and support points for families of analytic functions.* Notices Amer. Math. Soc. 23, January 1976, Abstract 731-30-11, p. A-100. [6, 88]
917. Makówka, B. *Quasi β-spiral starlike functions.* (Polish and Russian summaries) Zeszyty Nauk Politech Łódź. No. 208 Mat. No. 6(1975), 47–75. [6]; MR 52 #727
918. Malec, M.; Nikitch, N. *On a geometrical characteristic of analytic functions.* (Polish summary) Zeszyty Nauk. Akad. Gorn.-Hutniczej-Mat.-Fiz.-Chem. Zeszyt 13(1973), 133–135; MR 50 #10212. [35]
919. Malik, M. A. *On the derivative of a polynomial.* (Abstract), Notices, A. M. S., vol. 16, no. 1, issue no. 111, Jan. 1969, p. 290. [22, 32]
920. Marcus, M. *A radial averaging transformation, capacity and conformal radius.* Bull. Amer. Math. Soc. 78(1972), 456–460. [24]
921. Martynov, Ju. A. *The geometric properties of arcs of level curves under schlicht conformal mappings.* (Russian) Trudy Tomsk. Gos. Univ. Ser. Meh.-Mat. 210(1969), 53–61; MR 43 #5016. [2, 12, 29, 35]
922. Mathews, J. H. *Coefficients of uniformly normal-Bloch functions.* Yokohama Math. J. 21(1973), 29–31; MR 48 #4321.
923. Matsumoto, K. *Note on Schiffer's variation in the class of univalent functions in the unit disc.* Nagoya Math. J. 32(1968), 273–276; MR 38 #1247. [24, 42]
924. Mazur, R. *On the radius of β-convexity for certain family of functions meromorphic in the unit disc.* Demonstratio Math. 8(1975), 483–489. [2, 12]; MR 52 #8404
925. Mehrok, T. J. S.; Bajpai, S. K. *On the univalence of a certain integral.* Indian J. Pures Appl. Math. 5(1974), 186–191. [6, 10, 85]; MR 52 #3501
926. Merkes, E. P. *On convolutions and growth of typically real functions.* Mathematical Essays Dedicated to A. J. Macintyre, pp. 299–303. Ohio Univ. Press, Athens, Ohio, 1970; MR 42 #7880. [12, 13, 41, 47, 62]

927. Merkes, E. P.; Scott, W. T. *Covering theorems for univalent functions.* Michigan Math. J. 13(1966), 41-47; MR 32 #7729. [1, 6, 10, 19, 22, 30]
928. Merkes, E. P.; Wright, D. J. *On the univalence of a certain integral.* Proc. A. M. S. 27(1971), 97-100; MR 42 #4720. [2, 4, 5, 6, 10, 85]
929. Merkes, E. P.; Wright, D. J. *Typically-real polynomials of small degree.* Math. Student 39(1971), 205-206(1972); MR 48 #8762. [13, 32]
930. Miazga, J. *The radius of convexity for a class of regular functions.* (Polish and Russian summareis) Ann. Univ. Mariae Curie-Sklodowska Sect. A 26(1972), 31-35(1974); MR 50 #10226. [6, 12]
931. Miazga, J.; Stankiewicz, Jan; Stankiewicz, Zofia. Radii of convexity for some classes of close-to-convex functions. Ann. Univ. Mariae Curie-Sklodowska Sect. A 20(1966), 53-57(1971); MR 46 #3765. [5, 12, 70]
932. Michel, C. *Eine Bemerkung zu schlichten Polynomen.* (Loose Russian and English summaries) Bull. Acad. Polon. Sci. Ser. Sci. Math. Astronom. Phys. 18(1970), 513-519; MR 43 #6423. [32]
933. Mikka, V. P. *Sufficient conditions for the univalence of the solutions of inverse boundary value problems with corner points.* (Russian) Trudy Sem Kraev. Zadacam Vyp. 10(1973), 95-106; MR 50 #620. [4]
934. Mikolajczyk, L. *Certain properties of extremal univalent functions that are p-symmetric and bounded from below in the disc* $|z| > 1$. (Polish. English summary) Zeszyty Nauk. Univ. Łódźk Nauki Mat. Przyrod. Zeszyt 20 Matematyka (1966), 103-115; MR 50 #13483. [9, 24, 49, 58]
935. Mikolajczyk, L. *Theoreme sur la deformation pour les fonctions univalentes p-symetriques et bornees inferieurment dans le domaine* $|z| > 1$. (Russian summary) Bull. Acad. Polon. Sci. Sér. Sci. Math. Astronom. Phys. 14(1966), 245-250; MR 34 #330. [9, 24, 49, 58]
936. Mikolajczyk, L. *Le domaine de variablilite des coefficients A_2 et A_3 des fonctions univalentes bornees avec des coefficients reels.* (Russian summary) Bull. Acad. Polon. Sci. Sér. Sci. Math. Astronom. Phys. 14(1966), 251-254; MR 34 #331. [22, 29]
937. Mikolajczyk, L. *Domaine de variation des coefficients A_2 et A_3 des fonctions univalentes bornees a coefficients reels.* Ann. Polon. Math. 19(1967), 81-106; MR 35 #362. [22, 29, 39]
938. Mikolajczyk, L. *A theorem on distortion for univalent p-symmetrical functions bounded in the circle* $|z| > 1$. Prace Mat.

12(1968), 35-51; MR 38 #2294. [9, 22, 24]
939. Mikolajczyk, Leon. *Relation between the modulus of a schlicht function, p-symmetrical and bounded from below, and the modulus of its derivative.* (Polish. French summary) Zeszyty Nauk Univ. Lódźk. Nauki Mat. Przyrod. Sér. II Zeszyt 29(1968), 3-29; MR 39 #7082. [9, 49, 58]
940. Mikolajczyk, Leon. *A relation between the modulus of a schlicht function that is bounded below and the modulus of its derivative.* (Polish. English summary) Zeszyty Nauk Univ. Lódźk. Nauki Mat. Przyrod. Sér II Zeszyt 34(1969), 3-21; MR 41 #450. [9, 58]
941. Mikolajczyk, Leon. A certain solution of the equation $f(z) = pf(g)$ *in the class of α-spiral starlike funcns.* (Polish. English summary) Zeszyty Nauk Univ. Lodzk. Nauki Mat. Przyrod. Ser. II Zeszyt 39(Mat.)(1971), 53-68; MR 48 #11473. [6, 24]
942. Milcetich, John G. *On the extreme points of some sets of analytic functions.* Proc. Amer. Math. Soc. 45(1974), 223-228; MR 50 #4957. [1, 88]
943. Milcetich, J. G. *A general extremal problem for the class of close-to-convex functions.* Notices Amer. Math. Soc. 23, January 1976, Abstract 731-30-8, p. A-99. [5]
944. Milin, I. M. *On the coefficients of univalent functions.* (Russian) Dokl. Akad. Nauk SSSR 176(1967), 1015-1018; MR 36 #5328. {Soviet Math. Dokl. 8(1967), No. 5, pp. 1255-1258}. [9, 25, 49, 54]
945. Milin, I. M. *Adjacent coefficients of univalent functions.* (Russian) Dokl. Akad. Nauk SSSR 180(1968), 1294-1297; MR 38 #318. {This article has appeared in English translation [Soviet Math. Dokl. 9(1968), 762-765]}. [8]
946. Milin, I. M. *The method of areas for schlicht functions in finitely connected domains.* (Russian) Trudy Mat. Inst. Steklov. 94(1968), 90-122; MR 37 #2970. Translated from the Russian: Extremal Problems of the Geometric Theory of Functions (A. M. S. 1969). [4, 27, 29]
947. Milin, I. M. *Hayman's regularity theorem for the coefficients of univalent functions.* (Russian) Dokl. Akad. Nauk SSSR 192(1970), 738-741; MR 42 #3269. {Soviet Math. Dokl. 11(1970), 724-728.}. [62]
948. Milin, I. M. *Univalent Functions and Orthogonal Systems.* (Russian) Izdat. "Nauka", Moscow, 1971. 256 pp. {An English translation is to be published by Amer. Math. Soc., Providence, R.I.}; MR 51 #5916. [44]
949. Miller, Allen R. *A machine computation of the Grunsky coeffi-*

cients of Schlicht functions. NRL Report 7225; Naval Research Laboratory, Washington, D. C., 1971, ii + 164 pp.; MR 43 #5017. [53]
950. Miller, J. E. *Convex meromorphic mappings and related functions.* Proc. Amer. Math. Soc. 25(1970), 220–228; MR 41 #3740. [2, 5, 9, 12, 56, 58]
951. Miller, James. *Extremal functions for meromorphic univalent functions.* J. Analyse Math. 24(1971), 77–86; MR 44 #5443. [9, 24, 58, 68]
952. Miller, James E. *Starlike meromorphic functions.* Proc. Amer. Math. Soc. 31(1972), 446–452; MR 44 #5465. [6, 7, 9, 58]
953. Miller, J. E. *Sequences of quasi-subordinate functions.* Pacific J. Math. 43(1972), 437–442; MR 47 #7023. [1]
954. Miller, James. *On the maximum modulus for meromorphic univalent functions.* Proc. West Virginia Acad. Sci. 45(1973), 360–365; MR 50 #10237. [9, 24]
955. Miller, J. E. *Subordinating factor sequences for convex maps in C^n.* Notices A. M. S., vol. 21, November 1974, Abstract 717–B17, p. A–614. [1, 10, 47]
956. Miller, S. S. *The Hardy class of a Bazilevič function and its derivative.* Proc. Amer. Math. Soc. 30(1971), 125–132; MR 44 #5444. [3, 4, 16, 62, 70, 81]
957. Miller, Sanford S. *An arc-length problem for n-fold symmetric univalent functions.* Kodai Math. Sem. Rep. 24(1972), 195–202; MR 46 #9323. [5, 6, 7, 10, 35, 49]
958. Miller, S. S. *Distortion properties of alpha-starlike functions.* Proc. Amer. Math. Soc. 38(1973), 311–318; MR 46 #9324. [4, 19, 62, 69, 83]
959. Miller, Sanford S. *The H^p classes for alpha-convex functions.* II. Notices Amer. Math. Soc. 21, April 1974, Abstract 74T–B80, p. A–371. [69]
960. Miller, S. S. *The Hardy class of functions of bounded argument rotation.* Notices Amer. Math. Soc. 21, February 1974, Abstract 74T–B34, p. A–301. [62, 71]
961. Miller, S. S. *Differential inequalities and Carathéodory functions.* Bull. Amer. Math. Soc. 81(1975), 79–81; MR 50 #7533. [2, 69]
962. Miller, S. S. *A differential inequality implying boundedness.* The American Mathematical Monthly, Vol. 82, no. 5, May 1975 (Advanced Problems Department), p. 529. [22]
963. Miller, S. S. *A class of differential inequalities implying boundedness.* Notices Amer. Math. Soc., January 1975, Abstract 720–30–16, p. A–122. [22]

964. Miller, S. S. *Second order differential inequalities and some applications in univalent function theory.* Notices Amer. Math. Soc. 23, January 1976, Abstract 731-30-10, p. A-100. [2]
965. Miller, Sanford S.; Mocanu, Petru T. *The Hardy class of functions of bounded argument rotation.* J. Austral. Math. Soc. Ser. A 21 (1976), no. 1, 72-78. [71]
966. Miller, S. S.; Mocanu, P. T.; Reade, M. O. *All α-convex functions are starlike.* Rev. Roumaine Math. Pures Appl. 17(1972), 1395-1397; MR 48 #2364. [4, 6, 9, 12, 69, 84]
967. Miller, S. S.; Mocanu, P. T.; Reade, Maxwell O. *The radius of α-convexity of univalent functions, $1 \le \alpha \le \infty$.* (Preliminary report) Notices A. M. S. 20(Feb. 1974), Abstract 73T-B60. [12, 69]
968. Miller, S. S.; Mocanu, P.; Reade, M. O. *All α-convex functions are univalent and starlike.* Proc. Amer. Math. Soc. 37(1973), 553-554; MR 47 #2044. [6, 10, 69]
969. Miller, S. S.; Mocanu, P.; Reade, M. O. *Bazilevič functions and generalized convexity.* Rev. Roumaine Math. Pures Appl. 19(1974), 213-224; MR 49 #3105. [4, 12, 16, 68, 69, 70]
970. Miller, S. S.; Mocanu, Petru T.; Reade, M. O. *The Hardy classes for functions in the class $MV[\alpha, k]$.* J. Math. Anal. Appl. 51(1975), 33-42. [71]; MR 52 #728
971. Minsker, Steven. *A new proof of Pick's theorem.* J. Res. Nat. Bur. Standards Sect. B 78B(1974), 95-96; MR 49 #10873. [4, 22, 30]
972. Mirošničenko, Ja. S. *Certain extremal problems of the theory of univalent functions.* (Russian) Izv. Vyss. Ucebn. Zaved. Matematike 1965, no. 2(45), 104-109; MR 32 #2574. {Amer. Math. Soc. Transl. (2) vol. 88(1970)}. [7, 9, 35, 45]
973. Mirošničenko, Ja. s. *On the question of the curvature of level curves.* (Russian) Trudy Tomsk. Gos. Univ. Ser. Meh.-Mat. 210 (1969), 62-65; MR 44 #6950. [35]
974. Mitjuk, I. P. *Univalent conformal mappings of multiply connected domain.* (Russian) Problems of Math. Phys. and Theory of Functions, II (Russian), pp. 74-84. Naukova Dumka, Kiev, 1964; MR 32 #5856. [7, 18, 49]
975. Mitjuk, I. P. *Some properties of functions regular in a multiply connected region.* Dokl. Akad. Nauk SSSR 164(1965), 495-498; MR 32 #5862. {Soviet Math. Dokl. 6(1965), 1252-1255}. [24, 60]
976. Mitjuk, I. P. *The symmetrization principle for an annulus and certian of its applications.* (Russian) Sibirsk. Mat. Z. 6(1965), 1282-1291; MR 35 #4396. [14, 24, 33, 48]
977. Mitjuk, I. P. *The inner radius of a domain and certain of its properties.* (Russian) Ukrain. Mat. Z. 17(1965), no. 1, 117-122; MR 32

#7730. [14, 19]
978. Mitjuk, I. P. *Some theorems on functions regular in an annulus.* (Ukrainian. Russian and English summaries) Dopovidi Akad. Nauk Ukrain. RSR 1965, 160–163; MR 30 #4926. [14, 19, 33, 48]
979. Mitjuk, I. P. *The principle of symmetrization for multiply connected regions and certain of its applications.* (Russian) Ukrain. at. Z. 17(1965), no. 4, 46–54; MR 33 #7510. [6, 18]
980. Mitjuk, I. P. *Some covering theorems for functions regular in multiply-connected domains.* (Ukrainian. Russian and English summaries) Dopovidi Akad. Nauk Ukrain. RSR 1965, 550–554; MR 33 #2811. [19]
981. Mitjuk, I. P. *A certain subordination principle for regular functions.* (Russian) Trudy Tomsk. Gos. Univ. Ser. Meh.-Mat. 189 (1966), 161–167; MR 37 #4244. [1, 7, 18]
982. Mocanu, P. T. *Extremal domains in the class of univalent functions.* (Romanian. Russian and French summaries) Acad. R. P. Romine Fil. Cluj Stud. Cerc. Mat. 12(1961), 303–313; MR 30 #4939. [24, 29]
983. Mocanu, P. T. *On an extremal problem relative to univalent functions.* (Romanian. Russian and French summaries) Acad. R. P. Romine Fil Bluj Stud. Cerc. Mat. 14(1963), 85–91; MR 33 #5868. [24]
984. Mocanu, P. T. *On the radius of convexity of holomorphic functions.* (Romanian. Russian and French summaries) Studia Univ. Babes-Bolyai Ser. Math.-Phys. 9(1964), no. 2, 31–33; MR 30 #4918. [12, 31]
985. Mocanu, P. T. *On the equation $f(z) = \alpha f(\alpha)$ in the class of univalent functions.* Mathematica (Cluj)6(29)(1964), 63–79; MR 32 #7731. [24]
986. Mocanu, P. T. *Functions univalent on sectors.* (Romanian) Stud. Cerc. Mat. 17(1965), 925–931; MR 34 #4475. [4, 57]
987. Mocanu, P. T. *Generalized radii of starlikeness and convexity of analytic functions.* (Romanian and Russian summaries) Studia Univ. Babes-Bolyai Ser. Math.-Phys. 11 (1966), no. 2, 43–50; MR 35 #6810. [11, 12, 31]
988. Mocanu, Petru T. *Convexity and starlikeness of conformal mappings.* Mathematica (Cluj)8(31)(1966), 91–102; MR 35 #5600. [6, 10, 11, 12]
989. Mocanu, Petru T. *About the radius of starlikeness of the exponential function.* (Romanian and Russian summaries) Studia Univ. Babes-Bolyai Ser. Math.-Phys. 14(1969), no. 1, 35–40; MR 41 #1990. [6, 11, 28]

990. Mocanu, Petru T. *Une propriété de convexité généralisée dans la théorie de la représentation conforme.* Mathematica (Cluj)11(34) (1969), 127-133; MR 42 #7881. [6, 10, 12, 69]
991. Mocanu, Petru T. *Sur la géométrie de la représentation conforme.* Mathematica (Cluj)12(35)(1970), 299-308; MR 48 #6391. [6, 10, 35]
992. Mocanu, P. T. *An extremal problem for univalent functions associated with the Darboux formula.* Ann. Univ. Mariae Curie-Sklodowska, Sect. A, vols. 22/23/24(1968/1969/1970), p. 131-135; MR 49 #5335. [4, 16]
993. Mocanu, Petru T. *Sur deux notions de convexité généralisée dans la représentation conforme.* (Romanian and Russian summaries) Studia Univ. Babes-Bolyai Ser. Math.-Mech. 16(1971), fasc. 2, 13-19; MR 47 #3665. [10]
994. Mocanu, Petru T. *A generalized property of convexity in conformal mappings.* Rev. Roumaines Math. Pures Appl. 17(1972), 1391-1394; MR 49 #5327. [12, 35, 49]
995. Mocanu, Petru T. *Sur une propriété d'etoilement dans la théorie de la representation conforme.* (Russian and Romanian summaries) Studia Univ. Babes-Bolyai Ser. Math.-Mech. 17(1972), fasc. 2, 55-58; MR 50 #586. [6]
996. Mocanu, Petru T. *The radius of α-convexity for the class of starlike univalent functions, α real.* Notices A. M. S. 20(Jan. 1973), Abstract (Prelim. report) 701-30-34, pg. A-112. [12, 69]
997. Mocanu, P. T.; Moldovan, Gr.; Reade, M. O. *Numerical computation of the convex Koebe function.* (Romanian and Russian summaries) Studia Univ. Babes-Bolyai Ser. Math.-Mech. 19(1974), fasc. 1, 37-46; MR 49 #3106. [6, 10, 49, 69, 70]
998. Mocanu, P.; Reade, M. O. *The order of starlikeness of certain univalent functions.* Notices Amer. Math. Soc. 18(1971), 815. Abstract #71T-B182. [69]
999. Mocanu Petru.; Reade, M. O. *On generalized convexity in conformal mappings.* Rev. Roumaine Math. Pures Appl. 16(1971), 1541-1544; MR 46 #7495. [4, 5, 6, 10, 69]
1000. Mocanu, P.; Reade, M. O. *Multivalent α-convex functions.* I. Notices, A. M. S., vol. 19, no. 6, issue 140, October, 1972 (Abstract 72T-B280), p. A-702. [14, 69]
1001. Mocanu, P.; Reade, M. O. *On α-convex functions.* Notices, A. M. S., vol. 19, no. 2, issue 136, February, 1972 (Abstract 693-B13), p. A-394. [69]
1002. Mocanu, Petru T.; Reade, Maxwell O. *The radius of α-convexity of certain classes of starlike univalent functions, α real.* Proc. Amer. Math. Soc. 51(1975), 395-400; Rev. Roumaine Math.

Pures Appl. 20(1975), no. 5, 561-565; MR 51 #10604. [6, 9, 12, 29, 69]; MR 52 #3502

1003. Mocanu, P.; Reade, M. O.; Zlotkiewicz, E. *On criteria for the univalence of analytic functions.* Notices, A.M.S., vol. 18, no. 4, isue no. 130, June 1971 (Abstract 71T-B104), p. 637. [69, 70]

1004. Mocanu, P.; Reade, P. T.; Zlotkiewicz, E. *On Bazilevič functions.* Proc. Amer. Math. Soc. 39(1973), 173-174; MR 47 #2045. [4, 5, 70]

1005. Mocanu, Petru T.; Reade, Maxwell O.; Zlotkiewicz, Eligiusz. *On the functional $[f(z_1)/f'(z_2)]$, for typically-real functions.* Rev. Anal. Numer. Theorie Approximation 3(1974), no. 2, 209-214 (1975). [13, 66]; MR 52 #5972

1006. Mocanu, P.; Reade, M. O.; Zlotkiewicz, E. *Bazilevič functions and close-to-convex p-valent functions.* Notices, A. M. S., vol. 19, no. 5, issue 139, August, 1972 (Abstract 696-30-1), p. A-636. [5, 14, 70, 80, 81]

1007. Mogk, E. *Uber ein Variationslemma von M. Schiffer.* Mitt. Math. Sem. Giessen Heft 82(1969), ii + 42 pp.; MR 40 #329. [24]

1008. Mogk, Eberhard. *An external problem of a certain class of schlicht functions in an annulus.* MITT. Math. Sem. Giessen 92(1971), pp. 51-54; MR 45 #5357. [48]

1009. Mogra, M. L. *On a class of starlike functions in the unit disc.* Notices, A. M. S., vol. 22, no. 5, issue no. 163, August 1975 (Abstract 75T-B158), p. A-511. [6, 12, 36, 63]

1010. Mogra, M L. *Radii of convexity for certain classes of univalent functions.* Notices Amer. Math. Soc., October 1975, Abstract 75T-B190, p. A-625. [2, 12]

1011. Mozgovaja, L. I. *On a class of functions univalent in the disc $|z| < 1$.* (Russian. Romanian summary). Bull. Inst. Politehn. Iasi (N. S.)10(14)(1964), no. 3-4, 37-42; MR 32 #205. [4, 23]

1012. Moleda, Alicja. *Certain extremal problems in the classes P_m and S_m^* of functions holomorphic in the disc $|z| < 1$.* (Polish. English summary) Zeszyty Nauk Univ. Lodz Nauki Mat. Przyrod. Ser. II Zeszyt 52 Mat. (1973), 57-83; MR 49 #3117. [6, 23, 24]

1013. Montaldo, Oscar. *Sul comportamento delle funzione univalenti mell'intorno della funzione di Koebe.* (English summary) Boll. Un. Mat. Ital. (3)21(1966), 127-143; MR 34 #333. [24, 54]

1014. Moulis, E. J., Jr. *A generalization of univalent functions with bounded boundary rotation.* Trans. Amer. Math. Soc. 174(1972), 345-368; MR 47 #8835. [4, 6, 11, 12, 23, 31, 71, 76, 77]

1015. Mozgovaja, L. I.; Acilov, H. *On an inequality in the theory of special classes of analytic functions.* (Russian, Uzbek summary) Izv. Akad. Nauk UzSSR Ser. Fiz.-Mat. Nk 11(1967), no. 1, 3-8; MR 35 #4390. [2, 6, 12]

1016. Mozgovaja, L. I. *A certain generalization of the Marx-Cakalov class of functions.* (Russian) Ukrain. Mat. Z. 21(1969), 38-49; MR 39 #5784. [2, 11, 28, 43]

1017. Narayan, S. *Theory of functions of a complex variable.* S. Chand & Co., Delhi (distributed by Lawrence Verry, Inc.), Mystic, Conn. 1966 X + 379 pp.; MR 37 #1561. [44]

1018. Nasr, M. A. *On the radius of convexity of convex combinations of certian analytic functions.* (Serbo-Croatian summary). Glasnik Mat. Ser. III 10(30)(1975), no. 2, 257-262. [12, 82]

1019. Nehari, Zeev. *Some function-theoretic aspects of linear second-order differential equations.* J. D'analyse Math. 18(1967), 259-276; MR 35 #4391. [26, 31]

1020. Nehari, Zeev. *Inequalities for the coefficients of univalent functions.* Arch. Rational Mech. Anal. 34(1969), 301-330; MR 40 #330. [4, 16, 19, 27, 34, 42, 45, 53, 54]

1021. Nehari, Zeev. *On the coefficients of Bieberbach-Eilenberg functions.* J. Analyse Math. 23(1970), 297-303; MR 42 #7884. [3, 4, 19, 34, 87]

1022. Nehari, Zeev. *A proof of $|a_4| \le 4$ by Loewner's method.* Proceedings of the Symposium on Complex Analysis (Univ. Kent, Canterbury, 1973), pp. 107-110. London Math. Soc. Lecture Note Ser., No. 12, Cambridge Univ. Press, London, 1974. [54]

1023. Netanyahu, Elisha. *Un problème d'extremum concernant les fonctions univalentes.* C. R. Acad. Sci. Paris Ser. A-B 267(1968), A 261-A 263; MR 38 #2296. [19]

1024. Netanyahu, E. *The minimal distance of the image boundary from the origin and the second coefficient of a univalent function in $|z| < 1$.* Arch. Rational Mech. Anal. 32(1969), 100-112; MR 38 #3422. [19, 29]

1025. Netanyahu, Elisha. *Sur les fonctions univalentes dans le disque de rayon un dont l'image contient un disque donné.* C. R. Acad. Sci. Paris Sér. A-B269(1969), A762-A765; MR 42 #487. [19, 24]

1026. Netanyahu, E. *On univalent functions in the unit disk whose image contains a given disk.* Journal D'analyse Mathematique 23 (1970), 305-322; MR 43 #6420. [19, 24, 29]

1027. Neuwirth, J.; Newman, D. J. *Positive $H^{1/2}$ functions are constants.* Proc. Amer. Math. Soc. 18(1967), p. 958. [3, 62]

1028. Newman, D. J. *Successive differences of bounded sequences.*

Proc. Amer. Math. Soc. 17(1966), 285-286.
1029. Nicholls, P. J.; Sons, L. R. *Minimum modulus and zeroes of functions in the unit disc.* Proc. London Math. Soc. (3)31(1975), 99-113. [62, 90]
1030. Nickel, Paul A. *A note on principal functions and multiply-valent canonical mappings.* Pacific J. Math. 20(1967), 283-288; MR 34 #4476. [14]
1031. Nikolaēva, R. V. *Certain properties of univalent functions in the disc $|z| < 1$.* (Russian) Izv. Vyss. Ucebn. Zaved. Matematika 1971, no. 7(110), 50-53; MR 45 #2156. [2, 4, 6, 10]
1032. Nikolaēva, R. V.; Repnina, L. G. *On the question of the Pólya-Schoenberg hypothesis.* (Russian) Mathematical Physics, No. 7(Russian), pp. 143-147. Naukova Dumka, Kiev. 1970; MR 42 #7882. [47]
1033. Nikolaēva, R. V.; Repnina, L. G. *A certain generalization of theorems ddue to Livingston.* (Russian) Ukrain. Mat. Z. 24(1972), 268-273; MR 45 #5336. [2, 11, 12, 73, 82]
1034. Nishimiya, Han. *Conformal mapping onto gear-like domains.* Sci. Rep. Saitama Univ. Ser. A 5, 1-4(1965); MR 32 #201. [23, 49]
1035. Noonan, James. *Meromorphic functions of bounded boundary rotation.* Michigan Math. J., 18(1971), no. 4, pp. 343-352; MR 45 #2154. [9, 23, 71, 76, 77]
1036. Noonan, James W. *Coefficients of functions with bounded boundary rotation.* Proc. Amer. Math. Soc. 29(1971), 307-312; MR 43 #498. [6, 10, 71, 77]
1037. Noonan, J. W. *Boundary bahavior of functions with bounded boundary rotation.* J. Math. Anal. Appl., 38(1972), no. 3, pp. 721- 734; MR 47 #473. [4, 61, 62]
1038. Noonan, James W. *Asymptotic behavior of functions with bounded boundary rotation.* Trans. Amer. Math. Soc. 164(1972), 397-410; MR 45 #3688. [3, 6, 10, 62, 71, 76, 77]
1039. Noonan, J. W. *On functions of bounded rotation.* Proc. Amer. Math. Soc. 32(1972), 91-101; MR 45 #527. [3, 6, 7, 10, 18, 23, 62, 71]
1040. Noonan, J. W. *On close-to-convex functions of order β.* Pacific J. Math. 44(1973), 263-280; MR 47 #3661. [3, 4, 5, 8, 10, 23, 45, 49, 70, 71]
1041. Noonan, J. W. *Curvature and radius of curvature for functions with bounded boundary rotation.* Canadian J. Math. 25(1973), 1015-1023; MR 48 #2365. [35, 71]
1042. Noonan, James W. *Coefficient differences and Handel deter-*

minants of areally mean p-valent functions. PRoc. Amer. Math. Soc. 46(1974), 29-37; MR 50 #4927. [14, 33]

1043. Noonan, J. W. *Coefficient behavior of a class of meromorphic functions.* Canad. J. Math. 27(1975), no. 5, 1157-1165(1976). [6, 9, 23, 25, 62, 71]

1044. Noonan, J. W. *Powers of p-valent functions.* Notices Amer. Math. Soc., January 1975, Abstract 720-30-13, p. A-121. [14, 33]

1045. Noonan, J. W. *Hankel determinants of mean-valent functions.* Notices Amer. Math. Soc. 23, January 1976, Abstract 731-30-1, p. A-98. [33]

1046. Noonan, J. W.; Thomas, D. K. *Hankel determinants of areally mean p-valent functions.* Proc. London Math. Soc. 25(1972), part 3, 503-524; MR 46 #5605. [33, 62]

1047. Noonan, J. W.; Thomas, D. K. *On successive coefficients of functions of bounded boundary rotation.* J. London Math. Soc. 5 (1972), 656-662; MR 47 #448. [8, 33, 71, 77]

1048. Noonan, J. W.; Thomas, D. K. *The integral means of regular functions.* J. London Math. Soc. (2)9(1974/75), 557-560; MR 51 #879. [3]

1049. Nosenko, A. S. [Nosenko, O. S.] *The ranges of values of two functionals that are defined on Carathéodory's C-functions.* (Russian) Ukrain. Mat. Z. 25(1973), 827-830, 863; MR 48 #6392 {English transl: Ukrainian Math. J. 25(1973), 689-691(1974)}. [2, 29]

1050. Nunokawa, M. *On the univalency and multivalency of certain analytic functions.* Math. Zeit. 104(1968), 394-404; MR 57 #2971. [2, 4, 10, 14, 22, 31, 43, 56, 61, 68, 85]

1051. Nunokawa, M. *On Bazilevič and convex functions.* Trans. Amer. Math. Soc. 143(1969), 337-341; MR 40 #2840. [3, 5, 6, 7, 10, 18, 33, 70, 80]

1052. Nunokawa, M. *On the univalence of a certain integral.* Trans. Amer. Math. Soc. 146(1969), 439-446; MR 40 #4436. [2, 4, 5, 6, 10, 22, 31, 43, 61, 85]

1053. Nunokawa, Mamoru. *On the univalence of a certain integral.* Proc. Japan Acad. 45(1969), 841-845; MR 42 #1991. [2, 4, 5, 6, 10, 85]

1054. Nunokawa, M. *A note on convex and Bazilevič functions.* Proc. Amer. Math. Soc. 24(1970), 332-335; MR 40 #4437. [4, 6, 7, 10, 16, 18, 70]

1055. Nunokawa, Mamoru. *On Bazilevič functions of bounded boundary rotation.* J. Math. Soc. Japan, 24(1972), no. 2, pp. 275-278; MR 47 #2046. [4, 5, 7, 10, 18, 70, 71]

1056. Obrock, A. *An inequality for certain schlicht functions.* Proc. Amer. Math. Soc. 17(1966), 1250-1253; MR #6075. [24, 39, 54]
1057. Obrock, Arthur E. *Grötzsch Domains and Teichmüller Inequalities.* Indiana Univ. Math. J. 20(1971), 739-751; MR 42 #4714
1058. Obrock, Arthur E. *On the use of Teichmüller's principle in conjunction with the continuity method.* J. Analyse Math. 25(1972), 75-105; MR 48 #8775. [24, 29]
1059. Ogawa, Shôtarô. *A note on close-to-convex functions.* I. J. Nara Gakugei Univ. 8(1959), no. 2, 9-10; MR 31 #3588. [5, 64, 74]
1060. Ogawa, Shôtarô. *A note on close-to-convex functions.* II. J. Nara Gakugei Univ. 8(1959), no. 2, 11-17; MR 31 #3589. [5, 20]
1061. Ogawa, Shôtarô. *On some extension of radius of convexity.* J. Nara Gakugei Univ. Natur. Sci. 12(1964), 1-4; MR 34 #2856. [5, 10, 12]
1062. O'Hara, P. J.; Rodriguez, R. S. *Some properties of self-inversive polynomials.* Proc. Amer. Math. Soc. 44(1974), 331-335; MR 50 #2460. [32]
1063. Ozaki, S.; Nunokawa, M. *The Schwarzian derivative and univalent functions.* Proc. Amer. Math. Soc., 33(1972), no. 2, pp. 392-394; MR 4 #8821. [4, 31]
1064. Ozawa, Mitsuru. *On the sixth coefficient of a univalent function.* Kodai Math. Sem Rep. 17(1965), 1-9; MR 31 #2394. [42, 53, 54]
1065. Ozawa, Mitsuru. *An elementary proof of local maximality for a_6.* Kodai Math. Sem. Rep. 20(1968), 437-439; MR 38 #4670. [42, 53, 54]
1066. Ozawa, Mitsuru. *On local maximality for the coefficients a_6 and a_8.* Kodai Math. Sem. Rep. 20(1968), 440-441; MR 38 #4671. [54]
1067. Ozawa, Mitsuru. *On the Bieberbach conjecture for the sixth coefficient.* Kodai Math. Sem. Rep. 21(1969), 97-128; MR 39 #432. [42, 53, 54]
1068. Ozawa, Mitsuru. *An elementary proof of the Bieberbach conjecture for the sixth coefficient.* Kodai Math. Sem. Rep. 21(1969), 129-132; MR 40 #333. [42, 53, 54]
1069. Ozawa, Mitsuru. *On an elementary proof of local maximality for the coefficient a_8.* Kodai Math. Sem. Rep. 21(1969), 459-462; MR 41 #7090. [42, 53, 54]
1070. Ozawa, Mitsuru. *A proof of the Bieberbach conjecture for the fourth coefficient.* Kodai Math. Sem. Rep. 24(1972), 506-512; MR 47 #7020. [42, 53, 54]
1071. Ozawa, Mitsuru. *Certain coefficient inequalities for univalent functions.* Kodai Math. Sem. Rep. 25(1973), 1-31; MR 47 #7021. [54]

1072. Ozawa, Mitsuru; Kubota, Yoshihisa. *On the eight coefficient of univalent functions.* J. Analyse Math. 23(1970), 323-352; MR 42 #6214. [54]

1073. Ozawa, Mitsuru; Kubota, Yoshihisa. *On the eight coefficient of univalent functions.* II. Kodai Math. Sem. Rep. 23(1971), 1-59; MR 44 #425. [42, 53, 54]

1074. Ozawa, Mitsuru; Kubota, Yoshihisa. *Bieberbach conjecture for the eight coefficient.* Kodai Math. Sem. Rep. 24(1972), 331-382; MR 47 #5246. [42, 53, 54]

1075. Ozawa, M.; Kubota, Y. *Bieberbach conjecture for the eight coefficient.* II. Kodai Math. Sem. Rep. 25(1973), 257-288; MR 48 #4298. [42, 53, 54]

1076. Padmanabhan, K. S. *On certain classes of meromorphic funcitons in the unit circle.* Math. Z. 89(1965), 98-107. [2, 5, 9, 11, 20, 43, 56, 73]

1077. Padmanabhan, K. S. *On the radius of univalence and starlikeness for certain analytic functions.* J. Indian Math. Soc. (N.S.)29 (1965), 71-80; MR 32 #4262. [2, 6, 10, 11, 43, 68]

1078. Padmanabhan, K. S. *On the radius of univalence and starlikeness for certain analytic functions.* II. J. Indian Math. Soc. (N.S.) 29 (1965), 201-208; MR 34 #4477. [2, 6, 9, 10, 11, 22, 43, 56, 61]

1079. Padmanabhan, K. S. *On the radius of convexity of a certain class of meromorphically starlike function in the unit circle.* Math. Z. 91(1966), 308-313, MR 32 #7732. [6, 9, 12, 16]

1080. Padmanabhan, K. S. *Coefficient estimates for a certain class of meromorphic starlike multivalent functions.* J. London Math. Soc., 42(1967), 201-207; MR 35 #1769. [2, 6, 9, 14, 23, 25]

1081. Padmanabhan, K. S. *On the radius of univalence and starlikeness for a certain class of meromorphic functions.* J. Indian Math. Soc. (N.S.)30(1966), 207-212(1967); MR 37 #6453. [6, 9, 11, 12, 43]

1082. Padmanabhan, K. S. *Developments in the theory of univalent functions.* Symposia on Theoretical Physics and Mathematics, Vol. 8(Symposium Madras, 1967), pp. 141-150. Plenum, New York, 1968; MR 38 #4666. [5, 44]

1083. Padmanabhan, K. S. *On certain classes of starlike functions in the unit disk.* J. Indian Math. Soc. (N. S.)32(1968), 89-103, MR 39 #2965. [6, 9, 12, 23, 63]

1084. Padmanabhan, K. S. *On the radius of univalence of certain classes of analytic functions.* J. London Math. Soc. (2)1(1969), 225-231; MR 40 #331. [2, 5, 6, 10, 11, 12, 73]

1085. Padmanabhan, K. S. *Estimates of growth for certain convex and close-to-convex functions in the unit disc.* J. Indian Math. Soc.

(N.S.)33(1969), 37-47; MR 41 1991. [5, 6, 10, 37, 62, 63, 64, 72, 74]

1086. Padmanabhan, K. S. *On a certain class of functions whose derivatives have a positive real part in the unit disc.* Ann. Polon. Math. 23(1970/71), 73-81; MR 41 #8648. [2, 4, 7, 12, 18, 21, 56]

1087. Padmanabhan, K. S. *On the partial sums of certain analytic functions in the unit disc.* Ann. Polon. Math. 23(1970/71), 83-92; MR 42 #488. [2, 5, 6, 10, 11, 14, 20, 21, 36, 43]

1088. Padmanabhan, K. S. *The radius of univalence and starlikeness of a certian class of analytic functions.* Ann. Polon. Math., 26(1972), pp. 147-156; MR 46 #337. [2, 4, 6, 23, 43, 56]

1089. Padmanabhan, K. S. *On the arithmetic mean of univalent convex functions.* (Serbo-Croatian summary). Glasnik Mat. III9(29) (1974), 65-68; MR 49 #9183. [10, 82]

1090. Padmanabhan, K. S.; Parvatham, R. *Properties of a class of functions with bounded boundary rotation.* Ann. Polon. Math. 31 (1975/76), no. 3, 311-323. [4, 6, 10, 11, 12, 31, 71, 77]

1091. Palka, Janina. *Sharp estiamtes of $|p(w)|$, arg $(p(w)/w)$, $|p'(w)|$, arg $p'(w)$) in a class of univalent polynomials.* Ann. Univ. Mariae Curie-Sklodowska Sect. A., vols. 22/23/24(1968/1969/1970); 137-146; MR 50 #10227. [32]

1092. Palka, J. *On the fourth order Grunsky functionals for bounded univalent functions.* Ann. Univ. Mariae Curie-Sklodowska Sect A 25(1971), 67-81(1973); MR 48 #6398. [22, 30, 53]

1093. Panasovic, V. A. *On the construction of the mapping function and Faber polynomials for certain polygons.* (Russian) First Republ. Math. Conf. of Young Researchers, Part II, (Russian), pp. 552-560. Akad. Nauk Ukrain. SSR Inst. Mat., Kiev, 1965; MR 33 #7511. [49, 53]

1094. Pawel, Todorow. *Uber den Radius des Schlichtheitskreises einer Klasse meromorpher Funktionen.* Acad. Roy. Belg. Bull. Cl. Sci. (5)51(1965), 869-876; MR 32 #4274. [9, 43]

1095. Pederson, R. N. *On unitary properties of Grunsky's matrix.* Arch. Rational Mech. Anal. 29(1968), 370-377; MR 37 #2972. [4, 42, 53, 54]

1096. Pederson, R. N. *An extension of Grunsky's inequality with application to the coefficient problem.* Notices, Amer. Math. Soc. (Asbtract 68T-213), vol. 15, no. 2, issue no. 104, February, 1968, p. 362. [53, 54]

1097. Pederson, R. N. *A proof of the Bieberbach conjecture for the sixth coefficient.* Arch Rational Mech. Anal. 31(1968/69), 331-351; MR 39 #431. [4, 42, 53, 54]

1098. Pederson, R. N. *A note on the local coefficient problem*. Proc. Amer. Math. Soc. 20(1969), 345-347; MR 38 #4667. [42]
1099. Pederson, R. N. *Weak limit directions of plane sets and the Schiffer fundamental lemma*. Notices Amer. Math. Soc. 23, January 1976, Abstract 731-30-34, p. A-105. [62]
1100. Pederson, R. N.; Schiffer, M. *Further generalizations of the Grunsky inequalities*. J. Analyse Math. 23(1970), 353-380; MR 42 #6210. [4, 34, 42, 53]
1101. Pederson, R.; Schiffer, M. *A proof of the Bieberbach conjecture for the fifth coefficient*. Arch. Rational Mech. Anal. 45(1972), 161-193; MR 47 #453. [42]
1102. Pehleckii, I. D. *Riemann surfaces with a univalently accessible boundary*. (Russian) Perm Gos. Univ. Ucen. Zap. 1966, no. 131, 27-32; MR 40 #5852. [62]
1103. Pethe, K. *Estimation du coefficient a_{p+3} de la fonction p-valente dans le cèrcle unité*. (English and Russian summaries) Bull. Acad. Polon. Sci. Sér. Sci. Math. Astronom. Phys. 20(1972), 219-220; MR 46 #5603. [14, 50]
1104. Petruska, G. *A contribution to Bloch's theorem*. Ann. Univ. Sci. Budapest. Eötvös Sect. Math. 12(1969), 39-42; MR 41 #7080. [19]
1105. Pfaltzgraff, J. A. *Extremal problems and coefficient regions for analytic functions represented by a Stieltjes integral*. Trans. Amer. Math. Soc. 115(1965), 270-282; MR 33 #7526. [2, 23, 24, 29]
1106. Pfaltzgraff, J. A. *Extremal problems for functions with bounded boundary rotation*. Notices, Amer. Math. Soc. (Abstract 655-87), vol. 15, no. 3, issue no. 105, April 1968, p. 491. [23, 35]
1107. Pfaltzgraff, J. A. *Variations with constraints for classes of analytic functions*. (Abstract), Notices, A. M. S., vol. 16, no. 1, issue no. 111, Jan. 1969, p. 135. [24]
1108. Pfaltzgraff, J. A. *Constrained extremal problems for functions with positive real part*. (Abstract) Notices, A. M. S. 17(1970), p. 263. [24]
1109. Pfaltzgraff, J. A. *On the Marx conjecture for a class of close-to-convex functions*. Michigan Math. J. 18(1971), 275-278; MR 44 #421. [1, 5, 6, 16, 82]
1110. Pfaltzgraff, John A. *Univalence and quasiconformal extension of holomorphic maps in C^n*. Notices A. M. S. 20 (Jan. 1973), Abstract 701-32-3, pg. A-114; Bulletin Amer. Math. Soc. 80 (1974), 543-544. [4, 68]
1111. Pfaltzgraff, J. *Univalence of the integral $(f'(z))^\lambda$*. Bull. London Math. Soc. 7(1975), no. 3, 254-256. [85]; MR 52 #8410
1112. Pfaltzgraff, J. A.; Pinchuk, B. *Constrained extremal problems*

for classes of meromorphic functions. Bull. Amer. Math. Soc., vol. 75, no. 2, March, 1969, pp. 379-384; MR 39 #444. [9, 24, 58]

1113. Pfaltzgraff John A.; Pinchuk, Bernard. *A variational method for classes of meromorphic functions.* J. Analyse Math. 24(1971), 101-150; MR 43 #7613. [2, 4, 6, 9, 10, 16, 23, 24, 58, 71]

1114. Pfluger, A. *The convexity of certain sections of n-bodies of coefficients of univalent functions.* (Russian) Certain Problems of Mathematics and Mechanics (Russian), pp. 233-241. Izdat. "Nauka", Leningrad, 1970; MR 45 #7040. [29]

1115. Pfluger, A. *Lineare Extremalprobleme bei schlichter Funktionen.* Ann. Acad. Sci. Fenn. Ser. AI No. 489(1971), 32 pp.; MR 45 #5337. [24]

1116. Phelps, Dean. *On a coefficient problem in univalent functions.* Trans. Amer. Math. Soc. 143(1969), 475-485; MR 40 #2844. [29]

1117. Piłat, Barbara. *Sur une classe de fonctions normées univalentes dans le cercle unité.* (Polish and Russian summaries) Ann. Univ. Mariae Curie-Sklodowska Sect. A 17(1963), 69-74(1965); MR 33 #2803. [32]

1118. Piłat, Barbara. *On typically real functions with Montel's normalization.* (Polish and Russian summaries) Ann. Univ. Mariae Curie-Sklodowska Sect. A 18(1964), 53 -72(1967); MR 38 #317. [13, 23, 66]

1119. Pinchuk, B. *Extremal problems in the class of close-to-convex functions.* Bull. Amer. Math. Soc. 72(1966), 1014-1017; MR 34 #332. [2, 5, 6, 10, 23, 56, 64]

1120. Pinchuk, B. *Extremal problems in the class of close-to-convex functions.* Trans. Amer. Math. Soc 129(1967), 466-478; MR 36 #370. [2, 5, 6, 10, 23, 24, 56, 64]

1121. Pinchuk, B. *A variational method for functions of bounded boundary rotation.* Trans. Amer. Math. Soc. 138(1969), 107-113; MR 38 #6042. [24, 71]

1122. Pinchuk, B. *On starlike and convex functions or order α.* Duke Math. J. 35(1968), 721-734; MR 37 #6454. [1, 6, 10, 24, 36, 63, 64, 38]

1123. Pinchuk, B. *Faber polynomials for starlike functions.* Journal of Math. and Mechanics 19(1970), 981-990; MR 41 #5613. [6, 23, 24, 53]

1124. Pinchuk, Bernard. *Functions of bounded boundary rotation.* Israel J. Math. 10(1971), 6-16; MR 46 #338. [2, 4, 5, 6, 10, 11, 12, 16, 19, 23, 62, 71, 73, 85]

1125. Pinchuk, B. *The Hardy class of functions of bounded boundary* rotation. Proc. Amer. Math. Soc. 38(1973), 355-360; MR 47 #2070. [3, 4, 23, 71, 76, 77]

1126. Pinchuk, B. *Integral means of analytic functions.* Israel J. Math. 17(1974), 105-107; MR 49 #10868. [1, 3, 19]

1127. Piranian, G. *The shape of level curves.* Proc. Amer. Math. Soc. 17(1966), 1276-1279; MR 33 #7498. [35]

1128. Piranian, G. *Bounded Functions with Large Circular Variation.* Proc. A. M. S. 19(1968), 1255-1257; MR #6464. [7, 22]

1129. Pirl, Udo. *Über die geometrische Gestalt eines Extremalkontinuums aus der Theorie der konformen Abbildung.* Math. Nachr. 39(1969), 297-312; MR 40 #7432.

1130. Plaskota, Wieslaw. *The range of the functional $K[F(\delta), \bar{F}(\delta), F'(\delta), \bar{F}'(\delta)]$ in the family of functions meromorphic and schlicht in the circle $|z| > 1$.* (Polish. Russian summary) Zeszyty Nauk Univ. Lódźk. Nauki Mat. Przyrod. Ser. II Zeszyt 29(1968), 63-88; MR 40 #334. [9, 24, 29]

1131. Plaskota, W. *Le domaine de variation de la fonctionnelle $K[F(\zeta), \bar{F}(\zeta), F'(\zeta), \bar{F}'(\zeta)]$ dans la famille des fonctions meromorphes et univalentes dans le cèrcle $|z| > 1$.* Ann. Polon. Math. 21(1969), 195-215; MR 39 #7083. [9, 24]

1132. Plaskota, W. *On the coefficients of some families of regular functions.* (Loose Russian summary) Bull. Acad. Polon. Sci. Ser. Sci. Math. Astronom. Phys. 17(1969), 715-718; MR 41 #457. [6, 36]

1133. Plaskota, W. *Limitation des coefficients dans une famille de fonctions holomorphes dans le cèrcle $|z| < 1$.* Ann. Polon. Math. 24 (1970/71), 65-70; MR 42 #6215. [6, 36]

1134. Plaskota, Wieslaw. *Sur quelques problèmes extrémaux dans les familles des fonctions générées par les fonctions de Carathéodory.* Ann. Polon. Math. 25(1971/72), 139-144; MR 46 #344. [6, 11, 36]

1135. Pletneva (Èzrohi), T. G. *Radii of convexity and curvature bounds for the level curves of certain classes of functions regular on the unit disc.* (Russian) Mathematical Physics, No. 6(1969) (Russian), pp. 157-168. Naukova Dumka, Kiev, 1969; MR 41 #7083. [2, 12, 35]

1136. Pletneva, T. G. *Certain extremal problems in classes of analytic functions in the disc $|z| < 1$ and in the annular domain $1 < |\zeta| < \infty$.* (Ukrainian. English and Russian summaries) Dopovidi Akad. Nauk Ukrain. RSR Ser A 1972, 137-141, 188; MR 46 #3766. [48]

1137. Pohilevič, V. A. *On a theorem of M. Biernacki.* (Ukrainian. Russian and English summaries) Dopovidi Akad. Nauk Ukrain. RSR

1965, 423-425; MR 32 #5863. [4, 16]
1138. Pohilevič, V. A. *On a theorem of M. Biernacki in the theory of univalent functions.* (Russian) Ukrain. Mat. Z. 17(1965), no. 4, 63-71; MR 33 #5869. [23, 24]
1139. Pohilevič, V. A. *Extremal properties of certain classes of univalent functions.* (Russian) Ukrain. Mat. Z. 19 (1967), no. 2, 49-59; MR 35 #4392. [2, 5, 10, 12, 68]
1140. Pohilevič, V. A. *On some new special classes of schlicht functions.* (Ukrainian. Russian and English summaries) Dopovidi Akad. Nauk Ukrain. RSR Ser. A 1968, 620-623; MR 38 #6044. [2, 6]
1141. Pohilevič, V. A. [Pohilevič, V. O.] *Meromorphic close-to-convex functions in the disc.* (Russian). Ukrain. Mat. Z. 21(1969), 50-59; MR 40 #337. [2, 9, 12, 25]
1142. Pohilevič, V. A. [Pohilevič, V. O.] *The equivalence of two classes of univalent functions.* (Russian) Teor. Funkcii Funkcional. Anal. i Prilozen. Vyp. 8(1969), 57-62; MR 41 #7088. [5]
1143. Pohilevič, V. O. *Bounds of the convexity and of the curvature of the vel of the curvature of the level curves in certain classes of symmetric analytic functions.* (Ukrainian. English and Russian summaries). Dopovidi Akad. Nauk Ukrain. RSR Ser. A 1971, 422- 427, 478; MR 45 #554. [35]
1144. Pohilevič, V. O. *Meromorphic almost convex functions in a disc.* (Ukrainian. English and Russian summaries). Dopovidi Akad. Nuak Ukrain. RSR Ser. A 1971, 518-522, 574; MR 45 #8845. [2, 12, 35]
1145. Palka, Janina. *Sharp estimates of $|p(w)|$, arg $(p(w)/w)$, $|p'(w)|$, arg $p'(w)$ in a class of univalent polynomials.* Ann. Univ. Mariae Curie-Sklodowska, Sect. A, vols. 22/23/24(1968/1969/1970), p. 137. [32]
1146. Polomosnova, R. S. *Variational problems for certain classes of functions, schlicht and non-schlicht in doubly connected domains.* (Russian) Trudy Tomsk. Gos. Univ. Ser. Meh.-Mat. 175(1964), Vyp. 2, 59-72; MR 34 #6063. [24, 60]
1147. Polomosnova, R. S. *Extremal properties of starlike mappings of an annulus.* (Russian). Trudy Tomsk. Gos. Univ. Ser. Meh.-Mat. 189(1966), 168-175; MR 37 #2973. [6, 48]
1148. Polomosnova, R. S. *Boundary functions for classes of typically real functions and functions with positive real part in an annulus.* (Russian) Trudy Tomsk. Gos. Univ. Ser. Meh.-Mat. 200(1968), 173-180; MR 41 #451. [2, 13, 29, 48]
1149. Pommerenke, Ch. *Über nahezu konvexe analytische Funktionen.*

Arch. Math. (Basel)16(1965), 344-347; MR 32 #7733. [5, 7, 18, 74, 75]
1150. Pommerenke, V. C. *Über die subordination analytischer Funktionen.* J. Reine Angew. Math. 218(1965), 159-173; MR 31 #4900. [1, 6]
1151. Pommerenke, C. *Konforme Abbildung und Fekete-Punkte* {Conformal mapping and extremal points}. Math. Z. 89(1965), 422-438; MR 34 #6056. [9, 53]
1152. Pommerenke, Ch. *On the Loewner differential equation.* Michigan Math. J. 13(1966), 435-443; MR 34 #6064. [1, 24]
1153. Pommerenke, Ch. *On the coefficients and Hankel determinants of univalent functions.* J. London Math. Soc. 41(1966), 111-122; MR 32 #2575. [6, 8, 16, 23, 27, 33]
1154. Pommerenke, Ch. *On the coefficients of univalent functions.* J. London Math. Soc., 42(1967), 471-474; MR 36 #5329. [54]
1155. Pommerenke, C. *Relations between the coefficients of a univalent function.* Invent. Math. 3(1967), 1-15; MR 36 #3972. [8, 9, 22, 30, 33, 39, 45, 54, 58, 68]
1156. Pommerenke, C. *On the Hankel determinants of univalent functions.* Mathematika 14(1967), 108-112; MR 35 #6811. [54]
1157. Pommerenke, Ch. *Über die Verteilung der Fekete-Punkte.* Math. Ann. 168(1967), 111-127; MR 34 #6057.
1158. Pommerenke, Ch. *On the logarithmic capacity and conformal mapping.* Duke Math. J. 35(1968), 321-325; MR 37 #4246. [9, 58]
1159. Pommerenke, Ch. *On the growth of univalent functions.* Mich. Math. J. 15(1968), 485-494; MR 38 #4668. [68]
1160. Pommerenke, Ch. *On the Grunsky inequalities for univalent functions.* Arch. Rational Mech. Anal. 35(1969), 234-244; MR 40 #2845. [53]
1161. Pommerenke, Ch. *On a variational method for univalent functions.* Michigan Math. J. 17(1970), 1-3; MR 41 #452. [24]
1162. Pommerenke, Ch. *Normal functions.* Proc. NRL Conf. on Classical Function Theory (Math. Res. Center, Naval Res. Lab., Washington, D. C., 1970), pp. 77-93. Math. Res. Center, Naval Res. Lab., Washington, D. C., 1970; MR 48 #4322. [19, 44]
1163. Pommerenke, Ch. *Estiamtes for normal meromorphic functions.* Ann. Acad. Sci. Fenn. Ser. A I No. 476(1970), 10 pp.; MR 44 #2928. [22, 43, 46, 61, 68]
1164. Pommerenke, Ch. *On Bloch Functions.* J. London Math. Soc. (2) 2(1970), 689-695; MR 44 #1799. [16, 19, 62]
1165. Pommerenke, Ch. *On the growth of the coefficients of analytic*

functions. J. London Math. Soc. (2)5(1972), 624-628; MR 47 #2050. [22, 61]
1166. Pommerenke, Ch. *Problems in complex function theory.* Bull. London Math. Soc. 4(1972), 354-366. [16]
1167. Pommerenke, Ch. *On the boundary behavior of normal functions.* Proceedings of the Symposium on Complex Analysis (Univ. Kent, Canterbury, 1973), pp. 113-114. London Math. Soc. Lecture Note Ser., No. 12, Cambridge Univ. Press, London, 1974. [44, 62]
1168. Pommerenke, C. *Univalent functions with a chapter on quadratic differentials by Gerd Jensen.* Studia Mathematical Mathematische Lehrbucher, Band XXV. Vandenhoeck & Ruprecht, Gottingen, 1975, 376 pp. [44]
1169. Ponomar'ov, S. P. *On the monogeneity of symmetrically differentiable functions.* (Ukrainian) Teoret. Prikl. Mat. Vip. 2(1963), 11-14; MR 33 #4250.
1170. Poole, J. T. *Coefficient extremal problems for schlicht functions.* Trans. Amer. Math. Soc. 121(1966), 455-474; MR 32 #993. [6, 9, 24, 25, 36, 54]
1171. Poole, J. T. *On starlike functions.* Proc. Amer. Math. Soc. 19 (1968), 495-500; MR 36 #6602. [6, 9, 16, 25, 36]
1172. Poole, J. T. *A note on the coefficients of univalent functions.* Ann. Polon. Math. 20(1968), 91-93; MR 36 #3973. [52]
1173. Popov, V. I. *On the method of parametric representations.* (Russian) Trudy Tomsk. Gos. Univ. Ser. Meh.-Mat. 175(1964), Vyp. 2, 73-77; MR 31 #326. [22, 24, 29, 39]
1174. Popov, V. I. *The range of a system of functionals on the class S.* (Russian) Trudy Tomsk. Gos. Univ. Ser. Meh.-Mat. 182(1965), vyp. 3, 106-132; MR 33 #7521. [29]
1175. Popov, V. I. *A certain variational formula for univalent functions.* (Russian) Trudy Tomsk. Gos. Univ. Ser. Mch.-Mat. 200 (1968), 181-183; MR 41 #453. [24]
1176. Popov, V. I. *L. S. Pontrjagin's maximum principle in the theory of univalent functions.* (Russian) Dokl. Akad. Nauk SSSR 188 (1969), 532-534; MR 40 #7435.
1177. Prohorov, D. V. *A certain geometric property of functions that are starlike of order α.* (Russian) Mat. Zametki 10(1971), 287-293; MR 45 #528 [English transl. Math. Notes 10(1971), 597-600]. [6, 49]
1178. Prohorov, D. V. *A generalization of a class of close-to-convex functions.* Mat. Zametki 11(1972), 509-516. [Math. Notes 11 (1972), 311-315]; MR 46 #9325. [5, 70, 74]

1179. Prohorov, D. V. *The functions of I. E. Bazilevič's class.* (Russian) Zap. Naucn. Sem. Leningrad. Otdel. Mat. Inst. Steklov. (LOMI) 44(1974), 127–130, 187; MR 51 #10605. [70]

1180. Prohorov, D. V. *The geometric characterization of certain classes of univalent functions.* (Russian. English summary) Vestnik Leningrad. Univ. No. 13 Mat. Meh. Astronom. Vyp. 3(1974), 51–55, 156; MR 50 #7500. [5, 7]

1181. Prohorov, D. V. *The geometric characcterization of functions from Bazilevic subclasses.* (Russian) Izv. Vyss. Ucebn. Zaved Matematika 1975, no. 2(153), 130–132. [70]

1182. Qiu Hua-ji [Ch'iu Hua-chi]. *Some questions on the mapping of surfaces by bounded analytic functions.* (Chinese) Shuxue Jinzhan 8(1965), 243–250; MR 38 #2293. [39]

1183. Quine, John R. *On univalent polynomials.* Notices Amer. Math. Soc. 20(1973), Abstract 709-B38, p. A-666. [32]

1184. Quine, J. R. *Some topological theorems relating to close-to-convex functions.* Notices Amer. Math. Soc. 23, January 1976, Abstract 731-30-35, p. A-105. [5]

1185. Rahman, Q. I. *On a property of rational functions.* II. Proc. Amer. Mtah. Soc. 40(1973), 143–145; MR 50 #10214. [32]

1186. Rahman, Q. I.; Mohammad, Q. G. *Remarks on Schwarz's lemma.* Pacific J. Math. 23(1967), 139–142; [32, 37]

1187. Rahman, Q. I.; Stankiewicz, J. *Differential inequalities and local valency.* Pacific J. Math. 54(1974), 165–181; MR 51 #10593. [2, 11, 12, 14, 21, 22, 30, 32, 89]

1188. Rahman, Q. I.; Turán, P. *On a property of rational functions.* Ann. Univ. Sci. Budapest. Eötvös Sect. Math. 16(1973), 37–45 (1974); MR 50 #10213.

1189. Ratti, J. S. *The radius of univalence of certain analytic functions.* Math. Z. 107(1968), 241–248; MR 38 #4669. [2, 6, 10, 11, 43]

1190. Ratti, J. S. *The radius of convexity of certain analytic functions.* Ind. J. Pure Appl. Math. 1(1970), no. 1, 30–36; MR 41 #1992. [6, 10, 12]

1191. Raymon, L.; Tepper, D. E. *Star center points of starlike functions.* J. Austral. Math. Soc. 19(1975), part 4, 505–510. [6]

1192. Read, G. A. *Univalent derivatives of entire functions.* J. London Math. Soc. (2)1(1969), 189–192; MR 40 #2860. [90]

1193. Reade, Maxwell O. *Generalizations of alpha convex functions.* I. Notices Amer. Math. Soc. 21, August 1974, Abstract 74T-B138, p. A-486 [69]

1194. Reade, Maxwell O. *Generalizations of Bazilevič functions.* I.

Notices Amer. Math. Soc. 21, June 1974, Abstract 74T-B111, p. A-438. [70]
1195. Reade, M. O. *On functions of bounded boundary rotation.* Preliminary report. Notices Amer. Math. Soc., January 1975, Abstract 720-30-7, p. A-119. [71]
1196. Reade, M. O.; Mocanu, P. *The radius of α-convexity of starlike functions.* Notices, A. M. S., vol. 19, no. 1, isssue no. 135, January 1972 (Abstract 691-30-1), p. A-110 [6, 12, 69]
1197. Reade, M. O.; Mocanu, P. T.; Zlotkiewicz, E. *On the funcitonal $[f(a)/f'(b)]$ in S_R.* Notices Amer. Math. Soc. 21, January 1974, Abstract 711-30-25, p. A-125. [29, 39]
1198. Reade, M. O.; Ogawa, S.; Sakaguchi, K. *The radius of convexity for a certain class of analytic functions.* J. Nara Gakugei Univ. 13 (1965), 1-3; MR 34 #1509. [2, 11, 12, 43, 56]
1199. Reade, M. O.; Umezawa, T. *Some criteria for the multivalence of certain analytic functions.* Colloq. Math. 16(1967), 23-26; MR 35 #4397. [4, 5, 14, 85]
1200. Reade, M. O.; Umezawa, T. *An inequality for univalent functions due to Dvorak.* (Czech summary) Casopis Pest. 96(1971), 265-267, 301; MR 45 #532a. [16, 42, 45, 68]
1201. Reade, M. O.; Złotkiewicz, E. J. *The Koebe constant for a class of bounded functions.* Notices, A. M. S., vol. 17, no. 7, issue no. 125, November, 1970 (Abstract 680-B1), p. 1045. [19, 22, 24]
1202. Reade, M. O.; Złotkiewicz, J. *On the equation $f(z) = pf(a)$ in certain classes of analytic functions.* Ann. Univ. Mariae Curie-Sklodowska Sect. A, vols. 22/23/24(1968/1969/1970), p. 151-153; MR 49 #5333; Mathematica (Cluj) 13(36)6(1971), 281-286; MR 48 #11474. [29]
1203. Reade, Maxwell O.; Złotkiewicz, Eligiusz, J. *On univalent functions with two preassigned values.* Proc. Amer. Math. Soc. 30 (1971), 539-544; MR 44 #422. [6, 19, 22, 41]
1204. Reade, M. O.; Złotkiewicz, E. J. *On a theorem of Kaczmarski concerning the equation $f(z) = pf(a)$.* Notices, A. M. S., vol. 18, no. 1, issue no. 127, January 1971 (Abstract 682-30-2, Preliminary report), p. 144. [6, 10, 13]
1205. Reade, M. O.; Złotkiewicz, E. J. *Koebe sets for univalent functions with two preassigned values.* Bull. A. M. S. 77(1971), 103-105; MR 42 #3267. [6. 19, 22, 24, 41]
1206. Reade, M. O.; Złotkiewicz, E. *On values omitted by univalent functions with two preassigned values.* Compositio Math. 24 (1972), 355-358; MR 47 #8831. [19, 24]

1207. Redding, R. W. *Typically real functions of order α*. Notices, A. M. S., vol. 19, no. 5, issue 139, August, 1972 (Abstract 696-30-3), p. A-636. [13, 23, 51, 66]

1208. Red'kov, M. I. *Concerning the domain of values of a certain functional in the class S*. (Russian) Trudy Tomsk. Gos. Univ. 163 (1963), 44-47. [29]

1209. Red'kov, M. I. *On the coefficients of bounded univalent functions*. (Russian) Izv. Vyss. Ucebn. Zaved. Matematika 1965, no. 1 (44), 114-122; MR 31 #329. {Amer. Math. Soc. Transl. (2) vol. 88 (1970)}. [22, 24, 29]

1210. Red'kov, M. I. *Extremal problems in the class $S_1(\beta)$*. (Russian) Trudy Tomsk. Gos. Univ. Ser. Meh.-Mat. 189(1966), 176-183; MR 37 #1582. [22, 29]

1211. Red'kov, M. I. *The range of values of a certain system of functionals for bounded schlicht funcitons*. (Russian) Trudy Tomsk. Gos. Univ. Ser. Meh.-Mat. 210(1969), 83-96; MR 43 #5018. [22, 29]

1212. Reich, E.; Schiffer, M. *Estimates for the transfinite diameter of a continuum*. Math. Z. 85(1964), 91-106; MR 30 #4921. [19, 24]

1213. Ren Fu-yao [Jen Fu-yao]. *Extensions of some theorems of Robinson and Goluzin*. Chinese Math.-Acta 8(1966), 780-787 (1967); MR 37 #398. [49, 68]

1214. Reshetnyak, Yu. G. *Extremal properties of mappings with bounded distortion*. Siberian Math. J. 10(1969), p. 1300 [71]

1215. Reshetnyak, Yu. G. *The local structure of mappings with bounded distortion*. Siberian Math. J. 10(1969), p. 1311. [71]

1216. Revjakov, M. I. *On values omitted by univalent functions*. (Russian). Trudy Mat. Inst. Steklov. 94(1968), 123-129; MR 37 #2974. {Translated from the Russian: Extremal problems of the Geometric Theory of Functions (A. M. S., 1969)}. [19]

1217. Revjakov, M. I. *Certain covering theorems for univalent functions*. (Russian). Izv. Vyss. Ucebn. Zaved. Matematika 1968, no. 7(74), 85-92; MR 37 #6455. [19]

1218. Robertson, J. M. *A local mean value theorem for the complex plane*. PRoc. Edinburgh Math. Soc. (2)16(1968/69), 329-331; MR 41 #3715. [68]

1219. Robertson, M. S. *The Generalized Bieberbach Conjecture for Subordinate Functions*. Michigan Math. J. 12(1965), 421-429; MR 32 #2576. [1, 5, 6, 10, 16, 42]

1220. Robertson, M. S. *A generalization of the Bieberbach coefficient problem for univalent functions*. Michigan Math. J. 13(1966),

185-192; MR 33 #269. [5, 8, 16, 38, 41, 42, 79]
1221. Robertson, M. S. *Univalent functions $f(z)$ for which $zf'(z)$ is spirallike.* Mich. Math. J. 16(1969), 97-101; MR 39 #5785. [2, 4, 6, 10, 16, 31, 56]
1222. Robertson, M. S. *Power series with multiply monotonic coefficients.* Michigan Math. J. 16(1969), 27-31; MR 39 #2966. [4, 5, 20]
1223. Robertson, M. S. *Coefficients of functions with bounded boundary rotation.* Canad. J. Math. 21(1969), 1477-1482; MR 41 #458. [4, 12, 41, 71, 76, 77, 79]
1224. Robertson, M. S. *Quasi-subordinate functions.* Mathematical Essays Dedicated to A. J. Macintyre, pp. 311-330. Ohio Univ. Press, Athens, Ohio, 1970; MR 42 #7885. [1, 3, 6, 13, 16]
1225. Robertson, M. S. *Variational formulae for several classes of analytic functions.* Math. Z. 118(1970), 311-319; MR 43 #7614. [4, 5, 6, 10, 24, 71]
1226. Robertson, M. S. *Quasi-Subordination and Coefficient Conjectures.* Bull., A. M. S. 76(1970), 1-9; MR 40 #4441. [1, 3, 6, 10, 14, 16, 42, 54]
1227. Robertson, M. S. *The sum of univalent functions.* Duke Math. J. 38(1970), 411-419; MR 41 #8649. [10, 11, 13, 43, 73, 82, 85]
1228. Robertson, M. S. *A distortion theorem for analytic functions.* Proc. Amer. Math. Soc. 28(1971), 551-556; MR 43 #7615. [2, 4, 16]
1229. Rosenbloom, P. C. *Conformal mapping of nearly circular domains and Loewner's differential equation.* Inequalities, III (Proc. Third Sympos., Univ. California, Los Angeles, Calif., 1969), pp. 301-310. Academic Press, New York, 1972; MR 49 #548. [22, 61]
1230. Ross, George G. *On the computation of some Grunsky coefficients relevant to the Bieberbach conjecture.* Math. Comp. 25 (1971), 733-741; addendum, ibid. 25(1971), no. 116, Loose Microfiche suppl. A14-B5; MR 46 #9329. [42, 53]
1231. Royster, W. C. *On the univalence of a certain integral.* Michigan Math. J. 12(1965), 385-387; MR 32, p. 232 [4, 5, 85]
1232. Royster, W. C. Proceedings of the Symposium on univalent functions and related topics, Department of mathematics, Univ. of Kentucky, May 18-21, 1967. [44]
1233. Royster, W. C. *On the derivative of bounded functions.* Indian J. Math. 11(1969), 141-143; MR 41 #2020. [22, 61]
1234. Royster, W. C. *Convex meromorphic functions.* Mathematical Essays Dedicated to A. J. Macintyre, pp. 331-339. Ohio Univ. Press, Athens, Ohio, 1970; MR 42 #7883. [2, 6, 7, 9, 10, 23, 25]
1235. Royster, W. C. *Univalent functions convex in one direction.*

Notices, A. M. S., vol. 19, no. 1, issue no. 135, January, 1972 (Abstract 691-30-31), p. A-119. [7, 12, 41, 78]

1236. Royster, W. C.; Suffridge, T. J. *Typically real polynomials.* Notices, A. M. S., vol. 16, no. 6, issue no. 116, October 1969, p. 968 (Abstract 69T-B173). [13, 32, 51]

1237. Royster, W. C.; Suffridge, T. J. *Typically real polynomials.* Ann. Univ. Mariae Curie-Sklodowska, Sect. A, vols. 22/23/24(1968/ 1969/1970), p. 161. Publ. Math. Debrecen 17(1970), 307-312 (1971); MR 46 #2065; MR 49 #5357. [13, 32, 51]

1238. Rubel, L. A. *Some applications of the Gauss-Lucas theorem.* Enseignement Math. (2)12(1966), 33-39; MR 34 #4457. [32]

1239. Rubinstein, Zalman. *On a problem of Ilyeff.* Pacific J. Mth. 26 (1968), 159-161; MR 38 #6034. [32]

1240. Rubinstein, Zalman. *On analytic functions satsifying the mean value theorem and a conjecture of W. G. Dotson.* Math. Mag. 42 (1969), 256-259; MR 41 #435.

1241. Rubinstein, Zalman. *On the multivalence of a class of meromorphic functions.* Pacific J. Math., 38(1971), No. 3, pp. 771-784; MR 46 #3773. [6, 9, 14, 16, 32, 43]

1242. Rubinstein, Z.; Shaffer, Dorothy Browne. *On the multivalence and order of starlikeness of a class of meromorphic functions.* Israel J. Math 17(1974), 302-314; MR 50 #10228. [6, 9, 11, 12, 16, 22, 32, 43]

1243. Rung, D. C. *Inequalities on holomorphic functions omitting one value.* Notices, Amer. Math. Soc. (Abstract 630-166), vol. 13, no. 1, issue no. 87, January, 1966, p. 108. [19, 68, 90]

1244. Ruscheweyh, S. *On the radius of univalence of the partial sums of convex functions.* Bull. London Math. Soc. 4(1972), 367-369; MR 47 #5244. [10, 20, 43]

1245. Ruscheweyh, Stephan. *Über die Faltung schlichter Funktionen.* Mat. Z. 128(1972), 85-92; MR 47 #7017. [5, 6, 10, 47]

1246. Ruscheweyh, Stephan. *Eine Invarianzeigenschaft der Basilevič-Funktionen.* Math. Z. 134(1973), 215-219; MR 48 #11476. [5, 6, 10, 47, 70]

1247. Ruscheweyh, Stephan. *Nichtlineare Extremal probleme für holomorphe Stieltjesintegrale.* Math. Z. 142(1975), 19-23; MR 51 #10606. [13, 51, 82]

1248. Ruschewehy, Stephan. *New criteria for univalent functions.* Proc. Amer. Math. Soc. 49(1975), 109-115; MR 51 #3418. [4, 6, 10, 16, 36, 38, 63, 64, 68]

1249. Ruscheweyh, Stephan. *Duality for Hadamard products with ap-*

plications to extremal problems for functions regular in the unit disc. Trans. Amer. Math. Soc. 210(1975), 63-74. [2, 5, 6, 12, 20, 22, 42, 43, 47, 56, 74, 88]; MR 52 #3508

1250. Ruscheweyh, S.; Sheil-Small, T. *Hadamard products of schlicht functions and the Polya-Schoenberg conjecture.* Comment. Math. Helv. 48(1973), 119-135; MR 48 #6393. [1, 2, 4, 5, 6, 10, 20, 29, 47, 63, 64]

1251. Ruscheweyh, S.; Wirths, K.-J. *Über die Faltung schlichter Funktionen.* II. Math. Z. 131(1973), 11-23; MR 47 #8836. [11, 12, 22, 30, 47]

1252. Ruscheweyh, St.; Wirths, K.-J. *Über die Koeffizienten spezieller schlichter Polynome.* Ann. Polon. Math. 28(1973), 341-355; MR 48 #11475. [32]

1253. Ryan, F. B. *On the Poisson-Stieltjes representation for functions with bounded real part.* Michigan Math. J. 17(1970), 301-310; MR 43 #2221. [2, 22, 62]

1254. Ryff, J. V. *Subordinate H^p Functions.* Duke Math. J. 33(1966), 347-354; MR 33 #289. [1, 3, 16, 18, 37, 49]

1255. Sakaguchi, Koichi. *On certain multivalent functions.* J. Nara Gakugei Univ. 8(1959), no. 2, 19-23; MR 31 #3590. [14, 49, 65]

1256. Sakaguchi, K. *A variational method for functions with positive real part.* J. Math. Soc. Japan 16(1964), 287-297; MR 31 #1375. [2, 21, 23, 24]

1257. Sakaguchi, Koichi. *The radius of convexity for a certain class of regular functions.* J. Nara Gakugei Univ. Natur. Sci. 12(1964), 5-8; MR 34 #1510. [12]

1258. Sakaguchi, Koichi. *On extremal problems in the classes of functions with positive real part and typicaly real ones.* I, II. Bull. Nara. Univ. Ed. Natur. Sci. 17(1969), no. 2, 1-12; ibid. 18(1969), no. 2, 1-6; MR 44 #5471. [2, 13, 72]

1259. Sakaguchi, Koichi. *A property of convex functions and an application to criteria for univalence.* Bull. Nara Univ. Ed. Natur. Sci. 22(1973), no. 2, 1-5; MR 50 #13486. [4, 5, 10, 29]

1260. Sakaguchi, K.; Watanabe, S. *On close-to-convex functions.* J. Nara Gakugei Univ. Natur. Sci. 14(1966), 7-12; MR 34 #6065. [4, 5, 12, 74]

1261. Sakai, M. *On basic domains of extremal functions.* Kodai Math. Sem. Re. 24(1972), 251-258; MR 47 #8838. [9]

1262. Sanders, R. W. *The starlike radius for classes of regular bounded functions.* Proc. Amer. Math. Soc. 54(1976), 217-220. [11, 19, 22, 61]

1263. Sansone, Giovanni; Gerretsen, Johan. *Lectures on the theory of functions of a complex variable*. II: Geometric theory. Wolters-Noordhoff Publishing, Groningen, 1969. X + 700 pp.; MR 41 #3714. [44]
1264. Sato, Tsuneo. *Some variations of univalence radius*. J. College Arts Sci. Chiba Univ. 4(1965), no. 3, 241–243; MR 36 #6603. [22, 43, 82]
1265. Schiffer, M. *Univalent functions whose n first coefficients are real*. J. D'analyse Math. 18(1967), 329–349; MR 35 #5603. [39, 42]
1266. Schiffer, M. *On the coefficient problem for univalent functions*. Amer. Math. Soc. Trans. 134(1968), 95–101; MR 37 #4249. [24]
1267. Schiffer, M. *Some distortion theorems in the theory of conformal mapping*. Accademia Nazionale Dei Lincei. Rendiconti 10(1971), 1–20. [68]; MR 43 #3435
1268. Schiffer, M. *Inequalities in the theory of univalent functions*. Inequalities, III(Proc. Third Sympos., Univ. California, Los Angeles, Calif., 1969), 311–319. Academic Press, New York, 1972; MR 49 #558. [44]
1269. Schiffer, M.; Schmidt, H. G. *A new set of coefficient inequalities for univalent functions*. Arch. Rational Mech. Anal. 42(1971), 346–368; MR 48 #11479. [24, 53]
1270. Schiffer, M.; Schober, G. *Coefficient problems and generalized Grunsky inequalities for Schlicht functions with quasiconformal extensions*. Arch. Rational Mech. Anal. 60(1975/76), no. 3, 205–228. [53, 54]
1271. Schiffer, M.; Tammi, O. *The fourth coefficient of a bounded real univalent function*. Ann. Acad. Sci. Fenn. Ser. A I No. 354(1965), 32 pp.; MR 35 #4394. [22, 24, 30, 39]
1272. Schiffer, M.; Tammi, O. *On the fourth coefficient of bounded univalent functions*. Trans. Amer. Math. Soc. 119(1965), 67–78; MR 32 # [22, 30, 53, 54]
1273. Schiffer, M.; Tammi, O. *A method of variations for functions with bounded boundary rotation*. J. Analyse Math. 17(1966), 109–144; MR 35 #5601. [6, 10, 23, 24, 35, 36, 38, 71, 76, 77]
1274. Schiffer, M.; Tammi, O. *On the fourth coefficient of univalent functions with bounded boundary rotation*. Ann. Acad. Sci. Fenn. Ser. A I no. 396(1967), 26 pp.; MR 35 #3052. [4, 24, 71, 77]
1275. Schiffer, M.; Tammi, O. *On bounded univalent functions which are close to identity*. Ann. Acad. Sci. Fenn. Ser. A I No. 435(1968), 26 pp.; MR 39 #4372. [22, 30, 53]
1276. Schiffer, M.; Tammi, O. *On the coefficient problem for bounded*

univalent functions. Trans. Amer. Math. Soc. 140(1969), 461–474; MR 39 #7088. [4, 16, 22, 24, 30, 53]
1277. Schiffer, M.; Tammi, Olli. *A Green's inequality for the power matrix.* Ann. Acad. Sci. Fenn. Ser. A I No. 501(1971), 15 pp.; MR 45 #5339. [22, 30, 53]
1278. Schnack, D. H. *A coefficient problem for a class of meromorphic univalent functions.* (Abstract), Notices A. M. S., vol. 16, no. 1, issue no. 111, Jan. 1969, p. 215. [9, 25, 39]
1279. Schneider, W. J. *A uniqueness theorem for conformal maps.* Notices A. M. S. (Abstract 626-27), vol. 12, no. 6, issue no. 84, October, 1965, p. 698.
1280. Schober, Glenn. *Univalent functions-selected topics.* Lecture notes in mathematics, vol. 478. Springer-Verlag, Berlin-New York, 1975. V + 200 pp. [44]
1281. Schoenberg, I. J. *Extrema for gap power series of positive real part.* J. Analyse Math. 14(1965), 379–391. [2, 21]
1282. Scilard, Karl [Szilard, Karl]. *On a known generalization of the concept of convexity for plane regions and its application to distortion theorems in the theory of conformal mappings.* (Russian). Contemporary Problems in Theory Anal. Functions (Internat. Conf., Erevan, 1965) (Russian), pp. 277–281. Izdat. "Nauk", Moscow, 1966; MR 34 #6051. [10, 68]
1283. Srivastava, R. S. L. *Univalent spiral functions.* Topics in analysis (Colloq. Math. Anal., Jyvaskyla, 1970), pp. 327–341. Lecture Notes in Math., Vol. 419, Springer, Berlin, 1974. [6, 44]; MR 52 #3505
1284. Selljahova, T. N.; Sobolev, V. V. *The mutual growth of the coefficients of functions that are univalent in the half-plane.* (Russian) Dokl. Akad. Nauk SSSR 218(1974), 768–770; MR 50 #7503. [57]
1285. Sevcenko, V. I. *On the question of the univalence of polynomial mappings.* (Russian. Geogian and English summaries) Gamoqeneb. Math. Inst. Sem. Mohsen. Anotacie. Vyp. 8(1973), 5–8; MR 49 #9170. [32]
1286. Shaffer, Dorothy Browne. *Distortion theorems for lemniscates and level loci of Green's functions.* J. Analyse Math. 17(1966), 59–70; MR 36, p. 82 [35]
1287. Shaffer, Dorothy Browne. *The curvature of level curves of lacunary polynomials and Green's function.* Notices A. M. S., vol. 17, no. 2, issue no. 120, Feb. (1970). Abstract, pg. 393. [35]
1288. Shaffer, D. B. *The curvature of level curves.* Trans. Amer. Math. Soc. 158(1971), 143–150. [35]; MR 43 #3428

1289. Shaffer, Dorothy Browne. *Analytic Functions $f(z)$ with $\operatorname{Re} f(z) > 1/2$ and applications.* Notices, A. M. S., vol. 19, no. 6, issue 140, October, 1972 (Abstract 72T-B287), p. A-704. [2, 9, 11, 12]
1290. Shaffer, Dorothy Browne. *On bounds for the derivative of analytic functions.* Proc. Amer. Math. Soc. 37(1973), 517-520; MR 46 #9357. [2, 22, 56, 61]
1291. Shaffer, Dorothy Browne. *The radius of convexity for a special class of meromorphic functions.* Bull. Amer. Math. Soc. 79 (1973), 224-225. [2, 9, 12, 56]
1292. Shaffer, Dorothy Browne. *Distortion theorems for a special class of analytic functions.* Proc. Amer. Math. Soc. 39(1973), 281-287; MR 47 #3662. [2, 12, 19, 56]
1293. Shaffer, Dorothy B. *The order of starlikeness and convexity for analytic and meromorphic functions.* Notices A. M. S. 20, August 1973, Abstract 706-30-6, p. A-522. [2, 11]
1294. Shaffer, Dorothy B. *Radii of starlikeness and convexity for special classes of analytic functions.* J. Math. Anal. Appl. 45 (1974), 73-80; MR 48 #8772. [2, 6, 10, 11, 12, 14, 56]
1295. Shaffer, Dorothy Browne. *Analytic functions in class $H_{1/2,n}$ and their applications.* Notices Amer. Math. Soc., January 1975), Abstract 720-30-22, p. A-123. [2, 9, 11, 12, 56]
1296. Shaffer, Dorothy Browne. *In equalities for a special class of bounded analytic functions.* Notices, A. M. S., vol. 22, no. 5, issue no. 163, August 1975 (Abstract 726-30-2), p. A-556. [2, 9, 12, 22]
1297. Shah, G. M. *On some starlike and convex functions.* Trans. A. M. S. 154(1971), 83-91; MR 42 #4721. [4, 6, 10, 11, 12, 40, 41, 63, 85]
1298. Shah. G. M. *On the univalence of some analytic functions.* Pacific J. Math. 43(1972), 239-250; MR 47 #2047. [2, 11, 22, 43, 56, 61]
1299. Shah, G. M. *On multivalency of some analytic functions.* J. Natur. Sci. and Math. 12(1972), 113-121; MR 48 #11477. [2, 11]
1300. Shah, G. M. *On the univalence of certain rational functions.* J. Natur. Sci. and Math. 12(1972), 131-149; MR 49 #10869. [28]
1301. Shah, G. M. *On holomorphic functions convex in one direction.* J. Indian Math. Soc. (N.S.) 37(1973), 257-276; MR 50 #13487. [4, 5, 19, 36, 40, 41, 63, 78, 79, 85]
1302. Shah, S. M. *Univalent derivatives of entire functions of slow growth.* Archive for Rational Mechanics and Analysis, Vol. 35, no. 4, 1969, p. 259-266; MR 40 #352. [90]
1303. Shah, S. M. *Holomorphic functions with mean p-valent derivatives.* Math. Ann. 192(1971), 176-182; MR 45 #2160. [33]

1304. Shah, S. M. *Analytic functions with univalent derivatives and entire functions of exponential type.* Bull. Amer. Math. Soc. 78 (1972), 154–172; MR 45 #2157. [4, 5, 6, 16, 31, 33, 42, 43, 44, 90]
1305. Shah, S. M. *Univalence of derivatives of functions defined by Gap Power Series III.* Notices Amer. Math. Soc. 23, January 1976, Abstract 731-30-29, p. A-104. [39, 43]
1306. Shah, S. M.; Trimble, S. Y. *Univalent functions with univalent derivatives, I.* Bull. Amer. Math. Soc. 75(1969), 153–157, 888; MR 433 #4373. [4, 22, 30, 54, 90]
1307. Shah, S. M.; Trimble, S. Y. *Univalent functions with univalent derivatives, II.* Trans. A. M. S. 144(1969), 313–320; MR 40 #2841. [4, 43, 90]
1308. Shah, S. M.; Trimble, S. Y. *Univalent functions with univalent derivatives, III.* Journal of Math. and Mechanics, vol. 19, no. 5, November, 1969, pp. 451–460; MR 40 #4438. [4, 90]
1309. Shah, S. M.; Trimble, S. Y. *Univalent functions with univalent derivatives.* Notices A. M. S. (Abstract), vol. 17, no. 4, issue no. 122, June, 1970, p. 661. [90]
1310. Shah, S. M.; Trimble, S. Y. *Entire functions with univalent derivatives.* J. Math. Anal. Appl. 33(1971), 220–229. MR 43 #6435. [4, 90]
1311. Shah, S. M.; Trimble, S. Y. *Relations between the univalence of derivatives of power series and entire functions.* Notices, A. M. S., vol. 19, no. 1, issue no. 135, January, 1972 (Abstract 691-30-27), p. A-118. [90]
1312. Shah, S. M.; Trimble, S. Y. *Univalence of derivatives of an even entire function.* Notices, A. M. S., vol. 19, no. 6, issue 140, October, 1972 (Abstract 72T-B284), p. A-703. [90]
1313. Shah, S. M.; Trimble, S. Y. *Univalence of derivatives of a functions defined by a gap series.* Notices A. M. S. 20(Jan. 1973), Abstract 701-30-13, pg. A-106. [39]
1314. Shah, S. M.; Trimble, S. Y. *Univalence of derivatives of gap power series.* Notices A. M. S., vol. 20, no. 3, April 1973 (Abstract 703-B24, A-371). [39, 43]
1315. Shah, S. M.; Trimble, S. Y. *Entire functions with some derivatives univalent.* Canad. J. Math. 26(1974), 207–213; MR 50 #2493. [42, 90]
1316. Shah, S. M.; Trimble, S. Y. *The order of an entire function with some derivatives univalent.* J. Math. Anal. Appl. 46(1974), 395–409; MR 49 #9209. [4, 90]
1317. Shah, S. M.; Trimble, S. Y. *Univalence of derivatives of functions*

defined by gap power series. J. London Math. Soc. (2)9(1975), 501-512; MR 50 #13520. [4, 32, 43, 90]

1318. Shah, S. M.; Trimble, S. Y. *Univalence of derivatives of functions defined by gap power series II*. Notices, A. M. S., vol. 22, no. 5, issue no. 163, August 1975 (Abstract 75T-B179), p. A-517. [12, 39, 43]

1319. Sheil-Small, T. *On convex univalent functions*. J. London Math. Soc. (2)1(1969), 483-492; MR 40 #5844. [6, 10, 36, 38, 63, 64]

1320. Sheil-Small, T. *Some conformal mapping inequalities for starlike and convex functions*. J. London Math. Soc. (2)1(1969), 577-587; MR 40 #2842. [2, 3, 6, 7, 10, 16]

1321. Sheil-Small, T. *A note on the partial sums of convex schlicht functions*. Bull. London Math. Soc. 2(1970), 165-168; MR 42 #485. [4, 5, 6, 10, 20, 64]

1322. Sheil-Small, T. *Starlike univalent functions*. Proc. London Math. Soc. (3)21(1970), 577-613; MR 43 #2207. [1, 16, 62, 68]

1323. Sheil-Small, T. *On linear accessibility and the conformal mapping of convex domains*. J. Analyse Math. 25(1972), 259-276; MR 47 #444. [5, 6, 10, 16, 20, 63, 64, 74]

1324. Sheil-Small, T. *On Bazilevič functions*. Quart. J. Math. Oxford Ser., 23(1972), no. 90, pp. 135-142; MR 45 #8847. [4, 70]

1325. Sheil-Small, T. *On the convolution of analytic functions*. J. Reine Angew. Math. 258(1973), 137- . [1, 6, 16, 42, 45, 47]

1326. Sheil-Small, T. B. *Some linear operators in function theory*. Proceedings of the Symposium on Complex Analysis (Univ. Kent, Canterbury, 1973), pp. 119-123. London Math. Soc. Lecture Note Ser., No. 12, Cambridge Univ. Press, London, 1974. [44]

1327. Sheil-Small, T. *On linearly accessible univalent functions*. J. London Math. Soc. (2)6(1973), 385-398; MR 48 #4299. [4, 5, 6, 16, 23, 74, 75]

1328. Siewierski, L. *The local solution of the coefficient problem for bounded schlicht functions*. Societas Scientiarum Lodziensis, Sect. III, Nr. 68; Panstwowe Wydawnictwo Naukowe, Lodz, 1960. 124 pp. MR 32 #7735. [16, 22, 30]

1329. Siewierski, L. *Sharp estimation of the coefficients of bounded univalent functions near the identity*. (loose Russian summary). Bull. Acad. Polon. Sci. Ser. Sci. Math. Astronom. Phys. 16 (1968), 575-576; MR 38 #1246. [22, 30]

1330. Siewierski, L. *On the maximum value of the functional $|a_3 - \alpha a_2^2|$ in the classes of quasi-starlike functions*. (Loose Russian summary). Bull. Acad. Polon. Sci. Ser. Sci. Math. Astronom. Phys.

16(1968), 573-574; MR 38 #1245. [6, 36]
1331. Siewierski, L. *On the maximum of the functional* $|a_3 - \alpha a_2^2|$ *in the classes of quasi-starlike functions*. Ann. Polon. Math. 22 (1969/70), 243-253; MR 40 #4442. [6, 36]
1332. Siewierski, L.; Smialkowna, H. *On the coefficients of meromorphic quasi-convex functions*. Ann. Univ. Mariae Curie-Sklodowska, Sect. A., vols. 22/23/24(1968/1969/1970), 167-170; MR 51 #10607. [9, 10, 25]
1333. Silverman, H. *On a class of close-to-convex functions*. Proc. Amer. Math. Soc., vol. 36, no. 2, December 1972, 477-484; MR 47 #2048. [5, 10, 12, 19, 74, 75]
1334. Silverman, Herb. *Linear combinations of convex mappings*. Rocky Mountain J. Math. 5(1975), 629-631 [82]; MR 52 #3504
1335. Silverman, H. *Univalent functions with negative coefficients*. Proc. Amer. Math. Soc. 51(1975), 109-116; MR 51 #5910. [4, 6, 10, 11, 12, 16, 19, 23, 32, 36, 38, 39, 63, 64, 88]
1336. Silverman, H. *Extreme points of univalent functions with two fixed points*. Notices Amer. Math. Soc. 23, January 1976, Abstract 731-30-7, p. A-99. [6, 10, 88]
1337. Silverman, Herb; Silvia, Evelyn Marie. *On the order of convexity of α-starlike functions*. Notices A. M. S. 20(Jan. 1973), Abstract 701-30-9, pg. A-106. [69]
1338. Silverman, H. W.; Silvia, Evelyn M. *On linear combinations of convex functions of order β*. Preliminary report. Notices Amer. Math. Soc. 22, January 1975, Abstract 720-30-2, p. A-118 [10, 11, 12, 82]
1339. Silverman, H.; Telage, D. N. *Spiral-like functions and related classes with fixed second coefficient*. Notices, A. M. S., vol. 22, no. 4, issue no. 162, June 1975 (Abstract 75T-B124), p. A-462. [6, 10, 45]
1340. Silverman, H; Telage, D. N. *Extreme points of a subclass of close-to-convex functions*. Notices Amer. Math. Soc. 23, January 1976, Abstract 731-30-14, p. A-101. [5,, 74, 75, 88]
1341. Silvia, Evelyn Marie. *A variational method on certain classes of functions of bounded boundary rotation*. Notices Amer. Math. Soc. 20(1973), Abstract 73T-B304, p. A-634. [24, 71]
1342. Silvia, Evelyn Marie. *On a subclass of spirallike functions*. Proc. Amer. Math. Soc. 44(1974), 411-420; MR 49 #7433. [6, 23, 69, 83]
1343. Silvia, Evelyn M. *P-valent classes related to functions of bounded boundary rotation*. Notices Amer. Math. Soc. 23, January 1976, Abstract 731-30-3, p. A-98. [14, 23, 71, 76]

1344. Sinelnikova, N. I. *Certain chord distortion type problems on classes of schlicht functions.* (Russian). Trudy Tomsk. Gos. Univ. Ser. Meh.-Mat. 189(1966), 184-193; MR 37 #2975. [6, 24, 29]
1345. Singh, P.; Chandra, S. *On the coefficients of functions regular in the unit circle.* Notices Amer. Math. Soc. 20 (August 1973), Abstract 73T-B193, p. A-482. [71, 77]
1346. Singh, Ham. *A sort of converse of a theorem of Strohhäcker.* Ganita 18(1967), 13-16; MR 39 #428. [12]
1347. Singh, Ram. *A covering theorem for bounded convex schlicht functions.* Ganita 18(1967), no. 2, 67-70; MR 39 #5786. [10, 19, 22]
1348. Singh, Ram. *On a class of star-like functions.* Composito Math. 19(1967), 78-82; correction of an error which appears in same J. 21(1969), 230-231; MR 39 #7084. [6, 12, 19, 23, 36, 63]
1349. Singh, Ram. *On a class of starlike functions.* II. Ganita 19(1968), no. 2, 103-110; MR 41 #3741. [6, 12, 36, 63]
1350. Singh, Ram. *A note on sprial-like funcitons.* J. Indian Math. Soc. (N.S.) 33(1969), 49-55; MR 41 #454. [6, 11, 12, 16, 29, 63]
1351. Singh, Ram. *Meromorphic close-to-convex functions.* J. Indian Math. Soc. (N.S.) 33(1969), 13-20; MR 41 #3742. [2, 5, 9, 12, 16, 56]
1352. Singh, Ram. *Radius of convexity of partial sums of a certain power series.* J. Austral. Math. Soc. 11(1970), 407-410; MR 43 #7616. [2, 11, 12, 20, 21, 56, 73]
1353. Singh, Ram. *Some classes of regular univalent functions.* Rev. Mat. Hisp.-Amer. (4) 30(1970), 109-114; MR 43 #2208. [4, 5, 6, 10, 64, 73, 85]
1354. Singh, Ram. *A theorem on bounded analytic functions.* Publ. Math. Debrecen 18(1971), 183-185(1972); MR 46 #7518. [2, 22]
1355. Singh, Ram. *On Bazilevič Functions.* Proc. Amer. Math. Soc. 38 (1973), 261-271; MR 47 #449. [29, 70, 82, 85]
1356. Singh, R.; Singh, V. *Coefficient estimates for a certain class of bounded starlike functions.* Notices, A. M. S., vol. 18, no. 6, issue no. 132, October 1971 (Abstract 71T-B214), p. 954. [6, 22, 36]
1357. Singh, V.; Goel, R. M. *On radii of convexity and starlikeness of some classes of functions.* J. Math. Soc. Japan 23(1971), 323-339; MR 43 #7617. [2, 6, 9, 11, 12, 22, 56, 61, 72, 82]
1358. Sirokova, E. A. *Certain questions on the univalence of functions of class Σ.* (Russian). Mat. Zametki 18(1975), no. 3, 403-410. [9]
1359. Sirokov, N. A. *Hayman's regularity theorem.* (Russian) Zap. Naucn. Sem. Leningrad. Otdel. Mat. Inst. Steklov. (LOMI) 24

(1972), 182-200; MR 46 #3774. [39]

1360. Sitarski, R. *On some variation formulas for quasi-starlike functions.* (Polish and Russian summaries) Zeszyty Nauk. Politech. Lodz. No. 208 Mat. No. 6(1975), 107-116. [6]; MR 52 #729

1361. Sižuk, P. I. *On a certain result of Libera and Livingston.* (Russian) Sibirsk. Mat. Z. 16(1975), 98-102, 196; MR 51 #5911. [11, 82]

1362. Sižuk, P. I. *Regular functions $f(z)$ for which $zf'(z)$ is θ-spiral-shaped of order α.* (Russian) Sibirsk. Math. Z. 16(1975), no. 6, 1286-1290, 1371. [6]

1363. Sižuk, P. I.; Černikov, V. V. *Certain properties of univalent functions.* (Russian) Mat. Zametki 17(1975), 563-569.

1364. Skotnikova, G. A.; Goldina, N. M. *Ranges of values of certain coefficient systems of functions typically real in a ring.* (Russian) Trudy Mat. Inst. Steklov. 94(1968), 130-142; MR 37 #403. {Translated from the Russian: Extremal Problems of the Geometric Theory of Functions, (A. M. S. 1969)}. [13, 48]

1365. Skupien, Z. *On the problem of coefficients of univalent functions.* Zeszyty Nauk. Univ. Jagiello. Prace Mat. No. 10(1965), 89-91; MR 33 #5877. [54]

1366. Sladkowska, Janina. *Polynomes quasi-univalents et univalents. Variations elementaries et polynomes extremaux.* Societas Scientiarum Lodziensis, Sect. III, Nr. 64. Panstwowe Wydawnictwo Naukowe, Lodz, 1960; MR 33 #258. [32]

1367. Sladkowska, J. *Sur les conditions de Grunsky-Nehari pour les fonctions univalentes bornees dans le cercle unite.* (English and Russian summaries) Bull. Acad. Polon. Sci. Ser. Sci. Math. Astronom. Phys. 21(1973), 307-311; MR 50 #4928. [4, 22, 53]

1368. Sladkowska, Janina. *Coefficient inequalities for Shah's functions.* Demonstratio Math. 5(1973), 171-192; MR 49 #5334. [22, 27, 30, 53]

1369. Sladkowska, Janina. *Les polynômes de Faber dans le théorie de fonctions univalentes bornées.* Demonstratio Math. 8(1975), 99-112; MR 51 #5918. [22, 53]

1370. Sljusarova, K. M. *The domain of the values of C-functions with fixed real coefficients.* (Ukrainian. English and Russian summaries) Dopovidi Akad. Nauk Ukrain. RSR Ser A 1974, 222-225, 286. [29]

1371. Sljusarova, K. M. *Certain estiamtes of C-functions and starlike functions with fixed Taylor coefficients.* (Ukrainian. English and Russian summaries). Visnik Kiiv. Univ. Ser. Mat. Meh. No. 16 (1974), 105-108, 176. [29]

1372. Sljusarova, K. M.; Saran. L. A. *Estimation of the curvature of level curves for functions that are convex with respoect to the direction of the imaginary axis.* (Ukrainian. English and Russian summaries) Dopovidi Akad. Nauk Ukrain. RSR Ser. A 1972, 141–144, 189; MR 46 #3767. [35]

1373. Slyk, V. A. *Certain properties of functions that are weakly p-valent in an annulus.* (Russian) Kuban. Gos. Univ. Naucn. Trudy Vyp. 148 Mat. Anal. (1971). 75–80; MR 49 #555. [14, 48, 65]

1374. Slyk, V. A. *Certain estimates for functions that are regular and univalent in an annulus.* (Russian) Kuban. Gos. Univ. Naucn. Trudy Vyp. 148 Mat. Anal. (1971), 81–84; MR 48 #11483. [48]

1375. Slyk, V. A. *The Gronwall problem in the class of bounded functions that are weakly univalent in the mean with respect to a circle.* (Russian) Metric questions of the theory of functions and mappings. No. IV (Russian), pp. 165–171. Izdat. "Naukova Dumka", Kiev, 1973; MR 49 #10870. [14, 22]

1376. Slyk, V. A. *Certain covering theorems for weakly univalent, meromorphic functions in an annulus.* (Russian) Metric questions of the theory of functions and mappings, No. V (Russian), pp. 171–178. Izdat "Naukova Dumka", Kiev, 1974. [19, 33]

1377. Šmiałkówna, H. *On some properties of Schroeder's univalent functions.* (Russian summary). Bull. Acad. Polon. Sci. Ser. Sci. Math. Astronom. Phys. 23(1975), no. 9, 947–949.

1378. Smoluk, A.; Zamorski, J. *On conditions for extremal generalized spiral functions, involving their coefficients.* (Polish, Russian and English summaries) Prace Mat. 7(1962), 119–125; MR 32 #4263. [6]

1379. Sobolev, V. V. *On the problem of the relative extremum of a certain functional on the class of functions that are univalent in the half-plane.* (Russian) Trudy Tomsk. Gos. Univ. Ser. Meh.-Mat. 200(1968), 184–188; MR 41 #1988. [29, 57]

1380. Sone, N. *Sufficient conditions for p-valence of regular functions.* J. Math. Soc. Japan 16(1964), 406–418; MR 33 #1440. [4, 14]

1381. Sone, N. *Some radii of multivalently starlikeness or convexity.* Math. Japon. 10(1965), 30–34; MR 32 #7736. [2, 11, 12, 22, 43, 56, 61]

1382. Stankiewicz, Jan. *On some classes of close-to-convex functions.* Ann. Univ. Mariae Curie-Sklodowska Sect. A 20(1966), 77–83 (1971); MR 46 #3768. [5, 23, 29, 64]

1383. Stankiewicz, Jan. *Quelques problèmes extrêmaux dans les classes des fonctions α-angulairement étoilées.* Ann. Univ. Mariae Curie-Sklodowska Sect. A 20(1966), 59–75(1971); MR 41 #8650, MR 47 #3663. [6, 24]

1384. Stankiewicz, J. *Some remarks concerning starlike functions.*

(Loose Russian summary) Bull. Acad. Polon. Sci. Ser. Sci. Math. Astronom. Phys. 18(1970), 143-146; MR 41 #8650. [6, 19]

1385. Stankiewicz, Jan. *Some remarks on functions starlike with respect to symmetric points.* (Polish and Russian summaries) Ann. Univ. Mariae Curie-Sklodowska Sect. A 19(1965), 53-59(1970); MR 41 #5612. [1, 6, 10, 23]

1386. Stankiewicz, J. *On a family of starlike functions.* Ann. Univ. Mariae Curie-Sklodowska, Sect. A 22-24(1968/70), 175-181 (1972); MR 50 #590. [6, 10, 11, 12, 19, 23, 63]

1387. Stankiewicz, Jan. *Some extremal problems for the class S_α.* Ann. Univ. Mariae Curie-Sklodowska Sect. A 25(1971), 101-107(1973); MR 48 #6394. [2, 6, 11, 12, 23, 36, 63]

1388. Stankiewicz, Jan. *The influence of coefficients on some properties of regular functions.* (Polish and Russian summaries) Ann. Univ. Mariae Curie-Sklodowska Sect. A. 27(1973), 99-107(1975). [2, 43, 71]

1389. Stankiewicz, Zofia. *Sur la subordination en domaine de certain operateurs dans les classes $S(\alpha, \beta)$.* (Polish and Russian summaries). Ann. Univ. Mariae Curie-Sklodowska Sect. A 27(1973), 109-119(1975). [1]; MR 52 #8413

1390. Stankiewicz, Z. *An integral operator and inclusion of domains.* Bull. Acad. Polon. Sci. Ser. Sci. Math. Astronom. Phys. 22(1974), 1201-1207; MR 51 #5912. [1]

1391. Starkov, V. V. *A disproof of a certain conjecture on subordination for starlike functions.* (Russian) Dokl. Akad. Nauk SSSR 214 (1974), 52-55; MR 49 #3107. [1, 6, 16]

1392. Stepanova, O. V. *A property of level lines in univalent conformal mappings.* (Russian) Dokl. Akad. Nauk SSSR 163(1965), 1330; MR 32 #   . {Math. Dokl. 6(1965), 1124-1125}. [35]

1393. Stepanova, O. V. *Certain properties of the level curves of univalent conformal mappings.* (Russian) Mat. Sb. (N.S.)72(114) (1967), 72-96; MR 34 #6066. [35]

1394. Stepanova, O. V. *The range of values of a certain functional in the class of schlicht functions.* (Russian) Trudy Tomsk. Gos. Univ. Ser. Meh.-Mat. 210(1969), 97-110; MR 43 #7618. [9, 29]

1395. Strocik, T. V. *A remark on the conformal mapping of half-strips.* (Russian) Ukrain. Mat. Z. 26(1974), 686-690, 718; MR 50 #4920. [35]

1396. Stump, Robert K. *Linear combinations of univalent functions with complex coefficients.* Canad. J. Math. 23(1971), 712-717; MR 44 #2919. [10, 11, 12, 56, 82]

1397. Styer, David. *Close-to-convex multivalent functions with respect to weakly starlike functions.* Trans. Amer. Math. Soc. 169(1972), 105-112; MR 47 #3666. [2, 5, 6, 14, 23]
1398. Styer, David. *On weakly starlike multivalent functions.* J. Analyse Math. 18(1973), 217-233; MR 48 #6395. [6, 14]
1399. Styer, David. *Multivalent convex functions.* J. Analyse Math. 38 (1975), 60-77. [10, 14]
1400. Styer, D.; Wright, D. *On the valence of the sum of two convex functions.* Proc. Amer. Math. Soc. 37(1973), 511-516; MR 47 #2049. [5, 6, 10, 14, 16, 23, 45, 82]
1401. Suffridge, T. J. *Convolutions of Convex Functions.* J. of Math. and Mech. 15, May, 1966, 795-804; MR 33 #5871. [4, 5, 6, 10, 22, 23, 47]
1402. Suffridge, T. J. *A coefficient problem for a class of univalent functions.* Michigan Math. J. 16(1969), 33-42; MR 39 #1646. [6, 16, 17, 19, 22, 30, 36, 54]
1403. Suffridge, T. J. *On univalent polynomials.* J. London Math. Soc. 44(1969), 496-504; MR 38 #3419. [32]
1404. Suffridge, T. J. *The principle of subordination applied to functions of several variables.* Pacific J. Math. 33(1970), 241-248. [1, 6, 10]
1405. Suffridge, T. J. *Some remarks on convex maps of the unit disk.* Duke Math. J. 37(1970), 775-777; MR 42 #4722. [1, 10]
1406. Suffridge, T. J. *Extreme points in a class of polynomials having univalent sequential limits.* Trans. Amer. Math. Soc. 163(1972), 225-238; MR 45 #3679. [4, 32]
1407. Suffridge, T. J. *Starlike and convex maps in Banach spaces.* Pacific J. Math. 46(1973), 575-590. [1, 4, 6, 10]
1408. Suffridge, T. J. *Starlike functions as limits of polynomials.* Notices Amer. Math. Soc., January 1975, Abstract 720-30-4, p. A-118. [6, 32]
1409. Suffridge, T. J. *On univalent polynomials.* Notices Amer. Math. Soc. 22, April 1975, Abstract 723-B37, p. A-410. [5, 6, 32, 47]
1410. Suffridge, T. J. *Functions starlike of order α.* Notices Amer. Math. Soc. 23, January 1976, Abstract 731-30-23, p. A-103. [6]
1411. Szilard, K. *Über die Verzerrungseigenschaften der konformen Abbildung des Einheitskreises auf "ϱ_0-Konvexe" Gebiete. I.* Studia Sci. Math. Hungar. 1(1966), 133-136; MR 34 #6067. [19]
1412. Szilard, K. *Über die Verzerrungseigenschaften der konformen Abbildung des Einheitskreises auf "ϱ_0-konvexe" Gebiete. II.* Studia Sci. Math. Hungar. 2(1967), 49-51; MR 35 #4393. [10]

1413. Szilard, Karl. *Über die Koebesche Konstant 1/4.* Aequationes Math. 2(1969), 227–232; MR 39 #7077. [14, 19]
1414. Szilard, K. *On the Wolff-Noshiro-Warschawski-theorem in the theory of univalent functions.* Matematikai Lapok 21(1970), 45–50; MR 46 #7499. [4]
1415. Szynal, Anna; Szynal, Jan; Zygmunt, J. *On the coefficients of functions whose real part is bounded.* (Polish and Russian summaries). Ann. Univ. Mariae Curie-Sklodowska Sect. A 26(1972), 63–70(1974); MR 50 #2476. [2, 6, 21, 36, 56]
1416. Szynal, Jan. *On a certain class of regular functions.* Ann. Univ. Mariae Curie-Sklodowska Sect. A 25(1971), 109–120(1973); MR 48 #8773. [2, 6, 12, 36, 56, 63]
1417. Szynal, J. *Some remarks on coefficients inequality for α-convex functions.* Bull. Acad. Polon. Sci. Ser. Sci. Math. Astronom. Phys., vol. 20, no. 11(1972), 917–920; MR 47 #454. [4, 36, 38, 69, 84]
1418. Szynal, J.; Wajler, S. *On the fourth coefficient for α-convex functions.* Rev. Roumaine Math. Pures Appl. 19 (1974), 1153–1157; MR 50 #10229. [6, 10, 22, 30, 36, 38, 69, 84]
1419. Takatsuka, T. *Multivalent functions star-like in one direction.* Trans. Amer. Math. Soc. 120(1965), 72–82; MR 32 #206. [2, 6, 9, 23, 25, 40, 41, 56, 58, 63]
1420. Takatsuka, T. *The theory of multivalent functions.* Duke Math. J. 33(1966), 583–593; MR 33 #7522. [2, 4, 6, 9, 10, 14, 56]
1421. Talbot, A. *Some theorems on positive functions.* IEEE Trans. Circuit Theory CT-12(1965), 607–608; MR 41 #2021. [2, 26, 32, 57]
1422. Tammi, O. *On the use of the Grunsky-Nehari inequality for estimating the fourth coefficient of bounded univalent functions.* Colloq. Math. 16(1967), 35–42; MR 36 #1637. [22, 30]
1423. Tammi, Olli. *Grunsky type inequalities, and determination of the totality of the extremal functions.* Ann. Acad. Sci. Fenn. Ser. A I No. 443(1969), 20 pp.; MR 40 #4443. [22, 27, 30, 53]
1424. Tammi, Olli. *On Green's inequalities for the third coefficient of bounded univalent functions.* Ann. Acad. Sci. Fenn. Ser. A I No. 481(1970), pp. 3–10; MR 43 #5021. [22, 30]
1425. Tammi, Olli. *On optimizing parameters of the power inequality for a_4 in the class of bounded univalent functions.* Ann. Acad. Fenn. Ser. A I No. 560(1973), 24 pp. [22, 30]
1426. Tammi, Olli. *On Green's inequalities.* Topics in analysis (Colloq. Math. Anal., Jyvaskyla, 1970), pp. 370–375. Lecture Notes in Mah., vol. 419, Springer, Berlin, 1974; MR 51 #8396. [22, 30]

1427. Tamrazov, P. M. *The theory of conformal univalent mappings of domains of arbitrary connectivity onto domains with slits.* (Russian) Problems of Math. Phys. and Theory of Functions, II (Russian), pp. 110–117. Naukova Dumka, Kiev, 1964; MR 33 #5861. [60, 22, 61]

1428. Tamrazov, P. M. *A conformally metric theory of doubly-connected regions and a generalized Blaschke product.* (Russian) Dokl. Akad. Nauk SSSR 161(1965), 308–311; MR 31 #327. [48]

1429. Tamrozov, P. M. *Theorems on the covering of lines under a conformal mapping.* (Russian) Mat. Sb. (N.S.)66(108)(1965), 502–524; MR 31 #2392. [19]

1430. Tamrazov, P. M. *Certain extremal problems of the theory of univalent conformal mappings.* (Russian) Mat. Sb. (N.S.) 67(109) (1965), 329–337; MR 33 #7523.

1431. Tamrazov, P. M. *Some extremal problems of conformal mapping of doubly and multiply connected domains.* Dokl. Akad. Nauk SSSR 170(1966), 530–532; MR 35 #3047. {Soviet Math. Dokl. 7(1966), 1247–1249}. [60]

1432. Tamrazov, P. M. *Extremal conformal mappings and poles of quadratic differentials.* (Russian) Izv. Akad. Nauk SSSR Ser. Mat. 32(1968), 1033–1043; MR 38 #3417. [24]

1433. Tamrazov, P. M. *Supplement to the paper "On certain extremal problems of conformal mapping".* (Russian) Trud. Tomsk. Gos. Univ. Ser. Meh.-Mat. 210(1969), 111–118; MR 44 #423. [60]

1434. Tamrazov, P. M. *The angular displacement of the boundary in a certain class of univalent conformal mappings.* (Russian) Ukrain. Mat. Z. 22(1970), 563–566; MR 42 #4723. [48, 62]

1435. Tamura, J.; Oikawa, K.; Yamazaki, K. *Examples of minimal parallel slit domains.* Proc. Amer. Math. Soc. 17(1966), 283–284. [49]

1436. Tang, Sung-shih. *Two theorems on univalent functions.* (Chinese) Shuxue Jinzhan 7(1964), 431–432; MR 38 #2295. [9, 58, 68]

1437. Tepper, D. E. *On the radius of convexity and boundary distortion of schlicht functions.* Trans. Amer. Math. Soc. 150(1970), 519–528; MR 42 #3268. [2, 6, 12, 49, 62, 68, 72]

1438. Tepper, D. E. *Integral means of schlicht functions.* Notices, A. M. S., vol. 19, no. 6, issue 140, October, 1972 (Abstract 72T-B274), p. A-700. [3]

1439. Tepper, D. E. *Univalent functions with real coefficients.* Notices Amer. Math. Soc. 23, January 1976, Abstract 731-30-27, p. A-104. [39, 68]

1440. Thakare, N. K. *A note on the application of Schwarz's lemma.* J. Shivaji Univ. 4(1971), no. 8 (Science), 109; MR 50 #4931. [2, 56]
1441. Thomas, D. K. *On starlike and close-to-convex univalent functions.* J. London Math. Soc. 42(1967), 427–435; MR 35 #6812. [5, 6, 7, 18]
1442. Thomas, D. K. *A note on starlike functions.* J. London Math. Soc., 43(1968), 703–706; MR 37 #4248. [6, 7]
1443. Thomas, D. K. *On Bazilevič Functions.* Amer. Math. Soc. Trans. 132(1968), 353–361; MR 36 #5330. [4, 22, 70, 81]
1444. Thomas, D. K. *On an extremal problem in univalent functions.* Math. Z. 109(1969), 344–348; MR 39 #4374. [3, 4, 6, 16, 42, 70]
1445. Thomas, D. K. *On the coefficients of functions with bounded boundary rotation.* Proc. Amer. Math. Soc. 36, no. 1, November 1972, 123–129; MR 46 #7498. [7, 9, 25, 71, 76, 77]
1446. Thomas, D. K. *On the coefficients of meromorphic univalent functions.* Proc. Amer. Math. Soc. 47(1975), 161–166; MR 50 #10233. [9, 16, 25, 70, 81]
1447. Todorov, P. *The univalency of a class of meromorphic functions.* (Russian) Dokl. Akad. Nauk SSSR 162(1965), 285–286; MR 31 #2393. {Soviet Math. Dokl. 6(1965), 674–675}. [9, 43]
1448. Todorov, P. G. *On the radius of univalence of a certain class of meromorphic functions.* (Russian) Ukrain. Mat. Z. 17(1965), no. 3, 135–137; MR 32 #5864. [43]
1449. Todorov, P. G. *On the radius of univalence for a certain class of meromorphic functions.* (Russian) Uspehi Mat. Nauk 20(1965), no. 6(126), 162–163; MR 33 #4252. [9, 43]
1450. Todorov, P. G. *Über den Radius des Schlichtheitskreises einer Klasse meromorpher Functionen.* (Italian and English summaries) Rend. Accad. Sci. Fis. Mat. Napoli (4)32(1965), 108–115; MR 34 #6068. [9, 43]
1451. Todorov, P. *Über den Radius des Schlichtheitskreises einer Klasse rationaler Funktionen.* Österreich Akad. Wiss. Math.-Natur. Kl. S.-B. II 174(1965), 559–566; MR 34 #6069. [9, 43]
1452. Todorov, P. *The radius of univalence of a class of meromorphic functions.* (Russian) Mat. Vesnik 2(17)(1965), 197–199; MR 33 #5872. [9, 43]
1453. Todorov, P. *Über den Radius des Schlichtheitskreises einer Klasse meromorpher Funktionen.* Rev. Roumaine Math. Pures Appl. 11 (1966), 95–101; MR 33 #7524. [9, 43]
1454. Todorov, P. *Über den Radius des Schlichtheitskreises einer Klasse meromorpher Funktionen.* Math. Nachr. 31(1966), 379–386; MR 33 #5873. [9, 43]

1455. Todorov, P. *Über die Polstellen und den Radius des Schlichtheitskreises der meromorphen Klasse* Φ. Acad. Roy. Belg. Bull. Cl. Sci. (5)52(1966), 799–804; MR 34 #6070. [9, 43]
1456. Todorov, P. *On a class of schlicht meromorphic functions.* (Russian. Czech and English summaries) Casopis Pest. Mat. 91(1966), 77–79; MR 32 #7737. [9, 43]
1457. Todorov, P. *The poles and the radius of circular univalence in the meromorphic class* Φ. (Russian) Dokl. Akad. Nauk SSSR 168 (1966), 532–534; MR 33 #4253. {Soviet Math. Dokl. 7(1966), 693–695}. [9, 43]
1458. Todorov, P. *Ueber die Moeglichkeiten einer schlichten konformen Abbildung des Einheitskreises durch die hypergeometrische Funktion von Gauss.* (English summary). Acad. Roy. Belg. Bull. Cl. Sci. (5)53(1967), 432–441; MR 36 #6604. [28]
1459. Todorov, Pavel. *On the theory of schlicht mappings by means of the meromorphic class* φ. (Russian) Mat. Vesnik 4(19)(1967), 393–398; MR 37 #3000. [9, 43]
1460. Todorov, Pavel. *Zur Theorie der analytischen positiven und schlichten konformen Abbildungen durch zusammengefasste ganze und meromorphe Funktionen mit einer endlichen Anzahl Nullstellen und Polstellen.* Acad. Roy. Belg. Bull. Cl. Sci. (5) 54 (1968), 512–526; MR 38 #6043. [43]
1461. Todorov, P. G. [Todorov, Pavel] *On the theory of univalent conformal mappings that are realizable by meromorphic functions wit simple poles and positive residues.* (Russian) Ukrain. Mat. Z. 22 (1970), 416–422; MR 42 #1992. {See also MR 45 #7037; MR 43 #1696}. [4, 9, 43]
1462. Todorov, P. G. *On the theory of schlicht conformal mappings that are realized by certain Nevanlinna classes of analytic functions.* (Russian). Ukrain. Mat. Z. 23(1971), 118–122; MR 43 #6421. [43]
1463. Todorov, P. G. *The theory of analytic positive and schlicht conformal mappings by means of generalized entire and meromorphic functions with a finite number of zeroes and poles.* (Russian) Publ. Inst. Math. (Beograd) (N.S.)11(25)(1971), 5–15; MR 46 #2028. [2, 43]
1464. Todorov, P. G. *Zur Theorie der schlichten Abbildungen.* Publ. Inst. Math. (Beograd)(N.S.)11(25)(1971), 17–18; MR 46 #2031. [4]
1465. Todorov, P. G. *Maximal univalent mappings with bounded rotation that can be realized by a class of integrals of Schwarz-Christoffel type.* (Russian) Mathematica (Cluj) 15(38)(1973), 307–329. [71]

1466. Todorov, P. G. *Maximal domains of univalence for certain classes of meromorphic functions.* (Russian) Aequationes Math. 10 (1974), 177–188; MR 49 #10871. [9, 28]
1467. Tonti, Norman E.; Trahan, Donald H. *Analytic functions whose real parts are bounded below.* Math. Z. 115(1970), 252–258; MR 41 #7116. [1, 2, 6, 11, 12, 21, 36, 48, 56]
1468. Toppila, S. *A remark on Bloch's constant for schlicht functions.* Ann. Acad. Sci. Fenn. Ser. A I No. 423(1968), 4 pp.; MR 38 #1248. [19]
1469. Trimble, S. Y. *The convex sum of convex functions.* Math. Z. 109 (1969), 112–114; MR 39 #7085. [4, 5, 6, 10, 82]
1470. Trimble, S. Y. *A coefficient inequality for convex univalent functions.* Proc. Amer. Math. Soc. 48(1975), 266–267; MR 50 #7504. [10, 38]
1471. Turzynski, A. *Problems related to the class V^α of holomorphic functions of restricted rotation.* (Polish and Russian summaries) Zeszyty Nauk Politech. Lodz. No. 186 Mat. No. 5(1974), 61–80; MR 50 #13485. [12, 35, 71, 77]
1472. Turzynski, A. *On some problems in the class p^α and some classes generated by this class.* (Polish and Russian summaries) Zeszyty Nauk Politech. Lodz. No. 186 Mat. No. 5 (1974), 43–59; MR 50 #13484. [2, 6, 9, 11, 12, 24]
1473. Twomey, J. B. *An integral mean problem for bounded starlike functions.* (Abstract), Notices, A. M. S., vol. 16, no. 1, issue no. 111, Jan. 1969, p. 269. [3, 6, 22, 62]
1474. Twomey, J. B. *On starlike functions.* Proc. Amer. Math. Soc. 24(1970), 95–97; MR 40 #2843. [6, 16, 62, 63]
1475. Twomey, J. B. *On the derivative of a starlike function.* J. London Math. Soc. (2), 2(1970), 99–110; MR 40 #7436. [6, 9, 58, 62, 63]
1476. Twomey, J. B. *On meromorphic starlike functions.* J. London Math. Soc., 4(1971), 231–239; MR 44 #5445. [3, 6, 9, 58, 62]
1477. Twomey, J. B. *On bounded starlike functions.* J. Analyse Math. 24(1971), 191–204; MR 45 #3689. [2, 3, 6, 7, 22, 56]
1478. Ulina, G. V. *The range of values of a certain system of functionals in the class of schlicht functions with real coefficients.* (Russian) Baskir. Gos. Univ. Ucen. Zap. Vyp. 31(1968), Ser. Mat. no. 3, 342–350; MR 43 #7619. [24, 29, 39]
1479. Ullman, J. L. *The location of the zeroes of the derivatives of Faber polynomials.* Proc. Amer. Math. Soc. 34(1972), 422–424; MR 45 #8809. [9, 16, 53]
1480. Ullman, J. L. *An area theorem for schlicht functions.* Amer.

Math. Monthly, vol. 80, no. 2, Feb. 1973, 184-186; MR 47#3667. [27]
1481. Uluçay, C. *A general formulation of an extremalization problem and its appliction to analytic functions with positive real part.* (Turkish summary) Comm. Fac. Sci. Univ. Ankara Ser. A 19 (1970), 11-22(1971); MR 46 #3775. [2]
1482. Uluçay, C. *On the constant a.* Comm. Fac. Sci. Univ. Ankara Ser. A 21(1972), 77-87; MR 48 #11460. [19]
1483. Uluçay, C. *Proof of Bieberbach's conjecture.* (Turkish summary) Comm. Fac. Sci. Univ. Ankara Ser. A 22(1973), 1-20. [42]
1484. Umezawa, T. *Some theorems on multivalent functions.* Gumma Univ. Sci. Report, vol. 1, no. 1, November 1950. [14]
1485. Umezawa, T. *On some systems of regular functions.* J. Math. Soc. Japan, 7(1955), 139-165. [14, 43]
1486. Umezawa, T. *Some criteria of multivalence.* Gumma Univ. Sci. Report, vol. 10, no. 2, (1961), pp. 1-5. [4, 14, 31]
1487. Umezawa, T. *A class of multivalent functions.* Gumma Univ. Sci. Report, vol. 13(1965), pp. 19-23. [6, 7, 11, 12, 14, 18, 20]
1488. Ungar, Peter. *Bloch circles of univalent functions.* Comm. Pure Appl. Math. 18(1965), 313-318; MR 30 #4927. [19]
1489. Varjuhin, V. A.; Kas'janjuk, S. A.; Finogenova, V. G. *A problem of constrained extremum for a class of functions representable by a Stieltjes integral.* (Russian) Izv. Vyss. Ucebn. Zaved. Matematika 1966, no. 6(55), 40-49; MR 34 #6115.
1490. Veech, William A. *A second course in complex analysis.* W. A. Benjamin, Inc., New York-Amsterdam, 1967, IX + 246 pp.; MR 36 #3955. [44]
1491. Viswanath, G. R. *On convex univalent functions.* Notices Amer. Math. Soc. 22, February 1975, Abstract 75T-B65, p. A-317. [10, 23, 38, 64]
1492. Volk, B. *On the maximum n-th diameter.* Bulletin Amer. Math. Soc. 80, May 1974, 446-448; MR 48 #11487. [19]
1493. Volynec, I. A. *Distortion theorems for mappings of class BL.* (Russian) Dokl. Akad. Nauk SSSR 213(1973), 16-18. [34, 68]
1494. Volynec, I. A. *A certain conjecture in the theory of analytic functions.* (Russian) Dokl. Akad. Nauk SSSR 209(1973), 1261-1263; MR 48 #524. [31]
1495. Vorobiov, M. M. *Approximation of analytic functions by partial sums of the Faber series.* (Ukrainian. Russian and English summaries) Dopovidi Akad. Nauk Ukrain. RSR 1966, 138-141; MR 34 #7813. [53]

1496. Vowels, R. E. *A note on positive real functions.* IEEE Trans. Circuit Theory CT-18(1971), 383-386; MR 43 #3429. [2]
1497. Waadeland, H. *On an inequality of Bernardi.* I. Norske Vid. Selsk. Forh. (Trondheim)39(1966), 27-33; MR 35 #3053. [19]
1498. Waadeland, H. *On an inequality of Bernardi.* II. Norske Vid. Selsk. Forh. (Trondheim) 39(1966), 34-40; MR 35 #3054. [19]
1499. Waadeland, H. *Playing with a lemma.* (Norwegian. English summary) Nordisk Mat. Tidskr. 19(1971), 30-38, 60; MR 48 #516. [2]
1500. Wajler, Stanislaw. *Sur certaines sous-classes de la classe de fonctions quasi-étoilées.* Demonstratio Math. 8(1975), 89-97; MR 51 #5913. [6, 63]
1501. Walczak, S. *Estimation of $|f''(z_0)|$, $|f'''(z_0)|$, $|f^{(4)}(z_0)|$ for functions of class S.* (Polish. Russian summaries) Zeszyty Nauk. Univ. Lodzk. Nauki Mat. Przyrod. Ser. II Zeszyt 39 Mat. (1971), 69-73; MR 48 #8776. [68]
1502. Walczak, S. *Extremal problems in the class of close-to-convex functions.* Ann. Polon. Math. 25(1971), 23-38; MR 45 #2158. [5, 23, 24, 74, 75]
1503. Walczak, S. *The radius of conformity of some classes of regular functions.* Ann. Polon. Math. 27(1973), 189-195; MR 46 #9330. [43]
1504. Walczak, S. *Typical real functions in the exterior of the unit circle.* Comment. Math. Prace Mat. 17(1973/74), 513-531; MR 50 #7534. [13, 24]
1505. Warren, J. *Holomorphic functions bounded on a spiral.* (Abstract), Notices, A. M. S., vol. 16, no. 1, issue no. 111, Jan. 1969, p. 273.
1506. Warren, J. *Rate of growth of sprial functions.* Notices, Amer. Math. Soc., January 1975, Abstract 720-30-14, p. A-121. [62]
1507. Warschawski, S. E. *Remarks on the angular derivative.* Nagoya Math. J. 41(1971), 19-32; MR 43 #493. [46]
1508. Watson, Martha. *On functions that are bivalent in the unit circle.* J. Analyse Math. 17(1966), 383-409; MR 35 #6815. [14, 50]
1509. Wesolowski, A. *Relations entre la subordination et l'inègalité des modules dans le cas des majorantes appartenant a la classe $N_k(p, o:q)$.* Ann. Univ. Mariae Curie-Sklodowska Sect. A 20 (1966), 85-94(1971); MR 46 #5607. [1]
1510. Wesolowski, A. *Communiqué des formules variationnelles et des certains resultats recus dans les sous-classes de fonctions étoilées.* (Polish and Russian summaries). Ann. Univ. Mariae Curie-Sklodowska Sect. A 22-24(1968/70), 193-199(1972); MR 50 #591. {See also MR 46 #339}. [6, 23, 24, 36]

1511. Wesolowski, A. *Certains resultats concernant la classe $S^*(\alpha, \beta)$.* Ann. Univ. Mariae Curie-Sklodowska Sect. A 25(1971), 121-130(1973); MR 48 #8774. [6, 11, 23, 29, 36, 63]

1512. White, William L. *A subclass of Bazilevic Functions.* Notices, A. M. S., vol. 22, no. 4, issue no. 162, June 1975 (Abstract 75T-B120), p. A-461. [6, 10, 70]

1513. Whiteman, R. A. *A concise proof of Borel's theorem on coefficient bounds.* The Amer. Math. Monthly, vol. 74, no. 9, November 1967, p. 1094. [2, 21]

1514. Whiteman, R. A. *Compactness and Certain Subclasses of Schlicht Functions.* Amer. Math. Monthly 76, no. 6, June-July 1969, pp. 643-647; MR 39 #2967. [5, 6, 10]

1515. Whitney, E. L. *Loewner's equation and Grunsky's inequalities.* J. D'analyse Math. 18(1967), 377-388; MR 37 #1586. [24, 53]

1516. Wiatrowski, P. *Limitation exacte du module de la dérivée logarithmique dans la famille des fonctions univalentes bornées.* Panstwowe Wydawnictwo Naukowe, Lodz, 1960. 60 pp.; MR 33 #4273. [22, 24, 61]

1517. Wiatrowski, P. *A sharp estimate of the functional $|z^\beta f'(z)^\alpha / f(z)^\beta|$ in the class of univalent functions.* (Polish. French summary) Zeszyty Nauk. Univ. Lodzk. Nauki Mat. Przyrod. Ser. II Zeszyt 20 Matematyka (1966), 117-146; MR 50 #13492. [24, 29]

1518. Wiatrowski, P. *The coefficients of a certain family of holomorphic functions.* (Polish. English summary) Zeszyty Nauk. Univ. Lodz. Nauki Mat. Przyrod. Ser II Zeszyt 39 Mat. (1971), 75-85; MR 49 #3115. [54]

1519. Wiatrowski, P. *On the radius of convexity of some family of functions regular in the ring $0 < |z| < 1$.* Ann. Polon. Math. 25(1971), 85-98; MR 44 #4201. [2, 9, 12, 58]

1520. Wiatrowski, Pawel. *Extremal problems in some class of regular functions defined in the unit circle and applications.* Ann. Polon. Math., 26(1972), Fasc. 1, pp. 13-29; MR 46 #340. [2, 9, 12, 23]

1521. Wieczorek, Zbigniew. *The range of values of the functionals $\log (f(z)/z)$ and $\log [\{f(z_2)/z_2\}/\{f(z_1)/z_1\}]$ in certain classes of starlike functions.* (Polish. French summary) Zeszyty Nauk Univ. Lodzk Nauki Mat. Przyrod. Ser. II Zeszyt 34(1969), 73-86; MR 41 #8651. [6, 24, 29]

1522. Wieczorek, Z. *On the coefficients of starlike functions of some classes.* Comment. Math. Prace Mat. 18(1974/75), 113-119; MR 50 #13488. [6, 9, 25, 36]

1523. Wilken, D. R. *The integral means of close-to-convex functions.* Michigan Math. J. 19(1972), no. 4, pp. 377-380; MR 47 #450. [3, 5]

1524. Williams, R. K. *An elementary consequence of the area theorem.* Delta (Waukesha)3(1972/73), no. 3, 27-29; MR 47 #7030. [27]
1525. Wirths, Karl-Joachim. *Uber Potenzreihen mit monotonen Koeffizientenfolgen.* Math. Nachr. 66(1975), 43-55; MR 51 #10599. [39]
1526. Wright, D. J. *A note on the perturbation of starlike functions.* Notices, Amer. Math. Soc. (Abstract 648-93), vol. 14, no. 3, issue no. 99, August, 1967, p. 657. [1, 6, 11, 82]
1527. Wright, D. J. *On a class of starlike functions.* Composito Math. 21(1969), 122-124; MR 39 #7086. [6, 12, 19, 23, 36, 63]
1528. Wright, Donald J. *The radius of p-valent starlikeness for certain classes of analytic functions.* Ann. Polon. Math. 23(1970/71), 49-55; MR 41 #8653. [1, 2, 6, 11, 14, 22, 43, 56, 61]
1529. Yamaguchi, K. *On functions satisfying $Re\{f(z)/z\} > 0$.* Proc. Amer. Math. Soc. 17(1966), 588-591; MR 33 #268. [1, 2, 4, 20, 43, 56]
1530. Yamashita, S. *The derivative of a holomorphic function in the disk.* Michigan Math. J. 21(1974), 129-134. [16, 62]
1531. Yamashita, S. *Banach spaces of locally Schlicht functions with the Hornich operations.* Manuscripta Math. 16(1975), fasc. 3, 261-275.
1532. Yang, Chung Chun. *On the coefficients of a bounded univalent analytic function.* Proc. Indian Acad. Sci. Sect. A 82(1975), no. 2, 37-40. [22, 30]
1533. Yoshikawa, Hirosaku. *On a subclass of spiral-like functions.* Mem. Fac. Sci. Kyushu Univ. Ser. A, Math., vol. 25, no. 2, 1971. 271-279; MR 46 #9320. [6, 11, 12, 63]
1534. Yu, Cheng Shu; Chen, Ming Po. *The radius of convexity of some regular functions.* Kyungpook Math. J. 15(1975), 83-87; MR 51 #3419. [2, 6, 10, 12, 56]
1535. Zac'ko, V. N. *The boundaries of convexity of the classes S_α^* (m) and Σ_α^* (m).* Russian. Ukrain. Mat. Z. 21(1969), 386-392; MR 39 #4375. [9, 12, 49]
1536. Zalcman, L. *Hadamard Products of Schlicht Funcitons.* Proc. Amer. Math. Soc. 19, no. 3, June 1968, pp. 544-548; MR 37 #399. [4, 9, 14, 37, 82]
1537. Zawadzki, R. *On some theorems on distortion and rotation in the class of symmetric starlike functions of order α.* (Loose Russian summary) Bull. Acad. Polon. Sci. Ser. Sci. Math. Astronom. Phys. 17(1969), 639-645; MR 41 #7089; Ann. Polon. Math. 24 (1970/71), 169-185; MR 44 #5446. [6, 9, 49, 63]
1538. Zawadzki, R. *On the radius of convexity of some class of analytic*

k-symmetrical functions. (Polish and Russian summaries) Ann. Univ. Mariae Curie-Sklodowska Sect. A 26(1972), 79-101(1974); MR 49 #9184. [6, 12]

1539. Zedek, M. *The linear measures of n-podes in conformal maps of the unit disk.* J. London Math. Soc. (2)6(1973), 301-306; MR 47 #8832. [4, 19]

1540. Zhang, Kai-ming. *Typically real functions with starlike images.* Acta Math. Sinica 8(1958), 12-22 (Chinese); translated as Chinese Math.-Acta 8(1966), 599-607; MR 36 #2796. [6, 13, 23, 66]

1541. Ziegler, M. R. *Some integrals of univalent functions.* Indian J. of Math., vol. 11, no. 3, Sept. 1969, 145-151; MR 41 #3743. [2, 4, 6, 10, 56, 85]

1542. Ziegler, M. R. *The radius of starlikeness of certain analytic functions.* Math. Z. 122(1971), 351-354. [2, 6, 11, 56, 82]

1543. Ziegler, M. R. *A class of regular functions containing spirallike and close-to-convex functions.* Trans. Amer. math. Soc., 166 (1972), pp. 59-70; MR 45 #529. [2, 4, 5, 6, 8, 12, 31, 36, 63, 73]

1544. Ziegler, M. R.; Karunakaran, V. *The radius of starlikeness for a class of functions regular in the unit disc.* Notices Amer. Math. Soc. 23, February 1976, Abstract 76T-B29, p. A-277. [11, 85]

1545. Zinterhof, Peter. *Konstruktion von schlichten Funktionen mit unendlich vielen Fixpunkten.* (Italian and English summaries) Rend. Ist. Mat. Univ. Trieste 3(1971), 125-134(1972); MR 46 #9331. [10]; MR 52 #8405

1546. Złotkiewicz E. *Sur les domaines des valeurs de certaines fonctionelles dans la classe U(p).* (Polish and Russian summaries). Ann. Univ. Mariae Curie-Sklodowska Sect. A 18(1964), 73-83(1967); MR 39 #4376. [9, 12, 24, 58]

1547. Złotkiewicz, E. *Some remarks concerning meromorphic univalent functions.* Ann. Univ. Mariae Curie-Sklodowska Sect. A 21(1967), nos. 1-8, pp. 53-61; MR 49 #3118. [9, 24, 58]

1548. Złotkiewicz, E. *Some remarks concerning close-to-convex functions.* Ann. Univ. Mariae Curie-Sklodowska Sect. A, vol. 21 (1967), 47-51(1972); MR 48 #4296. [4, 5, 12, 23, 24, 29, 45, 82, 83, 84, 85]

1549. Złotkiewicz, E. *The region of variability of the ratio $f(b)/f(c)$ within the class of meromorphic and univalent functions in the unit disc.* Ann. Univ. Mariae Curie-Sklodowska, Sect. A, vols. 22/23/24(1968/1969/1970), p. 201-208; MR 49 #5336. [9, 24, 29]

1550. Zmorovič, V. A. *On bounds of convexity for starlike functions of order α in the circle $|z| < 1$ and in the circular region $0 < |z| <$ 1.* Mat. Sb. 68(110)(1965), 518-526; MR 33 #5875. {Amer. Math.

Soc. Transl. (2)80(1969), 203-213}. [6, 9, 12]
1551. Zmorovič, V. A. *On a class of extremal problems associated with regular functions with positive real part in the circle* $|z| < 1$. Ukrain. Mat. Z. 17(1965), no. 4, 12-21; MR 33 #5901. {Amer. Math. Soc. Transl. (2)80(1969), 215-226.}. [2, 6, 35, 49]
1552. Zmorovič, V. A. *On certain theorems of the theory of extremal estiamtes in special classes of analytic functions.* (Ukrainian. Russian and English summaries) Dopovidi Akad. Nauk Ukrain. RSR 1965, 980-984; MR 33 #5874. [6, 9, 12]
1553. Zmorovič, V. A. *On the radius of θ-spirality of θ-spiral functions in the circle* $|z| < 1$. (Ukrainian. Russian and English summaries) Dopovidi Akad. Nauk Ukrain. RSR 1965, 1262-1265; MR 32 #7738. [6, 11]
1554. Zmorovič, V. A. *Rotation theorems for the classes $S_\alpha^*(m)$ and $S_\gamma (m; \alpha; n)$ for functions univalent in* $|z| < 1$. (Ukrainian. Russian and English summaries) Dopovidi Akad. Nauk Ukrain. RSR 1966, 1117-1121; MR 34 #2857. [2, 6, 10, 23, 49]
1555. Zmorovič, V. A. *On the bounds of starlikeness and of univalence in certain classes of functions regular in the circle* $|z| < 1$. Ukrain. Mat. Z. 18(1966), no. 3, 28-39; MR 33 #7525. {Amer. Math. Soc. Transl. (2)80(1969), 227-242}. [6, 11, 43]
1556. Zmorovič, V. A.; Acilov, H. *On a class of univalent mappings of a circular ring.* (Russian. Uzbek summary) Izv. Akad. Nauk UzSSR Ser. Fiz.-Mat. Nauk 11(1967), no. 3, 17-22; MR 35 #5602. [48]
1557. Zmorovič, V. A.; Gudz, L. A. *Some theorems that are connected with A. Marx's conjecture.* (Ukrainian. English and Russian summaries) Dopovidi Akad. Nauk Ukrain. RSR Ser. A 1974, 687-691, 764; MR 50 #7501. [6, 13, 16]
1558. Zmorovič, V. A.; Jakubenko, O. A. *On certain theorems from the theory of special classes of functions analytic in a disc.* (Ukrainian. English and Russian summaries) Dopovidi Akad. Nauk Ukrain. RSR Ser. A 1972, 404-407, 476; MR 46 #5599. [6, 10, 63]
1559. Zmorovič, V. A.; Jakubenko, O. A. *On the theory of extremal properties of special classes of conformal mappings.* (Ukrainian. English and Russian summaries) Dopovidi Akad. Nauk Ukrain. RSR Ser. A (1972), 597-600, 669; MR 46 #9326. [5, 35, 49]
1560. Zmorovič, V. A.; Jakubenko, O. A. *Bounded starlike functions in the disc* $|z| < 1$. (Ukrainian. English and Russian summaries) Dopovidi Akad. Nauk Ukrain. RSR Ser. A(1972), 691-693, 764; MR 46 #9327. [6, 22, 23]
1561. Zmorovič, V. A.; Korobkova, I. K. *A certain generalization of*

Bernardi's integral transform. (Ukrainian. English and Russian summaries) Dopovidi Akad. Nauk. Ukrain. RSR Ser. A 1974, 595-597, 669; MR 50 #10230. [10, 85]

1562. Zmorovič, V. A.; Korobkova, I. K. *On the theory of analytic functions, with positive real part in the disc.* (Russian) Ukrain. Mat. Z. 26(1974), 545-549, 575; MR 50 #13535. [2, 56]

1563. Zmorovič, V. A.; Nikolaeva, R. V. *An application of Abel's integral equation in the theory of univalent functions.* (Ukrainian. English and Russian summaries) Dopovidi Akad. Nauk Ukrain. RSR Ser. A 1975, no. 5, 397-399, 473; MR 52 #8406.

TOPIC REFERENCES

Following is a list of ninety topics dealing directly with or closely related to the theory of univalent and p-valent functions. Following each topic is a list of those references in this Bibliography which contain results pertaining to that topic. The topics are numbered T1, T2, T3, ..., T90. The three topics T15, T17, and T59 are not being used in this Bibliography.

T1. *The Principle of Subordination* (majorization, quasi-subordination, hull-subordination).

[4, 42, 43, 50, 78, 87, 99, 125, 133, 134, 135, 136, 137, 139, 148, 149, 154, 155, 156, 157, 158, 159, 161, 162, 164, 165, 176, 177, 183, 188, 202, 207, 208, 212, 213, 272, 307, 330, 335, 340, 354, 387, 399, 482, 542, 544, 545, 546, 548, 549, 550, 551, 552, 603, 613, 614, 615, 629, 686, 687, 694, 695, 704, 710, 729, 745, 754, 818, 825, 827, 829, 830, 831, 837, 839, 840, 841, 850, 859, 892, 903, 904, 906, 910, 912, 913,

915, 927, 942, 953, 955, 981, 1109, 1122, 1126, 1150, 1152, 1219, 1224, 1226, 1250, 1254, 1322, 1325, 1385, 1389, 1390, 1391, 1404, 1405, 1407, 1467, 1509, 1526, 1528, 1529]

T2. *Functions of Positive Real Part* (functionals of positive real part).

[15, 32, 34, 42, 43, 91, 107, 141, 142, 147, 176, 184, 195, 203, 204, 224, 247, 254, 255, 256, 257, 258, 267, 268, 278, 319, 352, 354, 376, 387, 437, 440, 446, 448, 449, 450, 452, 454, 455, 459, 460, 462, 463, 470, 479, 487, 493, 494, 522, 523, 542, 546, 551, 571, 583, 584, 585, 588, 627, 628, 631, 640, 642, 643, 644, 657, 664, 666, 667, 668, 672, 679, 680, 681, 684, 703, 716, 718, 723, 728, 824, 831, 833, 835, 856, 868, 869, 877, 883, 884, 885, 886, 887, 907, 912, 921, 924, 928, 950, 961, 964, 1010, 1015, 1016, 1031, 1033, 1049, 1050, 1052, 1053, 1076, 1077, 1078, 1080, 1084, 1086, 1087, 1088, 1105, 1108, 1113, 1119, 1120, 1124, 1135, 1139, 1140, 1141, 1144, 1148, 1187, 1189, 1198, 1221, 1228, 1234, 1249, 1250, 1253, 1256, 1258, 1281, 1289, 1290, 1291, 1292, 1293, 1294, 1295, 1296, 1298, 1299, 1320, 1351, 1352, 1354, 1357, 1381, 1387, 1388, 1397, 1415, 1416, 1419, 1420, 1421, 1437, 1440, 1463, 1467, 1472, 1477, 1481, 1496, 1499, 1513, 1519, 1520, 1528, 1529, 1534, 1541, 1542, 1543, 1551, 1554, 1562]

T3. *Integral Means.*

[78, 87, 126, 172, 274, 282, 354, 386, 387, 480, 517, 551, 567, 587, 706, 874, 956, 1021, 1027, 1038, 1039, 1040, 1048, 1051, 1125, 1126, 1224, 1226, 1254, 1320, 1438, 1444, 1473, 1476, 1477, 1523]

T4. *Sufficient (necessary) Conditions for Univalency (p-valency); Transformations Preserving Univalency.*

[5, 8, 26, 27, 28, 33, 64, 71, 72, 75, 76, 77, 78, 79, 80, 101, 102, 103, 108, 124, 126, 129, 130, 133, 143, 144, 145, 171, 174, 183, 185, 190, 202, 203, 214, 217, 220, 225, 227, 245, 247, 248, 252. 255, 268, 273, 281, 288, 289, 290, 293, 294, 296, 303, 334, 337, 352, 355, 356, 370, 393, 395, 407, 409, 410, 413, 424, 436, 437, 439, 451, 475, 478, 479, 480, 489, 495, 542, 549, 552, 554, 575, 601, 605, 608, 658, 687, 694, 696, 697, 707, 755, 756, 757, 762, 773, 802, 805, 815, 817, 835, 836, 844, 851, 853, 858, 868, 872, 888, 900, 903, 906, 908, 910, 914, 928, 933, 946, 956, 958, 966, 969, 971, 986, 992, 999, 1004, 1011, 1014, 1020, 1021, 1031, 1037, 1040, 1050, 1052, 1053, 1054, 1055, 1063, 1086, 1088, 1090, 1095, 1097, 1100, 1110, 1113, 1124, 1125, 1137, 1199, 1221, 1222, 1223, 1225, 1228, 1231, 1248, 1250, 1259, 1260,

1274, 1276, 1297, 1301, 1304, 1306, 1307, 1308, 1310, 1316, 1317, 1321, 1324, 1327, 1335, 1353, 1367, 1380, 1401, 1406, 1407, 1414, 1417, 1420, 1438, 1443, 1461, 1464, 1469, 1486, 1529, 1536, 1539, 1541, 1543, 1548]

T5. *Close-to-Convex (linearly accessible functions), and Generalizations.*

[34, 44, 78, 80, 102, 106, 113, 141, 142, 143, 144, 145, 181, 194, 195, 203, 214, 220, 221, 227, 248, 252, 273, 278, 281, 292, 340, 354, 386, 391, 436, 440, 442, 443, 446, 453, 456, 460, 480, 520, 524, 526, 542, 543, 546, 548, 550, 551, 572, 588, 616, 692, 693, 694, 698, 703, 743, 747, 748, 802, 803, 835, 836, 837, 838, 855, 866, 867, 868, 869, 906, 910, 912, 928, 931, 943, 950, 957, 999, 1004, 1006, 1040, 1051, 1052, 1053, 1055, 1059, 1060, 1061, 1076, 1082, 1084, 1085, 1087, 1109, 1119, 1120, 1124, 1139, 1142, 1149, 1178, 1180, 1184, 1199, 1219, 1220, 1222, 1225, 1231, 1245, 1246, 1249, 1250, 1259, 1260, 1301, 1304, 1321, 1323, 1327, 1333, 1340, 1351, 1353, 1382, 1397, 1400, 1401, 1409, 1441, 1469, 1502, 1514, 1523, 1543, 1548, 1559]

T6. *Starlike Functions, Spiral-like Functions, and Generalizations.*

[20, 30, 33, 34, 42, 43, 44, 51, 54, 66, 70, 91, 92, 93, 99, 101, 102, 103, 106, 112, 113, 114, 115, 117, 118, 119, 126, 131, 141, 142, 143, 144, 145, 146, 155, 160, 162, 163, 165, 166, 170, 176, 179, 181, 184, 185, 188, 191, 192, 194, 195, 198, 202, 203, 214, 220, 221, 225, 229, 230, 248, 253, 254, 257, 268, 269, 273, 281, 291, 293, 313, 314, 319, 326, 327, 330, 335, 340, 342, 343, 344, 345, 347, 348, 349, 351, 352, 354, 356, 357, 377, 379, 388, 391, 392, 394, 420, 437, 438, 439, 447, 453, 455, 457, 458, 459, 460, 466, 468, 475, 488, 489, 517, 525, 526, 527, 528, 540, 542, 543, 544, 545, 546, 550, 551, 552, 553, 581, 582, 583, 586, 587, 588, 599, 600, 601, 602, 603, 606, 614, 615, 616, 619, 629, 633, 635, 639, 641, 642, 643, 658, 665, 666, 688, 689, 692, 693, 695, 697, 703, 709, 712, 713, 726, 727, 745, 747, 748, 752, 775, 776, 777, 790, 791, 802, 807, 825, 826, 830, 834, 835, 837, 841, 844, 848, 850, 852, 853, 855, 857, 858, 860, 866, 868, 873, 874, 882, 884, 887, 888, 889, 891, 892, 893, 904, 906, 912, 913, 915, 916, 917, 925, 927, 928, 930, 941, 952, 957, 966, 968, 979, 988, 989, 990, 991, 995, 997, 999, 1002, 1009, 1012, 1014, 1015, 1031, 1036, 1038, 1039, 1043, 1051, 1052, 1053, 1054, 1077, 1078, 1079, 1080, 1081, 1083, 1084, 1085, 1087, 1088, 1090, 1109, 1113, 1119, 1120, 1122, 1123, 1124, 1132, 1133, 1134, 1147, 1149, 1150, 1153, 1170, 1171, 1177, 1189, 1190, 1191, 1196, 1203, 1204, 1205, 1219, 1221, 1224, 1225, 1226,

1234, 1241, 1242, 1245, 1246, 1248, 1249, 1250, 1273, 1283, 1294, 1297, 1304, 1319, 1320, 1321, 1322, 1323, 1325, 1327, 1330, 1331, 1335, 1336, 1339, 1342, 1344, 1348, 1349, 1350, 1353, 1356, 1357, 1360, 1362, 1378, 1383, 1384, 1385, 1386, 1387, 1391, 1392, 1397, 1398, 1400, 1401, 1402, 1404, 1407, 1408, 1409, 1410, 1415, 1416, 1418, 1419, 1420, 1437, 1441, 1442, 1444, 1467, 1469, 1472, 1473, 1474 1475, 1476, 1477, 1487, 1500, 1510, 1511, 1512, 1514, 1521, 1522, 1526, 1527, 1528, 1533, 1534, 1537, 1538, 1540, 1541, 1542, 1543, 1550, 1551, 1552, 1553, 1554, 1555, 1557, 1558, 1560]

T7. *Length of Image Curve.*

[51, 179, 204, 225, 278, 287, 310, 311, 324, 349, 350, 376, 480, 620, 650, 689, 849, 863, 874, 952, 957, 974, 981, 1039, 1051, 1054, 1055, 1086, 1128, 1149, 1180, 1234, 1235, 1320, 1392, 1441, 1442, 1445, 1477, 1487]

T8. *Coefficient Bounds for* $\|a_{n+1}| - |a_n\|$

[81, 359, 362, 376, 466, 468, 609, 649, 881, 945, 1040, 1047, 1153, 1155, 1220, 1543]

T9. *Meromorphic Univalent Functions (functionals), including Starlike, Convex, Close-to-Convex, etc.*

[1, 8, 27, 68, 71, 72, 78, 91, 96, 104, 128, 134, 136, 176, 230, 233, 243, 265, 279, 282, 296, 314, 315, 316, 317, 331, 333, 340, 421, 429, 441, 467, 478, 489, 490, 506, 524, 535, 583, 621, 647, 650, 652, 664, 688, 708, 709, 712, 713, 722, 726, 727, 749, 750, 751, 757, 759, 764, 771, 791, 796, 797, 814, 823, 842, 843, 857, 867, 877, 934, 935, 938, 939, 940, 944, 950, 951, 952, 954, 966, 1002, 1035, 1043, 1076, 1078, 1079, 1080, 1081, 1083, 1094, 1112, 1113, 1130, 1131, 1141, 1151, 1155, 1158, 1170, 1171, 1234, 1241, 1242, 1243, 1261, 1278, 1289, 1291, 1295, 1296, 1332, 1351, 1357, 1358, 1394, 1419, 1420, 1436, 1445, 1446, 1447, 1449, 1450, 1451, 1452, 1453, 1454, 1455, 1456, 1457, 1459, 1461, 1466, 1472, 1475, 1476, 1479, 1519, 1520, 1522, 1535, 1536, 1537, 1546, 1547, 1549, 1550, 1552]

T10. *Convex Functions* (and generalizations).

[4, 20, 30, 33, 34, 54, 71, 72, 91, 99, 102, 106, 125, 139, 141, 142, 143, 144, 145, 147, 155, 161, 162, 165, 166, 176, 179, 180, 183, 188, 194, 195, 202, 212, 214, 221, 227, 231, 232, 247, 248, 252, 254, 259,

261, 264, 266, 268, 269, 272, 273, 291, 293, 324, 326, 340, 342, 346, 350, 352, 354, 376, 378, 386, 388, 397, 399, 453, 455, 457, 460, 473, 475, 479, 518, 526, 542, 543, 544, 551, 552, 568, 570, 619, 633, 634, 687, 693, 695, 698, 703, 704, 712, 745, 748, 795, 817, 834, 836, 841, 855, 859, 866, 868, 891, 893, 895, 904, 905, 906, 910, 912, 913, 915, 925, 927, 928, 955, 957, 968, 988, 990, 991, 993, 997, 999, 1031, 1036, 1038, 1039, 1040, 1050, 1051, 1052, 1053, 1054, 1055, 1061, 1077, 1078, 1084, 1085, 1087, 1089, 1090, 1113, 1119, 1120, 1122, 1124, 1139, 1189, 1190, 1204, 1219, 1221, 1225, 1226, 1227, 1234, 1244, 1245, 1246, 1248, 1250, 1259, 1273, 1282, 1294, 1297, 1319, 1320, 1321, 1323, 1332, 1333, 1335, 1336, 1338, 1339, 1347, 1353, 1385, 1386, 1392, 1396, 1399, 1400, 1401, 1404, 1405, 1407, 1412, 1418, 1420, 1469, 1470, 1491, 1512, 1514, 1534, 1541, 1545, 1558, 1561]

T11. *Radius of Starlikeness* (of various classes of functions).

[30, 32, 34, 91, 92, 94, 95, 100, 103, 105, 106, 114, 119, 123, 146, 147, 205, 214, 224, 229, 247, 253, 254, 257, 267, 287, 298, 376, 437, 438, 439, 447, 449, 451, 453, 454, 457, 458, 460, 487, 497, 516, 522, 527, 619, 635, 639, 641, 642, 665, 666, 669, 699, 714, 728, 855, 868, 873, 882, 987, 988, 989, 1014, 1016, 1033, 1076, 1077, 1078, 1081, 1084, 1087, 1090, 1124, 1134, 1187, 1189, 1198, 1227, 1242, 1251, 1262, 1289, 1293, 1294, 1295, 1297, 1298, 1299, 1335, 1338, 1350, 1352, 1357, 1361, 1381, 1386, 1387, 1396, 1467, 1472, 1487, 1511, 1526, 1528, 1533, 1542, 1544, 1553, 1555]

T12. *Radius of Convexity* (of various classes of functions).

[32, 33, 35, 73, 91, 93, 100, 103, 106, 117, 123, 142, 211, 212, 214, 220, 225, 239, 247, 255, 267, 268, 288, 292, 330, 340, 348, 351, 374, 376, 377, 380, 381, 440, 441, 442, 443, 446, 449, 451, 453, 455, 460, 526, 527, 542, 544, 572, 616, 619, 635, 640, 642, 643, 644, 657, 658, 667, 668, 702, 726, 727, 765, 776, 795, 805, 855, 858, 868, 882, 884, 885, 887, 888, 889, 905, 921, 924, 926, 930, 931, 950, 966, 967, 969, 984, 987, 988, 990, 996, 1002, 1009, 1010, 1014, 1015, 1018, 1033, 1061, 1079, 1081, 1083, 1084, 1086, 1090, 1124, 1135, 1139, 1141, 1144, 1187, 1190, 1196, 1198, 1235, 1242, 1249, 1251, 1257, 1260, 1289, 1291, 1292, 1294, 1295, 1296, 1297, 1318, 1333, 1335, 1338, 1346, 1348, 1349, 1350, 1351, 1352, 1357, 1381, 1386, 1387, 1396, 1416, 1437, 1467, 1471, 1472, 1487, 1519, 1520, 1527, 1533, 1534, 1535, 1538, 1543, 1546, 1548, 1550, 1552]

T13. *Typically-real Frunctions* (and generalizations).

[42, 43, 91, 151, 178, 197, 267, 300, 301, 319, 325, 366, 382, 444, 465, 471, 521, 546, 551, 628, 642, 683, 702, 705, 719, 725, 819, 820, 821, 879, 904, 926, 929, 1005, 1118, 1148, 1204, 1207, 1224, 1227, 1236, 1237, 1247, 1258, 1364, 1504, 1540, 1557]

T14. *p-valent (weakly p-valent), and Generalizations.*

[1, 103, 246, 282, 283, 364, 367, 377, 427, 453, 466, 468, 475, 482, 537, 538, 547, 548, 549, 550, 558, 559, 562, 564, 567, 599, 600, 653, 713, 714, 797, 803, 804, 805, 807, 866, 867, 869, 878, 976, 977, 978, 1000, 1006, 1030, 1042, 1044, 1050, 1080, 1087, 1103, 1187, 1199, 1226, 1241, 1255, 1294, 1343, 1373, 1375, 1380, 1381, 1397, 1398, 1399, 1400, 1413, 1420, 1484, 1485, 1486, 1487, 1508, 1528, 1536]

T15. *Distortion Theorems* (This topic is not being used.)

T16. *Conjectures, Open Problems.*

[19, 78, 87, 102, 112, 117, 177, 187, 200, 214, 215, 252, 268, 278, 288, 330, 335, 336, 337, 344, 355, 395, 398, 405, 408, 459, 473, 474, 475, 476, 477, 482, 484, 524, 542, 546, 550, 557, 594, 603, 606, 613, 619, 620, 704, 745, 746, 801, 818, 819, 845, 868, 871, 874, 889, 904, 905, 906, 914, 956, 969, 992, 1020, 1054, 1079, 1109, 1113, 1124, 1137, 1153, 1164, 1166, 1171, 1200, 1219, 1220, 1221, 1224, 1226, 1228, 1241, 1242, 1248, 1254, 1276, 1304, 1320, 1323, 1325, 1327, 1328, 1335, 1350, 1351, 1391, 1400, 1402, 1444, 1446, 1474, 1479, 1530, 1557]

T17. *Coefficient Bounds* (This topic is not being used.)

T18. *Area of Image Domain.*

[63, 162, 172, 204, 225, 261, 311, 355, 376, 440, 446, 480, 583, 586, 786, 787, 789, 863, 873, 874, 974, 979, 981, 1039, 1051, 1054, 1055, 1086, 1149, 1254, 1441, 1487]

T19. *Covering Theorems* (missed values, Koebe sets, Bloch's constant).

[1, 18, 19, 20, 117, 143, 144, 145, 152, 153, 163, 178, 201, 212, 260, 262, 263, 266, 291, 305, 306, 313, 335, 376, 391, 440, 493, 507, 513, 571, 642, 647, 650, 685, 686, 687, 711, 712, 745, 747, 748, 790, 791, 793, 826, 832, 834, 883, 889, 891, 900, 903, 927, 958, 977, 978, 980,

1020, 1021, 1023, 1024, 1025, 1026, 1104, 1124, 1126, 1162, 1164, 1201, 1203, 1205, 1206, 1212, 1216, 1217, 1243, 1262, 1292, 1301, 1333, 1335, 1347, 1348, 1376, 1384, 1386, 1402, 1411, 1413, 1429, 1468, 1482, 1488, 1492, 1497, 1498, 1527, 1539]

T20. *Partial Sums* (Cesàro sums).

[3, 125, 139, 147, 193, 200, 204, 436, 442, 447, 448, 451, 452, 540, 635, 687, 690, 691, 720, 746, 753, 836, 910, 913, 1060, 1076, 1087, 1222, 1244, 1249, 1250, 1321, 1323, 1352, 1487, 1529]

T21. *Coefficient Bounds (relations) for Functions (functionals) of Positive Real Part of Topic 2.*

[15, 107, 125, 139, 147, 204, 436, 450, 459, 462, 540, 542, 571, 585, 588, 627, 628, 629, 635, 640, 642, 657, 746, 824, 869, 883, 885, 887, 907, 910, 913, 1086, 1087, 1187, 1222, 1249, 1250, 1256, 1281, 1323, 1352, 1415, 1467, 1513]

T22. *Bounded Functions* (functionals).

[4, 31, 32, 59, 91, 111, 115, 125, 127, 130, 154, 156, 166, 170, 179, 180, 213, 236, 237, 238, 265, 276, 281, 295, 297, 305, 306, 307, 308, 309, 323, 341, 351, 353, 358, 384, 385, 388, 418, 430, 431, 432, 433, 434, 438, 450, 453, 496, 497, 507, 522, 536, 588, 605, 621, 622, 624, 625, 626, 661, 693, 704, 732, 733, 734, 798, 800, 801, 846, 854, 856, 883, 891, 895, 904, 919, 927, 936, 937, 938, 962, 963, 971, 1050, 1052, 1078, 1092, 1128, 1155, 1163, 1165, 1173, 1187, 1201, 1203, 1205, 1209, 1210, 1211, 1229, 1233, 1242, 1249, 1251, 1253, 1262, 1264, 1271, 1272, 1275, 1276, 1277, 1290, 1296, 1298, 1306, 1328, 1329, 1347, 1354, 1356, 1357, 1367, 1368, 1369, 1375, 1381, 1401, 1402, 1418, 1422, 1423, 1424, 1425, 1426, 1427, 1443, 1473, 1477, 1516, 1528, 1532, 1560]

T23. *Representation Theorems* (formulas).

[44, 130, 151, 178, 184, 187, 191, 192, 273, 278, 288, 293, 319, 325, 340, 347, 352, 355, 382, 409, 445, 465, 469, 481, 483, 513, 516, 542, 546, 551, 581, 583, 585, 660, 681, 707, 779, 802, 808, 819, 872, 873, 874, 879, 892, 1011, 1012, 1014, 1034, 1035, 1039, 1040, 1043, 1080, 1083, 1088, 1105, 1106, 1113, 1118, 1119, 1120, 1123, 1124, 1125, 1138, 1153, 1207, 1234, 1256, 1273, 1327, 1335, 1342, 1343, 1348, 1382, 1385, 1386, 1387, 1397, 1400, 1401, 1419, 1491, 1502, 1510, 1511, 1520, 1527, 1540, 1548, 1554, 1560]

TOPIC REFERENCES   261

T24. *Variational (symmetrization) Methods.*

[18, 19, 37, 42, 44, 45, 46, 52, 53, 55, 56, 68, 82, 83, 84, 87, 89, 90, 111, 115, 131, 133, 134, 152, 167, 168, 169, 184, 195, 230, 235, 238, 250, 251, 287, 288, 306, 312, 313, 341, 342, 391, 392, 395, 408, 421, 422, 424, 425, 426, 432, 434, 485, 486, 498, 500, 530, 532, 569, 576, 604, 607, 612, 626, 633, 636, 638, 642, 645, 648, 654, 661, 662, 672, 676, 703, 707, 733, 734, 741, 742, 749, 752, 758, 760, 769, 792, 808, 813, 831, 833, 842, 843, 860, 875, 878, 890, 892, 893, 894, 895, 896, 897, 914, 920, 923, 934, 935, 938, 941, 951, 954, 975, 976, 982, 983, 985, 1007, 1012, 1013, 1025, 1026, 1056, 1058, 1105, 1107, 1108, 1112, 1113, 1115, 1120, 1121, 1122, 1123, 1130, 1131, 1138, 1146, 1150, 1152, 1161, 1170, 1173, 1175, 1201, 1205, 1206, 1209, 1212, 1225, 1256, 1266, 1269, 1271, 1273, 1274, 1276, 1341, 1344, 1383, 1432, 1472, 1478, 1502]

T25. *Coefficient Bounds (relations) for Meromorphic Univalent Functions of Topic 9.*

[176, 265, 296, 314, 315, 316, 331, 333, 506, 583, 647, 664, 688, 708, 712, 749, 750, 751, 764, 771, 843, 883, 944, 1043, 1080, 1141, 1170, 1171, 1234, 1278, 1332, 1419, 1445, 1146, 1522]

T26. *Unrelated (disjoint) Functions* (image domains have no common values).

[15, 59, 60, 88, 89, 90, 196, 308, 512, 514, 515, 646, 651, 731, 811, 816, 1019, 1421]

T27. *Area Principle* (and generalizations).

[9, 62, 265, 282, 296, 338, 481, 511, 515, 534, 583, 587, 589, 611, 647, 656, 810, 811, 814, 816, 823, 845, 946, 1020, 1153, 1368, 1423, 1480, 1524]

T28. *Univalence of Special Functions* (such as Bessel functions, or of functions having special forms).

[24, 25, 121, 322, 469, 675, 700, 701, 788, 989, 1016, 1300, 1447, 1457, 1458, 1466]

T29. *Region of Variability* (of coefficients, or of certain functionals).

[46, 47, 48, 49, 54, 68, 160, 171, 191, 192, 197, 237, 238, 240, 242, 244, 314, 319, 373, 429, 430, 465, 467, 470, 471, 485, 525, 528, 623,

636, 637, 638, 647, 661, 683, 725, 733, 734, 741, 749, 777, 838, 842, 886, 921, 936, 937, 946, 982, 1002, 1024, 1026, 1049, 1058, 1105, 1114, 1116, 1130, 1148, 1173, 1174, 1197, 1202, 1204, 1208, 1209, 1210, 1211, 1250, 1259, 1344, 1350, 1355, 1370, 1371, 1379, 1382, 1394, 1478, 1511, 1517, 1521, 1548, 1549]

T30. *Coefficient Bounds (relations) for Bounded Functions of Topic 22.*

[127, 170, 180, 281, 305, 306, 309, 358, 431, 450, 536, 588, 624, 625, 693, 732, 800, 801, 904, 927, 971, 1092, 1155, 1187, 1251, 1271, 1272, 1275, 1276, 1277, 1306, 1328, 1329, 1368, 1402, 1418, 1422, 1423, 1424, 1425, 1426, 1532]

T31. *Schwarzian Derivative, the Differential Equation* $w'' + \varrho w = 0$.

[8, 13, 72, 78, 135, 227, 228, 265, 268, 274, 334, 337, 370, 407, 495, 512, 555, 696, 762, 773, 817, 823, 858, 897, 984, 987, 1014, 1019, 1050, 1052, 1063, 1090, 1221, 1304, 1486, 1494, 1543]

T32. *Univalence of Polynomials* (Schild, Remak, Hurwitz).

[122, 123, 124, 171, 174, 249, 294, 372, 435, 473, 484, 580, 584, 593, 659, 746, 799, 830, 910, 919, 929, 932, 1062, 1091, 1117, 1145, 1183, 1185, 1186, 1187, 1236, 1237, 1238, 1239, 1241, 1242, 1252, 1285, 1317, 1335, 1366, 1403, 1406, 1408, 1409, 1421]

T33. *Mean-valent Functions* (and generalizations).

[1, 6, 7, 132, 210, 212, 282, 358, 362, 364, 366, 558, 560, 561, 562, 563, 565, 653, 654, 878, 880, 881, 976, 978, 1042, 1044, 1045, 1046, 1047, 1051, 1153, 1155, 1303, 1304, 1376]

T34. *Bieberbach-Eilenberg Functions* (generalizations and related classes; Guelfer functions; Aharonov pairs, etc.)

[9, 11, 15, 18, 88, 138, 262, 307, 308, 509, 511, 512, 515, 605, 608, 646, 651, 1020, 1021, 1100, 1493]

T35. *Curvature, Level Curves.*

[18, 54, 65, 235, 350, 374, 375, 378, 380, 398, 412, 473, 474, 508, 520, 525, 528, 539, 620, 659, 682, 895, 918, 921, 957, 972, 973, 991, 1041, 1106, 1127, 1135, 1143, 1144, 1273, 1286, 1287, 1288, 1372, 1393, 1395, 1471, 1551, 1559]

T36. *Coefficient Bounds (relations) for Starlike Functions of Topic 6.*

[70, 101, 118, 126, 163, 170, 176, 179, 188, 191, 192, 253, 281, 343, 345, 347, 348, 455, 458, 459, 488, 545, 546, 553, 586, 587, 601, 602, 629, 658, 688, 693, 712, 776, 852, 857, 887, 888, 904, 1009, 1087, 1122, 1132, 1133, 1134, 1170, 1171, 1248, 1273, 1301, 1319, 1330, 1331, 1335, 1348, 1349, 1356, 1387, 1402, 1415, 1416, 1417, 1418, 1467, 1510, 1511, 1522, 1527, 1543]

T37. *Schwarz's Lemma* (and generalizations).

[388, 482, 1085, 1186, 1254]

T38. *Coefficient Bounds (relations) for Convex Functions of Topic 10.*

[143, 144, 145, 281, 346, 693, 712, 867, 904, 1122, 1220, 1248, 1273, 1319, 1335, 1417, 1418, 1470, 1491]

T39. *Special Conditions on Coefficients* (such as gaps, realness, monotonicity, etc.).

[12, 53, 150, 201, 234, 236, 237, 238, 240, 241, 242, 244, 248, 364, 373, 376, 395, 409, 410, 485, 489, 490, 491, 492, 542, 558, 624, 625, 663, 684, 733, 734, 747, 751, 752, 783, 784, 790, 791, 805, 834, 851, 886, 908, 937, 1056, 1155, 1173, 1182, 1197, 1265, 1271, 1278, 1305, 1313, 1314, 1318, 1335, 1359, 1439, 1478, 1525]

T40. *Functions which are Starlike in One Direction.*

[524, 642, 853, 1297, 1301]

T41. *Functions which are Convex in One Direction.*

[113, 300, 513, 524, 542, 571, 572, 575, 628, 642, 747, 834, 853, 926, 1203, 1205, 1219, 1220, 1223, 1235, 1297, 1301, 1419]

T42. *Bieberbach Conjecture* ($|a_n| \leq n$).

[12, 14, 45, 87, 167, 168, 187, 333, 336, 338, 339, 340, 359, 360, 395, 405, 408, 421, 422, 423, 424, 425, 426, 485, 510, 550, 593, 594, 610, 649, 655, 724, 767, 904, 923, 1020, 1064, 1065, 1067, 1068, 1069, 1070, 1073, 1074, 1075, 1095, 1097, 1098, 1100, 1101, 1200, 1219, 1220, 1226, 1230, 1249, 1265, 1304, 1315, 1325, 1444, 1483, 1508]

T43. *Radius of Univalency (p-valency)* (of various classes of functions).

[30, 31, 32, 54, 103, 116, 119, 122, 142, 146, 178, 205, 206, 212, 214, 224, 228, 253, 256, 257, 288, 293, 340, 391, 428, 436, 437, 438, 448, 449, 454, 460, 475, 478, 482, 522, 544, 572, 699, 702, 707, 720, 746, 753, 828, 882, 910, 1016, 1050, 1052, 1076, 1077, 1078, 1081, 1087, 1088, 1094, 1163, 1189, 1198, 1227, 1241, 1242, 1244, 1249, 1264, 1298, 1304, 1305, 1307, 1314, 1317, 1318, 1381, 1388, 1447, 1448, 1449, 1450, 1451, 1452, 1453, 1454, 1455, 1456, 1457, 1459, 1460, 1461, 1462, 1463, 1485, 1503, 1528, 1529, 1555]

T44. *Survey Articles, Books, Collections of Various Papers, Symposiums, Bibliographies.*

[22, 38, 41, 57, 61, 85, 86, 140, 175, 214, 226, 280, 285, 333, 371, 411, 414, 464, 476, 519, 556, 557, 559, 563, 786, 787, 789, 794, 948, 1017, 1082, 1162, 1167, 1168, 1232, 1263, 1268, 1280, 1283, 1304, 1326, 1490]

T45. *Odd Univalent Functions* (class S or class Σ)

[128, 252, 269, 291, 309, 336, 339, 388, 405, 426, 449, 458, 524, 546, 747, 802, 896, 904, 972, 1020, 1040, 1155, 1200, 1325, 1339, 1400]

T46. *Invariant (angular, spherical) Derivative.*

[279, 507, 876, 894, 1163, 1507]

T47. *Convolution (Faltung) of Functions, Hadamard Product.*

[4, 102, 116, 139, 272, 300, 302, 444, 450, 573, 729, 906, 926, 955, 1032, 1245, 1246, 1249, 1250, 1251, 1325, 1401, 1409, 1536]

T48. *Schlicht (or other properties) in an Annulus.*

[2, 150, 327, 470, 471, 730, 770, 792, 894, 895, 976, 978, 1008, 1136, 1147, 1148, 1364, 1373, 1374, 1428, 1434, 1467, 1556]

T49. *Special Geometry of Map* (such as k-fold symmetry).

[40, 47, 54, 110, 153, 170, 199, 221, 245, 291, 295, 299, 308, 317, 340, 363, 391, 394, 397, 431, 432, 433, 442, 513, 516, 521, 531, 536, 622, 673, 694, 744, 745, 747, 774, 781, 782, 785, 838, 853, 865, 901, 903, 914, 934, 935, 939, 944, 957, 974, 997, 1034, 1040, 1093, 1177, 1213, 1254, 1255, 1435, 1437, 1535, 1537, 1551, 1554, 1559]

T50. *Coefficient Bounds (relations) for p-valent Functions of Topic* 14.

[282, 550, 670, 803, 807, 866, 869, 1103, 1508]

T51. *Coefficient Bounds (relations) for Typically-Real Functions of Topic* 13.

[151, 628, 819, 820, 821, 904, 1207, 1236, 1237, 1247]

T52. *Coefficient Bounds (relations) for Odd Univalent Functions of Topic* 45.

[405, 426, 904, 1172]

T53. *Faber Polynomials* (Grunsky coefficients).

[10, 13, 29, 136, 296, 304, 306, 307, 308, 309, 331, 393, 395, 396, 406, 421, 422, 424, 425, 483, 499, 589, 590, 605, 608, 611, 617, 618, 655, 676, 677, 721, 737, 749, 762, 768, 770, 798, 811, 812, 814, 816, 824, 845, 854, 897, 949, 1020, 1064, 1065, 1067, 1068, 1069, 1070, 1073, 1074, 1075, 1092, 1093, 1095, 1096, 1097, 1100, 1123, 1151, 1160, 1230, 1269, 1270, 1272, 1275, 1276, 1277, 1367, 1368, 1369, 1423, 1479, 1495, 1515]

T54. *Coefficient Bounds (relations) for Functions of the Class S:* $f(z) = z + a_2 z^2 + \ldots$ *Analytic and Univalent in* $|z| < 1$.

[13, 14, 21, 49, 67, 107, 108, 113, 129, 152, 167, 169, 175, 212, 243, 277, 281, 306, 308, 332, 333, 338, 359, 360, 391, 395, 396, 405, 419, 422, 425, 445, 485, 486, 491, 493, 498, 499, 500, 501, 502, 503, 504, 505, 506, 509, 554, 589, 594, 601, 611, 623, 625, 630, 647, 648, 649, 655, 656, 673, 674, 685, 693, 735, 749, 771, 811, 822, 829, 845, 865, 903, 904, 914, 944, 1020, 1022, 1056, 1064, 1065, 1066, 1067, 1068, 1069, 1070, 1071, 1072, 1073, 1074, 1075, 1095, 1096, 1097, 1154, 1155, 1156, 1170, 1226, 1270, 1272, 1306, 1365, 1402, 1518]

T55. *Continued Fractions* (applied to Schlicht Functions). [None]

T56. *Distortion Theorems for Functions of Positive Real Part of Topic* 2.

[33, 91, 147, 224, 256, 265, 267, 268, 376, 387, 450, 522, 523, 524, 542, 627, 631, 640, 643, 644, 657, 679, 680, 723, 831, 833, 883, 885, 887, 912, 950, 1050, 1076, 1078, 1086, 1088, 1119, 1120, 1198, 1221, 1290, 1291, 1292, 1294, 1295, 1298, 1351, 1352, 1357, 1381, 1396, 1415, 1416, 1419, 1420, 1440, 1467, 1477, 1528, 1529, 1534, 1541, 1542, 1562]

T57. *Univalence Over Regions other than the Unit Disc.*

[24, 25, 50, 55, 65, 120, 326, 382, 390, 477, 597, 760, 986, 1284, 1379, 1421]

T58. *Distortion Theorems for Meromorphic Functions of Topic 9.*

[71, 134, 490, 535, 595, 708, 934, 935, 939, 940, 950, 951, 952, 1112, 1113, 1155, 1158, 1243, 1419, 1436, 1475, 1476, 1519, 1546, 1547]

T59. *Related Results from Analytic Function Theory.*
[This topic is not being used.]

T60. *Multiply-Connected Regions.*

[56, 88, 121, 328, 379, 383, 975, 1146, 1431, 1433]

T61. *Distortion Theorems for Bounded Functions of Topic 22.*

[32, 91, 213, 236, 265, 295, 297, 308, 351, 388, 418, 431, 432, 433, 434, 453, 496, 806, 854, 883, 891, 895, 904, 1037, 1050, 1052, 1078, 1163, 1165, 1229, 1233, 1262, 1290, 1298, 1357, 1381, 1516, 1528]

T62. *Boundary Behavior* (rate of growth of coefficients or of functionals).

[26, 66, 74, 126, 129, 151, 162, 181, 189, 210, 212, 221, 234, 279, 287, 332, 349, 351, 352, 354, 361, 365, 367, 368, 369, 390, 391, 415, 416, 538, 553, 562, 566, 567, 581, 582, 587, 588, 590, 596, 598, 620, 638, 705, 797, 802, 805, 857, 861, 863, 870, 871, 872, 874, 876, 898, 899, 900, 902, 926, 947, 956, 958, 960, 1027, 1029, 1037, 1038, 1039, 1043, 1046, 1085, 1099, 1102, 1124, 1164, 1167, 1253, 1322, 1434, 1437, 1473, 1474, 1475, 1476, 1506, 1530]

T63. *Distortion Theorems for Starlike Functions of Topic 6.*

[30, 126, 131, 163, 191, 192, 194, 230, 253, 269, 347, 377, 388, 458, 459, 524, 543, 587, 616, 619, 643, 658, 775, 776, 802, 857, 858, 888, 912, 1009, 1083, 1085, 1122, 1248, 1250, 1297, 1301, 1319, 1323, 1335, 1348, 1349, 1350, 1386, 1387, 1416, 1419, 1474, 1475, 1500, 1511, 1527, 1533, 1537, 1543, 1558]

T64. *Distortion Theorems for Convex Functions of Topic 10.*

[30, 139, 143, 144, 145, 147, 194, 247, 252, 266, 388, 479, 543, 619,

795, 891, 905, 906, 912, 1059, 1085, 1119, 1120, 1122, 1248, 1250, 1319, 1321, 1323, 1335, 1353, 1382, 1491]

T65. *Distortion Theorems for p-valent Functions of Topic* 14.

[803, 807, 1373]

T66. *Distortion Theorems for Typically-Real Functions of Topic* 13.

[178, 820, 1005, 1118, 1207, 1540]

T67. *Distortion Theorems for Odd Univalent Functions of Topic* 45.

[252, 388, 458, 524]

T68. *Distortion Theorems for Functions of the Class S:* $f(z) = z + a_2 z^2 + \ldots$ *Analytic and Univalent in* $|z| < 1$.

[47, 74, 109, 134, 212, 222, 241, 265, 266, 271, 321, 336, 338, 339, 379, 383, 396, 437, 458, 490, 493, 525, 528, 529, 531, 533, 535, 554, 579, 595, 613, 622, 630, 671, 724, 738, 741, 829, 846, 864, 905, 914, 951, 969, 1050, 1077, 1110, 1139, 1155, 1159, 1163, 1200, 1213, 1218, 1243, 1248, 1267, 1282, 1322, 1436, 1437, 1439, 1493, 1501]

T69. α-*Convex,* α-*Starlike Functions (Mocanu functions) and Generalizations.*

[33, 35, 95, 97, 98, 100, 105, 239, 289, 290, 291, 293, 318, 353, 355, 358, 777, 778, 779, 780, 808, 835, 958, 959, 961, 966, 967, 968, 969, 990, 996, 997, 998, 999, 1000, 1001, 1002, 1003, 1193, 1196, 1337, 1342, 1417, 1418]

T70. *Bazilevič Functions* (and generalizations).

[2, 216, 219, 289, 290, 293, 356, 358, 694, 779, 780, 808, 847, 931, 956, 969, 997, 1003, 1004, 1006, 1040, 1051, 1054, 1055, 1178, 1179, 1181, 1194, 1246, 1324, 1355, 1443, 1444, 1446, 1512]

T71. *Functions of Bounded Boundary Rotation.*

[17, 78, 172, 173, 177, 179, 214, 286, 288, 292, 293, 340, 494, 704, 706, 707, 774, 802, 804, 805, 809, 863, 875, 882, 960, 965, 970, 1014, 1035, 1036, 1038, 1039, 1040, 1041, 1043, 1047, 1055, 1090, 1113, 1121, 1124, 1125, 1195, 1214, 1215, 1223, 1225, 1273, 1274, 1341, 1343, 1345, 1388, 1445, 1465, 1471]

T72. *Distortion Theorems Involving Coefficients* (various classes of functions).

[32, 91, 388, 444, 448, 449, 846, 883, 884, 887, 1085, 1258, 1357, 1437]

T73. *Radius of Close-to-Convexity* (of various classes of functions).

[91, 142, 214, 221, 222, 292, 340, 453, 460, 699, 746, 776, 802, 855, 858, 1033, 1076, 1084, 1124, 1227, 1352, 1353, 1543]

T74. *Distortion Theorems of Close-to-Convex Functions of Topic 5.*

[194, 252, 440, 442, 443, 446, 838, 912, 1085, 1149, 1178, 1249, 1260, 1323, 1327, 1333, 1340, 1502]

T75. *Coefficient Bounds (relations) for Close-to-Convex Functions of Topic 5.*

[113, 440, 443, 446, 456, 524, 546, 548, 588, 693, 867, 912, 1149, 1219, 1327, 1333, 1340, 1502]

T76. *Distortion Theorems for Functions of Bounded Boundary Rotation of Topic 71.*

[172, 288, 293, 340, 707, 802, 805, 1014, 1035, 1038, 1125, 1223, 1273, 1343, 1445]

T77. *Coefficient Bounds for Functions of Bounded Boundary Rotation of Topic 71.*

[172, 173, 177, 286, 288, 340, 706, 774, 802, 804, 805, 875, 1014, 1035, 1036, 1038, 1047, 1090, 1125, 1223, 1273, 1274, 1345, 1445, 1471]

T78. *Distortion Theorems for Functions Convex in One Direction of Topic 41.*

[524, 571, 572, 1235, 1301]

T79. *Coefficient Bounds for Functions Convex in One Direction of Topic 41.*

[113, 524, 571, 572, 628, 972, 1219, 1220, 1223, 1301]

T80. *Distortion Theorems for Bazilevic Functions of Topic 70.*

[847, 1006, 1051]

T81. *Coefficient Bounds (relations) for Bazilevic Functions of Topic* 70.

[358, 694, 956, 1006, 1443, 1446]

T82. *Linear Combinations, Products, of Univalent Functions.*

[32, 36, 91, 142, 205, 211, 214, 248, 269, 351, 440, 446, 451, 460, 468, 475, 477, 526, 552, 678, 715, 795, 837, 855, 859, 1018, 1033, 1089, 1109, 1227, 1247, 1264, 1334, 1338, 1355, 1357, 1361, 1396, 1400, 1469, 1526, 1536, 1542, 1548]

T83. *Distortion Theorems for α-Convex (α-starlike) Functions of Topic* 69.

[291, 293, 777, 958, 1342, 1548]

T84. *Coefficient Bounds (relations) for α-Convex (α-starlike) Functions of Topic* 69.

[291, 358, 777, 778, 779, 780, 966, 1417, 1418, 1548]

T85. *Univalence of Integrals.*

[34, 71, 72, 91, 94, 96, 99, 102, 105, 139, 141, 142, 203, 205, 214, 224, 228, 248, 269, 357, 527, 697, 698, 746, 828, 906, 925, 928, 1050, 1052, 1053, 1111, 1124, 1199, 1227, 1231, 1297, 1301, 1353, 1355, 1541, 1544, 1548, 1561]

T86. *Distortion Theorems for Bieberbach-Eilenberg (and related classes) Functions of Topic* 34.

[9, 11, 15, 138, 509, 511, 646, 651]

T87. *Coefficient Bounds (relations) for Bieberbach-Eilenberg (and related classes) Functions of Topic* 34.

[11, 15, 509, 511, 512, 608, 1021]

T88. *Extreme Point Theory.*

[182, 186, 187, 188, 189, 273, 284, 387, 500, 541, 542, 544, 545, 546, 547, 548, 550, 551, 574, 575, 576, 577, 578, 585, 709, 807, 912, 913, 916, 942, 1249, 1335, 1336, 1340]

T89. *Functions of Bounded Index.*

[401, 402, 403, 699, 1187]

T90. *Entire Functions.*

[400, 404, 406, 579, 699, 817, 1029, 1192, 1243, 1302, 1304, 1306, 1307, 1308, 1309, 1310, 1311, 1312, 1315, 1316, 1317]

TABLE 1

Following is a list of those references in this bibliography which were published prior to the year 1966 and which were not included in Bibliography I.

Year	References
1950	[1484]
1955	[1485]
1959	[1059, 1060, 1255]
1960	[341, 781, 1328, 1366, 1516]
1961	[782, 982, 1486]
1962	[149, 150, 372, 621, 645, 783, 877, 1378]
1963	[37, 81, 234, 300, 312, 498, 670, 731, 741, 758, 782, 784, 785, 983, 1117, 1169, 1208]
1964	[1, 38, 46, 148, 235, 310, 373, 414, 660, 676, 752, 759, 786, 787, 865, 878, 947, 984, 985, 1011, 1061, 1146, 1173, 1212, 1256, 1257, 1380, 1427, 1436, 1546]
1965	[4, 39, 42, 51, 58, 66, 127, 128, 236, 276, 311, 330, 374, 375, 417, 424, 436, 465, 499, 500, 513, 516, 615, 636, 646, 677, 681, 684, 716, 736, 753, 755, 788, 790, 825, 826, 827, 828, 836, 837, 842, 843, 851, 866, 867, 879, 903, 972, 975, 976, 977, 978, 979, 980, 986, 1034, 1064, 1076, 1077, 1078, 1093, 1094, 1105, 1137, 1138, 1149, 1150, 1151, 1174, 1182, 1198, 1209, 1219, 1231, 1264, 1271, 1272, 1279, 1281, 1365, 1381, 1392, 1419, 1421, 1428, 1429, 1430, 1447, 1448, 1449, 1450, 1451, 1452, 1487, 1488, 1550, 1551, 1552, 1553]

TABLE 2

Following is a list of those references in this bibliography which were published during the year 1976.

[116, 153, 216, 218, 272, 290, 333, 357, 397, 548, 549, 550, 578, 580, 594, 652, 680, 691, 708, 739, 822, 849, 874, 916, 943, 964, 965, 1045, 1099, 1184, 1262, 1305, 1336, 1340, 1343, 1410, 1439, 1544]

TABLE 3

Following is a listing of the total number of research papers in this bibliography which were published in each of the given years.

Year	Total Number of Papers
1966	123
1967	110
1968	107
1969	130
1970	146
1971	149
1972	160
1973	166
1974	144
1975	125

TABLE 4

Following is a list of Math. Review (MR) numbers (which had not been available at time of first publication) for some of the References in the Bibliography, Part II.

Ref.	MR#	Ref.	MR#	Ref.	MR#	Ref.	MR#
17	MR54 #2939	186	MR53 #5849	332	MR52 #8407	519	MR53 #5845
23	53 3297	194	53 788	345	52 11029	547	53 11036
36	53 3288	205	57 9950	359	53 3286	548	53 11036
37	54 528	222	54 7768	360	56 12250	594	53 792
76	53 801	234	52 11031	366	52 14259	632	54 2946
79	53 5848	258	53 3314	367	52 11030	678	52 14261
87	54 5456	264	54 5451	370	57 12837	707	52 14271
153	54 525	271	53 8427	400	52 11039	720	57 12839
166	52 14258	303	53 11031	461	57 3372	758	54 5457
175	54 537	320	52 11023	496	53 11043	776	53 11037
780	52 14266	965	52 11024	1262	52 11027	1376	53 5844
803	53 11038	1018	52 14263	1270	53 8409	1377	52 14265
805	52 14268	1022	53 5846	1280	58 22527	1388	53 3290
822	52 14267	1043	54 2943	1318	54 5465	1425	55 10662
835	53 3282	1090	52 11025	1326	54 5479	1465	53 13549

Ref.	MR#	Ref.	MR#	Ref.	MR#	Ref.	MR#
849	52 14262	1166	58 1101	1336	54 2945	1483	53 8410
859	54 13057	1181	52 11026	1358	53 11041	1530	53 8429
862	53 11040	1183	53 5847	1362	52 14264	1531	54 2948
896	55 653	1208	52 11032	1363	53 8408	1532	53 3287
963	54 554	1250	56 5862	1371	56 5857		

Corrections

Reference	same as	Reference
385	......	384
607	......	604
618	......	617
692	......	126

Notes (a) Reference 1094: author is Todorov, Pavel
(b) Ezrohi, T. G. is same author as Pletneva, T. G.

BIBLIOGRAPHY OF SCHLICHT FUNCTIONS Part III (1976-1981)

CONTENTS

Preface 275
Bibliography.. 276
Table 1. References Prior to Year 1976 351
Table 2. References in Year 1982 352
Table 3. Number of References Published Each Year 352
Corrections 353

PREFACE

Bibliography of Schlicht Functions, Part III, contains 1025 references to publications in the theory of analytic univalent (Schlicht) and multivalent functions. Part III covers the years 1976 through 1981 and is a continuation of Bibliography of Schlicht Functions, Parts I, II, described in earlier pages.

Some papers published prior to the year 1976 and which were not included in Parts I, II are now listed in Table I. Some papers published in the year 1982 are listed in Table II. Math. Review numbers (MR) are included for most references.

Part III differs from Parts I, II in two respects. Abstracts have been omitted and also no subtopic classification and cross-index listings are included. The omission of abstracts is easily justified because of their transient value. Classification of the various results in the theory of Schlicht functions into subtopics (68 subtopics are used in Part I, 90 subtopics are used in Part II) and cross-index listings between references and subtopics is a very valuable but monumental task requiring hundreds of hours of work—precious time not available prior to the publication deadline for Part III.

S. D. Bernardi
July 1982

BIBLIOGRAPHY OF SCHLICHT FUNCTIONS (Part III)

1. Abe, H. *On some analytic functions.* Mem. Ehime Univ. Sect. III Engrg. 7 (1973), no. 4, 287-292; MR 58 #28477.
2. Abe. H. *On multivalent functions in multiply connected domains.* I. Proc. Japan Acad. 53 (1977), no. 3, 116-119; MR 56 #594.
3. Abe, H. *On multivalent functions in multiply connected domains.* II. Proc. Japan Acad. Ser. A Math. Sci. 53 (1977), no. 2, 68-71; MR 58 #6210.
4. Abian, A. *The coefficients of the Laurent expansion of analytic functions.* Arch. Math. (Brno) 13 (1977), no. 2, 65-68.
5. Abian, A. *Hurwitz' theorem implies Rouché's theorem.* J. Math. Anal. Appl. 61 (1977), no. 1, 113-115.
6. Abian, A.; Johnston, E. H. *Zeros of partial sums of the Laurent series of analytic functions.* Kyungpook Math. J. 21 (1981), no. 1, 87-90.
7. Abu-Muhanna, Y.; MacGregor, T. H. *Variability regions for bounded analytic functions with applications to families defined by*

subordination. Proc. Amer. Math Soc. 80 (1980), 227-233; MR 81m:30022.
8. Abu-Muhanna, Y.; MacGregor, T. H. *Extreme points of families of analytic functions subordinate to convex mappings.* Math. Z. 176 (1981), no. 4, 511-519; MR 82d:30021.
9. Abu-Muhanna, Y.; MacGregor, T. H. *Families of real and symmetric analytic functions.* Trans. Amer. Math. Soc. 263 (1981), no. 1, 59-74; MR 82a:30011.
10. Acker, A. *An isoperimetric inequality involving conformal mapping.* Proc. Amer. Math. Soc. 65 (1977), no. 2, 230-234; MR 57#3364.
11. Aharonov, D. *Special topics in univalent functions.* Lecture Notes, University of Maryland (1971).
12. Aharonov, D.; Shapiro, H. S. *A minimal-area problem in conformal mapping.* (Abstract) Proc. of the Symposium on Complex Analysis Univ. Kent, Canterbury, 1973, pp. 1-5. London Math. Soc. Lecture Note Ser. No. 12, Cambridge Univ. Press, London (1974); MR 54 #526.
13. Aharonov, D.; Srebro, U. *A short proof of the Denjoy conjecture.* Bulletin (N. S.) Amer. Math. Soc. 4, no. 3 (May 1981).
14. Ahlfors, L. V. *An inequality between the coefficients a_3 and a_4 of a univalent function.* Amer. Math. Soc. Transl. (2) vol. 104 (1976), pp. 57-60.
15. Ahmad, I. *Certain classes of functions univalent in the unit disc.* Bull. Inst. Math. Acad. Sinica 5 (1977), no. 2, 379-389.
16. Ahuja, O. P.; Jain, P. K. *On starlike and convex functions with missing and negative coefficients.* Bull. Malaysian Math. Soc. (2) 3 (1980), no. 2, 95-101; MR 82d:30013.
17. Aksent'ev, L. A. *The univalent solvability of inverse boundary value problems.* (Russian) Trudy Sem. Kraev. Zadacam Vyp. 11 (1974), 9-18; MR 57 #652.
18. Aksent'ev, L. A. *Univalent change of polygonal domains.* (Russian) Trudy Sem. Kraev. Zadacam Vyp. 13 (1976), 30-39; MR 58 #22522.
19. Aksent'ev, L. A.; Gaiduk, V. N.; Mikka, V. P. *The univalent solvability of the inverse boundary value problem for a regular function in a doubly connected region.* (Russian) Trudy Sem. Kraev. Zadacam Vyp. 12 (1975), 3-8; MR 56 #15948.
20. Aksent'ev, L. A.; Kudrjasov, S. N. *Some conditions for the univalence of the solution of an inverse boundary value problem for a symmetric profile.* (Russian) Trudy Sem. Kraev. Zadacam Vyp. 6 (1969), 3-15. For a review of this item see Zbl 236 #76007; MR 58 #17113.

21. Al-Amiri, H. S. *Applications of the domain of variability of some functionals within the class of Caratheodory functions.* (Polish and Russian summaries) Ann. Univ. Mariae Curie-Sklodowska Sect. A 31 (1977), 5–14 (1979); MR 81e:30019.
22. Al-Amiri, H. S. *Certain analogy of the α-convex functions.* Rev. Roumaine Math. Pures Appl. 23 (1978), no. 10, 1449–1454; MR 80i:30017.
23. Al-Amiri, H. S. *Certain nth order differential inequalities in the complex plane.* Canad. Math. Bull. 21 (1978), no. 3, 273–277; MR 80m:30002.
24. Al-Amiri, H. S. *The domain of variability of a functional within the class of univalent starlike functions.* (Serbo-Croatian summary) Glas. Mat. Ser. III 14 (34) (1979), no. 1, 55–66; MR 81e:30018.
25. Al-Amiri, H. S. *Certain generalizations of prestarlike functions.* J. Austral. Math. Soc. Ser. A 28 (1979), no. 3, 325–334; MR 81b:30018.
26. Al-Amiri, H. S. *On Ruscheweyh derivatives.* Ann. Polon. Math. 38 (1980), no. 1, 88–94; MR 82c:30010.
27. Al-Amiri, H.; Mocanu, P. T. *Certain sufficient conditions for univalency of the Class C'.* J. Math. Anal. Appl. 80 (1981), no. 2, 387–392; MR82g:30033.
28. Al-Amiri, H.; Mocanu, P. *Spirallike nonanalytic functions.* Proc. Amer. Math. Soc. 82 (1981), 61–65; MR 82j:30028.
29. Aleksandrov, I. A. *Parametric continuations in the theory of univalent functions.* Izdat. "Nauka," Moscow (1976), 343 pp. 2.08r; MR 58 #1099.
30. Aleksandrov, I. A. *A case of integration of the Löwner equation.* (Russian) Sibirsk. Mat. Z. 22 (1981), no. 2, 207–209, 238; MR 82f:30017.
31. Aleksandrov, I. A.; Andreev, V. A. *Extremal problems for systems of functions without common values.* (Russian) Sibirsk. Mat. Z. 19 (1978), no. 5, 970–982, 1213; MR 80d:30018.
32. Aleksandrov, I. A.; Cvetkov, B. G. *Functions that conformally map the strip into itself.* (Russian) Sibirsk. Mat. Z. 21 (1980), no. 1, 4–25, 235.
33. Aleksandrov, I. A.; Mandik, V. P. *Extremal properties of simultaneously p-valent functions.* (Russian) Sibirsk. Mat. Z. 2 (1981), no. 4, 3–13, 229; MR 82i:30028.
34. Aleksandrov, I. A.; Zavozin, G. G.; Kopanev, S. A. *Optimal controls in coefficient problems for univalent functions.* (Russian) Differencial'nye Uravnenija 12 (1976), no. 4, 599–611, 771; MR 54 #536.

35. Alenicyn, J. E. *Inequalities for generalized areas in the case of multivalent conformal mappings of domains with circular cuts.* (Russian) Mat. Zametki 29 (1981), no. 3, 387-395, 479; MR 82g:30040.
36. Alenicyn, J. E. *On the least area of a form of a multiply connected domain in a class of p-sheeted conformal mappings.* (Russian) Mat. Zametki 30 (1981), no. 6, 807-812, 957.
37. Anderson, J. M.; Barth, K. F.; Brannan, D. A. *Research problems in complex analysis.* Bull. London Math. Soc. 9 (1977), no. 2, 129-162; MR 55 #12899.
38. Anderson, J. M.; Rubel, L. A. *Hypernormal meromorphic functions.* Houston J. Math. 4 (1978), no. 3, 301-309; MR 80b:30026.
39. Anderson, J. M.; Shields, A. L. *Coefficient multipliers of Bloch functions.* Trans. Amer. Math Soc. 224 (1976), no. 2, 255-265.
40. Andreev, V. A. *Extremal problems for a certain class of functions that are regular and bounded in the disc.* (Russian) Dokl. Akad. Nauk SSSR 228 (1976), no. 4, 769-771; MR 54 #13067.
41. Andreev, V. A. *Certain problems of nonoverlapping domains.* (Russian) Sibirsk. Mat. Z. 17 (1976), no. 3, 483-498, 715; MR 55 #3235.
42. Anh, V. V.; Tuan, P. D. *On starlikeness and convexity of certain analytic functions.* Pacific J. Math. 69 (1977), no. 1, 1-9, MR 55 #5848.
43. Anh, V. V.; Tuan, P. D. *On β-convexity of certain starlike univalent functions.* Rev. Roumaine Math. Pures Appl. 24 (1979), no. 10, 1413-1424; MR 81b:30019.
44. Astahov, V. N. (Astahov, V. M.) *An extremal problem for univalent analytic functions.* (Russian) Akad. Nauk Ukrain. SSR Inst. Kibernet. Preprint No. 11 Teor. Optimal. Processov (1977), 18-24; MR 58 #28472.
45. Astahov, V. M. *The range of values of a system of functionals in classes of univalent functions.* (Russian, English summary) Dokl. Akad. Nauk Ukrain. SSR Ser. A (1978), no. 3, 195-198, 284; MR 58 #1129.
46. Astahov, V. M. *The range of a functional on the class of univalent functions with real coefficients.* (Russian) Theory of functions and mappings (Russian), "Naukova Dumka," Kiev (1979), pp. 3-27, 174; MR 81d:30013.
47. Astahov, V. N. (Astahov, V. M.); Gutljans'kli, V. J. *Some extremal problems for univalent analytic functions.* (Russian) Metric questions of the theory of functions and mappings (Russian), "Naukova Dumka," Kiev (1977), pp. 3-19, 166; MR 58 #28475.
48. Atzmon, A. *Extremal functions for functionals on some classes of*

analytic functions. J. Math. Anal. Appl. 65 (1978), no. 2, 333–338; MR 80c:30008.
49. Aumann, G. *Distortion of a segment under conformal mapping and related problems.* Proceedings of the C. Carathéodory International Symposium (Athens, 1973), pp. 46–53. Greek Math. Soc., Athens, 1974; MR 57 #6406.
50. Avhadiev, F. G. *Application of the Schwarzian derivative for the study of the univalent solvability of inverse boundary value problems.* (Russian) Trudy Sem. Kraev. Zadacam Vyp. 7 (1970), 78–80; MR 58 #17117.
51. Avhadiev, F. G. *Some univalent mappings of the half-plane.* (Russian) Trudy Sem. Kraev. Zadacam Vyp. 11 (1974), 3–8; MR 57 #602.
52. Axler, S.; Shields, A. *Extreme points in VMO and BMO.* Indiana Univ. Math. J. 31 (1982), no. 1, 1–6.
53. Baernstein, A., II. *Univalence and bounded mean oscillation.* Michigan Math. J. 23 (1976), no. 3, 217–223 (1977); MR 56 #3281.
54. Baernstein, A., II. *How the *-function solves extremal problems.* Proceedings of the International Congress of Mathematicians (Helsinki, 1978), pp. 639–644, Acad. Sci. Fennica, Helsinki, 1980; MR 81b:30028.
55. Baernstein, A., II. *Some sharp inequalities for conjugate functions.* Indiana Univ. Math. J. 27 (1978), no. 5, 833–852; MR 80g:30022.
56. Baernstein, A., II.; Rochberg, R. *Means and coefficients of functions which omit a sequence of values.* Math. Proc. Cambridge Philos. Soc. 81 (1977), no. 1, 47–57.
57. Baernstein, A., II.; Schober, G. *Estimates for inverse coefficients of univalent functions from integral means.* Isarel J. Math. 36 (1980), no. 1, 75–82; MR 82a:30022.
58. Bahtin, A. K. *Coefficients of univalent functions.* (Russian) Akad. Nauk Ukrain. SSR Inst. Mat. Prepring No. 32 (1978), 8 pp.: MR 80c:30015.
59. Bahtin, A. K. *Functions of class S.* (Russian) Akad. Nauk Ukrain. SSR Inst. Mat. Preprint (1979), no. 12, Issled. po Teorii Funkcii i Topologii, 3–13; MR 81k:30014.
60. Bahtin, A. K. (Bahtin, O. K.) *Some properties of the coefficients of univalent functions.* (Russian) The theory of functions and its applications (Russian), pp. 3–8, 207, "Naukova Dumka," Kiev (1979); MR 81g:30024.
61. Bahtin, A. K. *On coefficients of functions of class S.* (Russian) Dokl. Akad. Nauk SSSR 254 (1980), no. 5, 1033–35.
62. Bahtin, A. K. *On functions of the Gel'fer class.* (Russian) Akad.

Nauk Ukrain. SSR Inst. Mat. Preprint (1980), no. 31, O Nekotor. Zadacah v Teor. Odnolist. Funkcii, 3-10; MR 82f:30022.

63. Bahtin, A. K. *Extrema of coefficients of univalent functions.* (Russian) Akad. Nauk Ukrain. SSR Inst. Mat. Preprint (1980), no. 30, 20 pp.

64. Bahtin, A. K. *On the coefficients of univalent functions.* (Russian) Questions of the theory of the approximation of functions (Russian), pp. 3-14, 193, Akad. Nauk Ukrain. SSR, Inst. Mat., Kiev (1980).

65. Bahtin, A. K. *Some properties of functions of the class S.* (Russian) Ukrain. Mat. Z. 33 (1981), no. 2, 154-159.

66. Bahtina, G. P.; Bahtin, A. K. *Some extremal problems of nonoverlapping domains.* (Russian) Akad. Nauk Ukrain. SSR Inst. Mat. Preprint (1980), no. 31, O Nekotor. Zadacah v Teor. Odnolist. Funkcii, 11-16; MR 82f:30023.

67. Bajpai, P. L.; Singh, P. *The radius of convexity of certain analytic functions in the unit disk.* Indian J. Pure Appl. Math. 5 (1974), no. 8, 701-707; MR 55 #640.

68. Bajpai, P. L.; Singh, P. *The radius of univalence of certain analytic functions.* Ann. Polon. Mth. 32 (1976), no. 2, 119-128; MR 54 #10591.

69. Bajpai, S. K. *On the univalence of some analytic functions.* Publ. Inst. Math. (Beograd) N. S.) 19(33) (1975), 25-31; MR 53 #13546.

70. Bajpai, S. K. *The coefficients of power series giving a coefficients characterization for type.* Proceedings of the Eleventh Brazilian Mathematical Colloquium (Pocos de Caldas, 1977), Vol. II (Portuguese), pp. 897-902. Inst. Mat. Pura Apl., Rio de Janeiro (1978).

71. Bajpai, S. K. *A note on a class of meromorphic univalent functions.* Rev. Roumaine Math. Pures Appl. 22 (1977), no. 3, 295-297; MR 56 #5856.

72. Bajpai, Shyam Kishore. *Influence of ½f'' (0) on the α-convexity of normalised starlike univalent analytic functions of order β.* Demonstratio Math. 11 (1978), no. 2, 301-330; MR 80j:30013.

73. Bajpai, S. K. *Special arithmetic and geometric means preserve Φ-like univalence.* Rev. Colombiana Mat. 12 (1978), no. 3-4, 83-90; MR 81h:30014.

74. Bajpai, S. K. *An analogue of R. J. Libera's result.* (Italian summary) Rend. Mat. (6) 12 (1979), no. 2, 285-289; MR 81c:30003.

75. Bajpai, S. K. *A critical investigation of various subclasses of functions whose real part is bounded.* Internat. J. Math. Math. Sci. 4 (1981), no. 1, 89-99; MR 82j:30016.

76. Bajpai, S. K. *Spiral like integral operators.* Internat. J. Math. Math. Sci. 4 (1981), no. 2, 337-351.

77. Bajpai, S. K. *The coefficients of power series giving a coefficients characterization for type.* (Italian summary) Rend. Mat. (7) 1 (1981), no. 2, 293–303.
78. Bajpai, S. K.; Dwivedi, S. P. *On the radii of starlikeness, convexity and close to convexity of p-valent functions.* Rev. Roumaine Math. Pures Appl. 23 (1978), no. 6, 839–842; MR 58 #17065.
79. Bajpai, S. K.; Dwivedi, S. P. *Certain classes of univalent analytic functions.* Rev. Colombiana Mat. 13 (1979), no. 3, 207–243; MR 81i:30016.
80. Bajpai, S. K.; Dwivedi, S. P. *Certain convexity theorems for univalent analytic functions.* Publ. Inst. Math. (Beograd) (N. S.) 28(42) (1980), 5–11.
81. Bajpai, S. K.; Mehrok, T. J. S. *On regions of α-convexity for subclasses of starlike functions.* Indian J. Pure Appl. Math. 5 (1974), no. 10, 902–908; MR 55 #5849.
82. Baranowicz, Jozef, *On the coefficients of bounded real univalent functions.* Ann. Polon. Math. 32 (1976), no. 2, 135–144. (Errata insert); MR 54 #10589.
83. Barnard, R. W. *On quasi-starlike functions.* Ann. Polon. Math. 32 (1976), no. 3, 303–308; MR 54 #529.
84. Barnard, R. W. *On bounded univalent functions whose ranges contain a fixed disk.* Trans. Amer. Math. Soc. 225 (1977), 123–144; MR 54 #10585.
85. Barnard, R. W. *A generalization of Study's theorem on convex maps.* Proc. Amer. Math. Soc. 72 (1978), no. 1, 127–134; MR 80k:30014.
86. Barnard, R. W. *On Robinson's 1/2 conjecture.* Proc. Amer. Math. Soc. 72 (1978), no. 1, 135–139; MR 80j:30014.
87. Barnard, R. W.; Kellogg, C. *Applications of convolution operators to problems in univalent function theory.* Michigan Math. J. 27 (1980), no. 1, 81–94; MR 81f:30006.
88. Barnard, R.; Lewis, J. L. *Subordination theorems for some classes of starlike functions.* Pacific J. Math. 56 (1975), no. 2, 333–366; MR 52 #721.
89. Barnard, R.; Lewis, J. L. Correction to: *"Subordination theorems for some classes of starlike functions"* (Pacific J. Math. 56 (1975), no. 2, 333–366). Pacific J. Math. 61 (1975), no. 2, 607; MR 53 #5841.
90. Barr, A.; Causey, W. M. *The radius of univalence of certain classes of analytic functions.* Indian J. Math. 17 (1975), no. 1, 33–39.
91. Barsegjan, G. A. *Geometric structure of the image of the circle in*

mappings of meromorphic functions. (Russian) Mat. Sb. (N. S.) 106(148) (1978), no. 1, 35-43, 143.
92. Başgöze, T. *Some results on spiral-like functions*. Canad. Math. Bull. 18 (1975), no. 5, 633-637; MR 53 #13538.
93. *Bauer, K. W.; Ruscheweyh, S. *Differential operators for partial differential equations and function theoretic applications*. Lecture Notes in Mathematics, 791. Springer-Verlag, Berlin-New York (1980). v + 259 pp. ISBN 3-540-09975-1.
94. Bavrin, I. I. *Generalization of regular functions in the disc that are close-to-convex to the case of several complex variables*. (Russian. English summary) Anal. Math. 2 (1976), no. 4, 235-248.
95. Beardon, A. F.; Gehring, F. W. *Schwarzian derivatives, the Poincaré metric and the kernel function*. Comment. Math. Helv. 55 (1980), no. 1, 50-64; MR 81c:30020.
96. Becker, J. *Some inequalities for univalent functions with quasiconformal extensions*. General inequalities, 2 (Proc. Second Internat. Conf., Oberwolfach, 1978), pp. 411-415, Birkhauser, Basel, 1980; MR 82f:30020.
97. Becker,, J. *Conformal mappings with quasiconformal extensions*. Aspects of contemporary complex analysis (Proc. NATO Adv. Study Inst. Univ. Durham, Durham, 1979), pp. 37-77, Academic Press, London, 1980; MR 82g:30034.
98. Becker, J. *A problem in univalent-function theory*. General inequalities, 2 (Proc. Second Internat. Conf., Oberwolfach, 1978), p. 462, Birkhauser, Basel (1980); MR 82c:30019.
99. Becker, J.; Henson, C. W.; Rubel, L. A. *First-order conformal invariants*. Ann of Math. (2) 112 (1980), no. 1, 123-178.
100. Becker, J.; Pommerenke, C. *Über die quasikonforme Fortsetzung schlichter Funktionen*. Math. Z. 161 (1978), no. 1, 69-80; MR 58 #22541.
101. *Begehr, H. *Topics in complex analysis*. Four lectures given at the University of Delaware in January and February 1977. Lecture Notes, No. 4. Institute for Mathematical Sciences, University of Delaware, Newark, Del. (1977). i + 49 pp.
102. Belikov, V. S. *An area theorem for a certain class of quasiconformal mappings*. (Russian) Sibirsk. Mat. Z. 17 (1976), no. 6, 1203-1219, 1437.
103. *Belinskii, P. P. *Some problems in modern function theory*. (Proc. Conf. Modern Problems of Geometric Theory of Functions, Inst. Math., Acad. Sci, USSR, Novosibirsk, 1976). Nekotorye voprosy sovremennoi teorii funkcii. Materialy konferencii. (Russian) [Some

problems in modern function theory. Proceedings of a Conference] Papers presented at the Conference on Modern Problems of the Geometric Theory of Functions held at the Institute of Mathematics, Siberian Branch, Academy of Sciences of the USSR, Novosibirsk (1976). Edited by P. P. Belinskii. Akad. Nauk SSSR Sibirsk. Otdel. Inst. Mat., Novosibirsk (1976), 187 pp. 1.00r.

104. Beller, E.; Pinchuk, B. *Minimal H^2 interpolation in the Caratheodory class.* Proc. Amer. Math. Soc. 72 (1978), no. 2, 289–293; MR 80b:30027.

105. Bennett, G.; Stegenga, D. A.; Timoney, R. M. *Coefficients of Bloch and Lipschitz functions.* Illinois J. Math. 25 (1981), no. 3, 520–531.

106. Beresniewicz-Rajca, O. *On the Schwarzian for the functions of Shah-Tao-Shing.* Demonstratio Math. 9 (1976), No. 3, 307–319; MR 55 #3232.

107. Beresniewicz-Rajca, O. *The coefficients of star-like symmetrical functions.* (Polish, Russian and English summaries) Zeszyty Nauk. Politech. Slask. No. 441 Mat.-Fiz. Zeszyt 26 (1976), 161–168; MR 56 #595.

108. Beresniewicz-Rajca, O. *Coefficients of the Grunsky-Shah function.* (Polish, Russian and English summaries). Zeszyty Nauk. Politech. Slask. Mat.-Fiz. No. 35 (1979), 3–6; MR 81k:30018.

109. *Bernardi, S. D. *Bibliography of Schlicht functions. Part II. 1966–1975.* New York University Technical Report, IMM-414. Courant Institute of Mathematical Sciences, New York University, New York (1977). x + 168 pp.; MR 80h:30017

110. Bharati, R. *Some radius of convexity problems.* Indian J. Pure Appl. Math. 9 (1978), no. 11, 1118–1130; MR 80a:30009.

111. Bharati, R. *On α-close-to-convex functions.* Proc. Indian Acad. Sci. Sect. A Math. Sci. 88 (1979), no. 2, 93–103; MR 81c:30021.

112. Bharati, R. *Some radius of convexity problems.* II. Indian J. Pure Appl. Math. 12 (1981), no. 9, 1133–1145.

113. Bhoosnurmath. S. S.; Swamy, S. R. *Analytic functions with negative coefficients.* Indian J. Pure Appl. Math. 12 (1981), no. 6, 738–742; MR 82g:30019.

114. Biernacki, M. *Deux remarques au sujet d'un théorème de M. Rogosinski.* Mathematica, vol. 10 (1934), pp. 46–48.

115. Blatter, C. *Ein Verzerrungssatz fur schlichte Funktionen.* Comment. Math. Helv. 53 (1978), no. 4, 651–659; MR 80d:30010.

116. Blevins, D. K. *Covering theorems for univalent functions mapping onto domains bounded by quasiconformal circles.* Canad. J. Math. 28 (1976), no. 3, 627–631.

117. Blezu, Dorin; Pascu, N. N. *A generalization of the class of close-to-convex functions.* (Romanian summary) Studia Univ. Babes-Bolyai Math. 25 (1980), no. 2, 32–38.
118. Blezu, D.; Pascu, N.; Rotaru, P. *Alpha-starlike functions of order β.* (Romanian, English summary) Bul. Univ. Brasov Ser. C 21 (1979), 41–44; MR 82j:30017.
119. Bogda, R. A.; Shankar, H. *Convolutions and growth numbers of analytic functions.* Rocky Mountain J. Math. 10 (1980), no. 3, 475–583; MR 82a:30002.
120. Bogucki, Z.; Zderkiewicz, J. *Le rayon d'univalence de certains fonctions analytiques.* (Polish and Russian summaries) Ann. Univ. Mariae Curie-Sklodowska Sect. A 31(1977), 27–33 (1979); MR 81d:30022.
121. Bogucki, Z.; Zderkiewicz, J. *Sur les majorantes convexes des fonctions analytiques.* (Polish and Russian summaries) Ann. Univ. Mariae Curie-Sklodowska Sec. A 31 (1977), 21–25 (1979); MR 81f:30024.
122. Bogucki, Z.; Zderkiewicz, J. *Inégalitiés entre les modules des dérivées de fonctions subordonnées dans le cas de majorantes convexes.* English and Russian summaries) Bull. Acad. Polon. Sci. Ser. Sci. Math. Astronom. Phys. 25 (1977), no. 11, 1093–1098; MR 58 #6200.
123. Bogucki, Z.; Zderkiewicz, J. *Sur un problème de Jenkins pour les fonctions convexes.* Ann. Polon. Math. 38 (1980), no. 3, 255–258; MR 82g:30020.
124. Bogucki, Z.; Zderkiewicz, J. *Sur une sous-classe de fonctions étoilées dont les valeurs recouvrent un cèrcle fixe.* Ann. Polon. Math. 38 (1980), no. 3, 245–253; MR 82c:30011.
125. Bojarska, M.; Wesolowska, M. *Les dependances entre la subordination et l'inégaltié des modules dans le cas des majorantes appartenantes aux classes S^* (α, β), S^* (α, β).* (Polish and Russian summaries) Ann. Univ. Mariae Curie-Sklodowska Sect. A 29 (1975), 7–13 (1977).
126. Bondar, A. V.; Diab, F. M. *The set of singular points of a univalent analytic function.* (Russian) Ukrain. Mat. Z. 28 (1976), no. 2, 147–158, 284; MR 54 #2936.
127. Boutellier, R. *Le théorème de deformation de la class Σ^*.* (English summary) C. R. Acad. Sci. Paris Ser. A-B 286 (1978), no. 1, A33–A35; MR 8le:30020.
128. Boutellier, R. *Curvature and functions with bounded boundary rotation.* Math. Z. 177 (1981), no. 3, 395–400.
129. Boutellier, R. *(with Pfluger, A.)* Some extremal problems for func-

tions of bounded boundary rotation. Israel J. Math. 39 (1981), no. 1-2, 46-62; MR 82g:30021.
130. Brandt, M. *Ein Existenzsatz für schlicht-knoforme Abbildungen endlich-vielfach zusammenhängender Gebiete bei hoherer Normierung.* (Polish summary) Bull. Soc. Sci. Lett. Łódź 29 (1979), no. 5, 13 pp.
131. Brannan, D. A. *The Grunsky coefficients of meromorphic starlike and convex functions.* (Polish and Russian summaries) Ann. Univ. Mariae Curie-Sklodowska Sect. A. 31 (1977), 45-48 (1979); MR 81c:30054.
132. Brannan, D. A. *The Löwner differential equations.* Aspects of contemporary complex analysis (Proc. NATO Adv. Study Inst., Univ. Durham, Durham, 1979), pp. 79-95, Academic Press, London, 1980; MR 82i:30029.
133. Brannan, D. A.; Brickman, L. *Coefficient regions for starlike polynomials.* (Polish and Russian summaries) Ann. Univ. Mariae Curie-Sklodowska Sect. A 29 (1975), 15-21 (1977); MR 56 #8819.
134. Brannan, D. A.; Clunie, J. G. **Aspects of contemporary complex analysis* (Proc. NATO Adv. Study Inst., Univ. Durham, Durham 1979). Academic Press, Inc., New York, 1980. xiii + 572 pp; MR 82f:30001.
135. Brannan, D. A.; Kirwan, W. E. *Some covering theorems for analytic functions.* J. London Math. Soc. (2) 19 (1979), no. 1, 93-101; MR 80d:30011.
136. Brennan, J. E. *The integrability of the derivative in conformal mapping.* J. London Math. Soc. (2) 18 (1978), no. 2, 261-272.
137. Brickman, L.; Wilken, D. *Subordination and insuperable elements.* Michigan Math. J. 23 (1976), no. 3, 225-233 (1977); MR 54 #13066, MR 81d:30041.
138. Brown, J. E. *Geometric properties of a class of support points of univalent functions.* Trans. Amer. Math. Soc. 256 (1979), 371-382; MR 80k:30019.
139. Brown, J. E. *Derivatives of close-to-convex functions, integral means and bounded mean oscillation.* Math. Z. 178 (1981), no. 3, 353-358; MR 82j:30046.
140. Brown, J. E. *Univalent functions maximizing Re $\{a_3 + \lambda a_2\}$.* Illinois J. Math. 25 (1981), no. 3, 446-454; MR 82j:30022.
141. Bshouty, D. H. *the Bieberbach conjecture for univalent functions with small second coefficients.* Math. Z. 149 (1976), no. 2, 183-187; MR 55 #8341.
142. Bshouty, D. *A note on Hadamard products of univalent functions.* Proc. Amer. Math. Soc. 80 (1980), 271-272; MR 81h:30017.

143. Bshouty, D.; Hengartner, W. *Asymptotic FitzGerald inequalities.* Comment. Math. Helv. 53 (1978), no. 2, 228–238; MR 80b:30010.
144. Bshouty, D.; Hengartner, W.; Schober, G. *Estimates for the Koebe constant and the second coefficient for some classes of univalent functions.* Canad. J. Math. 32 (1980), no. 6, 1311–1324; MR 82f:30009.
145. Bucka, C.; Ciozda, K. *Some estimations and problems of the majorization in the classes of functions Sk (α, β).* (Polish and Russian summaries) Ann. Univ. Mariae Curie-Sklodowska Sect. A 29 (1975), 29–41 (1977); MR 57 #3370.
146. Bucka, C.; Ciozda, K. *Sur l'interpretation geometrique de certains sous-classes de la classe S.* (Polish and Russian summaries) Ann. Univ. Mariae Curie-Sklodowska Sect. A. 29 (1975), 23–28 (1977); MR 58 #1117.
147. Buckholtz, J. D.; Shah, S. M. *Absolute starlike and absolute convex functions.* J. Analyse Math. 38 (1980), 113–143; MR 82b:30009.
148. Buckholtz, J. D.; Shah, S. M. *Analytic functions with constraints on coefficients.* Nonlinear Anal. 5 (1981), no. 5, 553–564; MR 82g:30022.
149. Buckholtz, J. D.; Suffridge, T. J. **Complex analysis.* Proceedings of the Conference held in honor of Professor Swarupchand M. Shah at the University of Kentucky, Lexington, Ky., May 18-22, 1976. Edited by James D. Buckholtz and Teddy J. Suffridge. Lecture Notes in Mathematics, Vol. 599. Springer-Verlag, Berlin-New York, 1977. x + 159 pp. ISBN 3-540-08343-X; MR 56 #3260.
150. Burbea, J. *The Schwarzian derivative and the Poincaré metric.* Pacific J. Math. 85 (1979), no. 2, 345–354; MR 81g:30013.
151. Burdick, G. R.; Keogh, F. R.; Merkes, E. P. *On a ratio of a univalent function.* J. Math. Anal. Appl. 53 (1976), no. 2, 221–224.
152. Burdick, G. R.; Merkes, E. P. *On ratios of certain analytic functions.* Rev. Colombiana Mat. 9 (1975), no. 1, 23–28; MR 53 #5842.
153. Busovaskaja, O. A.; Gorjainov, V. V. *On the theory of star-like functions.* (Russian, English summary) Dokl. Akad. Nauk Ukrain. SSR Ser. A (1981), no. 2, 6–8, 95; MR 82k:30011.
154. Byers, R. B. *On the derivative of a typically-real function.* Ann. Fac. Sci. Univ. Nat. Zaire (Kinshasa) Sect. Math.-Phys. 5 (1979), no. 1-2, 47–53; MR 82h:30008.
155. Byers, R. B. *On growth of typically-real functions and a related class.* Bull. Inst. Math. Acad. Sinica 9 (1981), no. 1, 39–45; MR 82f:30010.
156. Byers, R. B.; Merkes, E. P. *Subordinations for typically-real and*

related functions. Rocky Mountain J. Math. 11 (1981), no. 2, 297-304; MR 82g:30015.
157. Campbell, D. M. *The growth of eventually areally mean p-valent functions.* Bull. Calcutta Math. Soc. 68 (1976), no. 1, 31-36; MR 56 #15904.
158. Campbell, D. M. *Eventually p-valent functions.* Rocky Mountain J. Math. 7 (1977), no. 4, 639-658; MR 56 #3282.
159. Campbell, D. M. *A truly isolated univalent function.* Michigan Math. J. 27 (1980), 253-255; MR 8lh:30020.
160. Campbell, D. M. *Analytic functions all of whose level sets are of infinite length.* Houston J. of Math. 6, no. 1 (1980), 15-17.
161. Campbell, D.; Cima, J.; Stephenson, K. *A Bloch function in all H_p classes, but not in BMOA.* Proc. Amer. Math. Soc. 78 (1980), no. 2, 229-230; MR 80k:30033.
162. Campbell, D. M.; Clunie, J. G.; Hayman, W. K. *Research problems in complex analysis.* Aspects of contemporary complex analysis. (Proc. NATO Adv. Study Inst., Univ. Durham, 1979), pp. 527-572, Academic Press, London, 1980.
163. Campbell, D. M.; Pearce, K. *Generalized Bazilevič functions.* Rocky Mountain J. Math. 9 (1979), no. 2, 197-226; MR 80d:30005.
164. Campbell, D. M.; Pfaltzgraff, J. A. *Boundary behavior and linear invariant families.* J. D'analyse Math. 29 (1976), 67-92.
165. Campbell, D. M.; Singh, V. (Singh, Vikramaditya). *Valence properties of the solution of a differential equation.* Pacific J. Math. 84 (1979), no. 1, 29-33; MR 81a:30006.
166. Campbell, D. M.; Ziegler, M. R. *The argument of the derivative of linear-invariant families of finite order and the radius of close-to-convexity.* (Polish and Russian summaries) Ann. Univ. Mariae Curie-Sklodowska Sect. A 28 (1974), 5-22 (1976); MR 54 #7768.
167. Carlson, B. C.; Shaffer, D. B. *Starlike and prestarlike hypergeometric functions.* [Preprint available in Applied Mathematical Science, Ames Laboratory, Iowa State University].
168. Causey, W. M.; White, W. L. *Starlikeness of certain functions with integral representations.* J. Math. Anal. Appl. 64 (1978), no. 2, 458-466; MR 57 #16561.
169. Cazacu, C. A. **Romanian-Finnish Seminar on Complex Analysis. Proceedings of the Seminar held in Bucharest, June 27-July 2, 1976.* Edited by Cabiria Andreian Cazacu, Aurel Cornea, Martin Jurchescu and Ion Suciu. Lecture Notes in mathematics, 743. Springer-Verlag, Berlin-New York, 1979. xvi + 713 pp. ISBN 3-540-09550-0; MR 80i:30002.

170. Černikov, V. V. *Estimation of the curvature of level curves in a certain class.* (Russian) Mat. Zametki 19 (1976), no. 3, 381–388; MR 53 #13537.
171. Černikov, V. V. *Some extremal properties of typically-real functions.* (Russian) Sibirsk. Mat. Z. 21 (1980), no. 6, 208–209, 224; MR 82g:30023.
172. Černikov, V. V. *α-convexity of order β of the class of all regular functions that are univalent in the disc.* (Russian) Mat. Zametki 29 (1981), no. 6, 859–866, 956.
173. Černikov, V. V.; Sizuk, P. I. *Some properties of star-like univalent functions.* (Russian) Sibirsk. Mat. Z. 19 (1978), no. 1, 193–200, 239; MR 57 #603, MR 58 #22503.
174. Černikov, V. V.; Sizuk, P. I. *Some properties of spiral-shaped univalent functions.* (Russina) Metric questions of the theory of functions. (Russian), pp. 132–135, 163, "Naukova Dumka," Kiev, 1980; MR 82k:30012.
175. Chambers, B. F. *Analytic continuation via Hadamard's product.* SIAM J. Math. Anal. 9 (1978), no. 6, 1096–1104; MR 80f:30002.
176. Chand, R.; Singh, P. *On certain Schlicht mappings.* Indian J. Pure Appl. Math. 10 (1979), no. 9, 1167–1174; MR 80j:30015.
177. Chandra, S. *Some results on regular functions.* Indian J. Pure Appl. Math. 8 (1977), no. 8, 915–919.
178. Chandra, S. *Coefficient estimates for the regular p-valent functions.* Ann. Fac. Sci. Univ. Nat. Zaire (Kinshasa) Sect. Math.-Phys. 3 (1977), no. 1, 65–78; MR 57 #6399.
179. Chandra, S. *Coefficient estimates for certain classes of meromorphic functions.* Indian J. Pure Appl. Math. 8 (1977), no. 9, 987–991; MR 58 #28498.
180. Chandra, S. *Some results on regular functions.* Indian J. Pure Appl. Math. 8 (1977), no. 8, 915–919; MR 57 #9951.
181. Chandra, S.; Singh, P. *An integral transformation of some classes of regular univalent functions.* Indian J. Pure Appl. Math. 6 (1975), 1270–1275; MR 57 #3371.
182. Charzynski, Z.; Lawrynowicz, J. *A simple method of deriving the Löwner equation with the help of algebraic functions.* Proceedings of the First Finnish-Polish Summer School in Complex Analysis (Podlesice, 1977), Part I, pp 1–9, Univ. Lodz, Lodz, 1978; MR 81i:30033.
183. Chen, M. P. *The radius of univalence and starlikeness of some classes of analytic functions.* Comment. Math. Univ. St. Paul. 24 (1975/76), no. 2, 91–95; MR 53 #8404.

184. Chen, M. P. *The radius of univalence and starlikeness of some classes of analytic functions.* Chinese J. Math. 3 (1975), no. 1, 17–22; MR 54 #13053, MR 53 #8404.
185. Chen, M. P. *On a class of starlike functions.* Nanta Math. 8 (1975), no. 1, 79–82; MR 54 #13054.
186. Chen, M. P. *A class of univalent functions.* Soochow J. Math. 6 (1980), 49–57; MR 82h:30009.
187. Chichra, P. N. *New subclasses of the class of close-to-convex functions.* Proc. Amer. Math. Soc. 62 (1976), no. 1, 37–43; MR 54 #13055.
188. Chichra, P. N. *Convex sum of regular functions.* J. Indian Math. Soc. (N.S.) 39 (1975), 299–304 (1976); MR 54 #10586.
189. Chou, T. W. *On some classes of analytic functions.* Soochow J. Math. Natur. Sci. 3 (1977), 109–122; MR 58 #11348.
190. Cima, J. A. *Hadamard products of convex schlicht functions.* (Polish and Russian summaries) Ann. Univ. Mariae Curie-Sklodowska Sect. A 29 (1975), 49–60 (1977); MR 58 #17066.
191. Cima, J. A. *A note on log (f(z)/z) for f in S.* Monatsh. Math. 81 (1976), no. 2, 89–93.
192. Cima, J. A. *The basic properties of Bloch functions.* Internat. J. Math. Math. Sci. 2 (1979), no. 3, 369–413; MR 80k:30037.
193. Cima, J. A.; Stegbuchner, H. *On the duals of some spaces of locally schlicht functions.* Indiana Univ. Math. J. 27 (1978), no. 4, 539–550.
194. Ciozda, K. *Sur la classe des fonctions convexes vers l'axe reel negatif.* (English and Russian summaries) Bull. Acad. Polon. Sci. Ser. Sci. Math. 27 (1979), no. 3–4, 255–261; MR 80m:30011.
195. Ciozda, K. *Sur quelques problèmes extrêmaux dans les classes des fonctions convexes vers l'axe reel negatif.* (English summary) Ann. Polon. Math. 38 (1980), no. 3, 311–317; MR 82b:30010.
196. Clemens, C. H. *A scrapbook of complex curve theory.* The University Series in Mathematics. Plenum Press, New York-London (1980), ix + 186 pp. ISBN 0-306-40536-9.
197. Clunie, J. G. *Inverse coefficients for convex univalent functions.* J. D'analyse Math. 36 (1979), 31–35; MR 81i:30017.
198. Clunie, J. G. *Some remarks on extreme points in function theory.* Aspects of contemporary complex analysis (Proc. NATO Adv. Study Inst., Univ. Durham, Durham, 1979), pp. 137–146, Academic Press, London, 1980; MR 82g:30024.
199. Clunie, J. G.; Rubel, L. A. *Riemann-Lebesgue centers of plane domains.* Illinois J. Math. 22 (1978), no. 4, 682–692.

200. Cohrane, P. C.; MacGregor, T. H. *Frechet differentiable functionals and support points for families of analytic functions.* Transactions Amer. Math. Soc. 236 (1978), 75-92; MR 57 #604.
201. Coonce, H. B.; Eenigenburg, P. J.; Ziegler, M. R. *Functions with bounded Mocanu variation.* II. (Polish and Russian summaries) Ann. Univ. Mariae Curie-Sklodowska Sect. A 28 (1974), 23-28 (1976); MR 53 #11035.
202. Cvetkov, B. G. *Some results for functions that are univalent in a strip.* (Russian) Some problems in modern function theory (Proc. Conf. Modern Problems of Geometric Theory of Functions, Inst. Math., Acad. Sci. USSR, Novosibirsk, 1976) (Russian), pp. 169-173. Akad. Nauk SSSR Sibirsk. Otdel. Inst. Mat., Novosibirsk, 1976. For a review of this item see RZMat 1977 #11 B146. [For the entire collection see MR 57 #15813.]; MR 58 #22524.
203. Das, R. N.; Singh, P. *On properties of certain subclasses of close-to-convex functions.* (Polish and Russian summaries) Ann. Univ. Mariae Curie-Sklodowska Sect. A 30 (1976), 15-22 (1978); MR 58 #17068.
204. Das, R. N.; Singh, P. *On the coefficient bounds of a subclass of analytic functions in the unit disc.* J. Indian Math. Soc. (N.S.) 40 (1976), no. 1-4, 153-158 (1977); MR 56 #3277.
205. Das, R. N.; Singh, P. *On subclasses of schlicht mapping.* Indian J. Pure Appl. Math. 8 (1977), no. 8, 864-872; MR 58 #6201.
206. Das, R. N.; Singh, P. *Radius of convexity for a certain subclass of close to convex functions.* J. Indian Math. Soc. (N.S.) 41 (1977), no. 3-4, 363-369; MR 80a:30010.
207. Das, R. N.; Singh, P. *The radius of starlikeness of certain subclasses of analytic functions.* Publ. Math. Debrecen 25 (1978), no. 1-2, 97-106; MR 58 #11349.
208. de Branges, L. *Coefficient estimates.* J. Math. Anal. Appl. 82 (1981), no. 2, 420-450.
209. de Temple, D. W.; Jenkins, J. A. *A Loewner approach to a coefficient inequality for bounded univalent functions.* Proc. Amer. Math. Soc. 65 (1977), no. 1, 125-126; MR 56 #3278.
210. de Temple, D. W.; Oulton, D. B. *Formulas for the Nehari coefficients of bounded univalent functions.* Canad. J. Math. 29 (1977), no. 3, 587-605; MR 56 #3276.
211. Din-van-F'eu. *An extension of the variational formula of G. M. Goluzin to an annulus.* (Russian) Extremal problems of the geometric theory of functions of a complex variable, II. Zap. Naucn. Sem. Leningrad. Otdel. Mat. Inst. Steklov. (LOMI) 44

(1974), 131–178; 188; MR 56 #5863.
212. Doppel, K. *Über lokal schlichte Funktionen, deren erste Ableitung beschränktes Argument hat.* Ann. Acad. Sci. Fenn. Ser. A 1 Math. 3 (1977), no. 2, 317–325; MR 58 #22504.
213. Doppel, K.; Koditz, H.; Timmann, S. *Bemerkungen über Fixpunktmengen schlichter Funktionen.* (Italian and English summary) Rend. Ist. Mat. Univ. Trieste 8 (1976), no. 2, 162–166 (1977); MR 57 #3375.
214. Doppel, K.; Zinterhof, P. *Über die ϵ-Entropie der Menge der schlichten Funktionen.* Anz. Osterrreich. Akad. Wiss. Math.-Naturwiss. Kl. (1971), no. 3, 45–46; MR 52 #5953.
215. Dubinin, V. N. *A certain symmetrization method.* (Russian) Mathematical analysis, No. 2. Kuban. Gos. Univ. Naucn. Trudy Vyp. 180 (1974), 50–64, 160–161; MR 53 #5852.
216. Dubinin, V. N. *Covering of vertical segments in conformal mappings.* (Russian) Mat. Zametki 28 (1980), no. 1, 25–32, 167; MR 81m:30008.
217. Dundučenko, L. I. (Dundučenko, L. O.); Goncarenko, S. V. *Mapping an n-connected circular domain onto a plane with cuts along arcs of logarithmic spirals.* (Russian) Mat. Zametki 21 (1977), no. 3, 329–334; MR 57 #600.
218. Duren, P. L. *Asymptotic behavior of coefficients of univalent functions.* Advances in complex function theory (Proc. Sem., Univ. Maryland, College Park, Md., 1973-1974), pp. 17–23. Lecture Notes in Math., Vol. 505, Springer, Berlin, 1976; MR 56 #597
219. Duren, P. L. *Applications of the Garabedian-Schiffer inequality.* J. Analyse Math. 30 (1976), 141–149; MR 55 #10656.
220. Duren, P. Subordination. *Complex analysis* (Proc. Conf. Univ. Kentucky, Lexington, Ky., 1976), pp. 22–29. Lecture Notes in Math., Vol. 599, Springer, Berlin, 1977; MR 56 #12252.
221. Duren, P. L. *Coefficients of univalent functions.* Bull. Amer. Math. Soc. 83 (1977), no. 5, 891–911; MR 57 #9943.
222. Duren, P. L. *Extreme points of spaces of univalent functions.* Linear spaces and approximation (Proc. Conf., Math. Res. Inst., Oberwolfach, 1977), pp. 471–477. Internat. Ser. Number. Math., Vol. 40, Birkhauser, Basel, 1978; MR 58 #17072.
223. Duren, P. L. *Successive coefficients of univalent functions.* J. London Math. Soc. (2) 19 (1979), no. 3, 448–450; MR 80i:30030.
224. Duren, P. L. *Extremal problems for univalent functions.* Aspects of contemporary complex analysis (Proc. NATO Adv. Study Inst., Univ. Durham, Durham, 1979), pp. 181–208, Academic Press, London, 1980; MR82h:30013.

225. Duren, P. L. *Arcs omitted by support points of univalent functions.* Comment. Math. Helv. 56 (1981), no. 3, 352-365.
226. Duren, P.; Leung, Y. J. *Logarithmic coefficients of univalent functions.* J. D'analyse Math. 36 (1979), 36-43; MR 81i:30018.
227. Duren, P.; Schober, G. *Nonvanishing univalent functions.* Math. Z. 170 (1980), 195-216; MR 81e:30021.
228. Dwivedi, S. P.; Bhargava, G. P.; Shukla, S. L. *On some classes of meromorphic univalent functions.* Rev. Roumaine Math. Pures Appl. 25 (1980), no. 2, 209-215; MR 81g:30016.
229. Eenigenburg, P. J. *The integral means of analytic functions.* Quart. J. Math. Oxford Ser. (2) 32 (1981), no. 127, 313-322.
230. Eenigenburg, P.; Nelson, D. *On the order of starlikeness for alpha-convex functions.* Mathematica (Cluj) 18 (41) (1976), no. 2, 143-146; MR 58 #22505.
231. Eenigenburg, P. J.; Nelson, J. D. On area and subordination. Complex analysis (Proc. S.U.N.Y. Conf., Brockport, N.Y., 1976), pp. 53-60. Lecture Notes in Pure and Appl. Math., Vol. 36, Dekker, New York, 1978; MR 58 #6211.
232. Eenigenburg, P. J.; Waniurski, J. *An area inequality for quasi-subordinate analytic functions.* Ann. Polon. Math. 34 (1977), no. 1, 25-33. (errata insert); MR 55 #652.
233. Eenigenburg, P.; Yoshikawa, H. *An application of the method of Zmorovič in geometric function theory.* J. Math. Anal. Appl. 56 (1976), no. 3, 683-688; MR 54 #13056.
234. Egerland, W. O.; Ziegler, M. R. *Best radii of univalence.* Amer. Math. Monthly, vol. 78 (1971), pp. 1031-1032.
235. Eke, B. G. *On the Bieberbach conjecture for multivalent functions.* J. London Math. Soc. (2) 13 (1976), no. 1, 19-26; MR 53 #791.
236. Essen, M. *Boundary behavior of univalent functions satisfying a Hölder condition.* Proc. Amer. Math. Soc. 83(1981), no. 1, 83-84; MR 82i:30010.
237. Fait, M.; Krzyz, J. G.; Zygmunt, J. *Explicit quasiconformal extensions for some classes of univalent functions.* Comment. Math. Helv. 51 (1976), no. 2, 279-285; MR 54 #10587.
238. Fait, M.; Stankiewicz, J.; Zygmunt, J. *On some classes of polynomials.* (Polish and Russian summaries) Ann. Univ. Mariae Curie-Sklodowska Sect. A 29 (1975), 61-67 (1977); MR 56 #12236.
239. Fait, M.; Złotkiewicz, E. *Convex hulls of some classes of univalent functions.* (Polish and Russian summaries) Ann. Univ. Mariae Curie-Sklodowska Sect. A 30 (1976), 35-41 (1978); MR 80c:30009.
240. Fan, K. *Distortion of univalent functions.* J. Math. Anal. Appl. 66 (1978), no. 3, 626-631; MR 80a:30015.

241. Fedorov, S. I. *On the maximum of a conformal invariant in a problem on nonoverlapping domains.* (Russian) Analytic number theory and the theory of functions, 4. Zap. Naucn. Sem. Leningrad. Otdel. Mat. Inst. Steklov (LOMI) 112 (1981), 172–183, 202.
242. Fejér, L. *La convergence sur son cèrcle de convergence d'une série de puissance effectuant une représentation du cèrcle sur le plan simple.* Comptes Rendus, vol. 156 (1913), pp. 46–49.
243. Fejér, L. *Über gewisse durch die Fouriersche und Laplacesche Reihe definierten Mittelkurven und Mitteflächen.* Rendiconti del Circolo Mat. di Palermo, vol. 38 (1914), pp. 79–97.
244. Fejér, L. *Über die Grenzen der Abscnitte gewisser Potenzreihen.* Acta Szeged, vol. 4 (1928-29), pp. 14–20.
245. Fekete, O. *On some subclasses of Bazilevič functions.* Romanian-Finnish Seminar on Complex Analysis (Proc., Bucharest, 1976), pp. 268–273, Lecture Notes in Math., 743, Springer, Berlin, 1979; MR 80k:30015.
246. Fekete, O. *A certain class of Bazilevič functions.* (Romanian and English summary) Studia Univ. Babes-Bolyai Math. 22 (1977), no. 1, 26–34; MR 58 #22506.
247. Feng, J. *Extreme points and integral mean estimates for classes of analytic functions.* Ph.D. thesis, State University of New York at Albany (1974).
248. Feng, J.; Macgregor, T. H. *Estimates on integral means of the derivatives of univalent functions.* J. Analyse Math. 29 (1976), 203–231.
249. Fenton, P. C. *On conformal maps with starlike images.* Math. Scand. 39 (1976), no. 1, 102–112; MR 54 #13050.
250. FitzGerald, C. H. *Quadratic inequalities and analytic continuation.* J. Analyse Math. 31 (1977), 19–47; MR 58 #6189.
251. FitzGerald, C. H. *Quadratic inequalities and univalent functions.* Aspects of contemporary complex analysis (Proc. NATO Adv. Study Inst., Univ. Durham, Durham (1979), pp. 393–397, Academic Press, London, 1980; MR 82g:30001.
252. FitzGerald, C. H.; Horn, R. A. *On the structure of Hermitian-Symmetric Inequalities.* J. London Math. Soc. (2), 15 (1977), 419–430.
253. Frank, J. L. *Subordination and convex univalent polynomials.* J. Reine Angew. Math. 290 (1977), 63–69; MR 56 #3269.
254. Fricke, G. H.; Shah, S. M. *Note on a theorem on analytic functions with univalent derivatives.* Indian J. Math. 21 (1979), no. 1, 57–61; MR 82b:30011.
255. Friedland, S.; Schiffer, M. *Global results in control theory with ap-*

plications to univalent functions. Bull. Amer. Math. Soc. 82, no. 6 (November 1976), 913–915.
256. Friedland, S.; Schiffer, M. *On coefficient regions of univalent functions.* J. Analyse Math. 31 (1977), 125–168; MR 58 #22519.
257. Fukui, S. *On the radii of convexity and starlikeness of $(zf(z))'/2$.* Bull. Fac. Ed. Wakayama Univ. Natur. Sci., no. 27 (1978), 1–4; MR 80k:30016.
258. Fukui, S. *The radii of starlikeness and convexity for certain class of analytic functions.* Bull. Fac. Ed. Wakayama Univ. Natur. Sci., no. 30 (1981), 1–4.
259. Fukui, S.; Sakaguchi, K. *An extension of a theorem of S. Ruscheweyh.* Bull. Fac. Ed. Wakayama Univ. Natur. Sci., no. 29 (1980), 1–3; MR 81g:30017.
260. Gaiduk, V. N. *Application of typically-real functions to inverse boundary value problems.* (Russian) Trudy. Sem. Kraev. Zadacam Vyp. 6 (1969), 26–30; MR 58 #17124.
261. Gaiduk, V. N. *Some conditions for univalence of the solutions of inverse boundary value problems.* (Russian) Trudy Sem. Kraev. Zadacam Vyp. 7 (1970), 98–102; MR 58 #17125.
262. Gaiduk, V. N. *Some conditions for univalence of the solution of an inverse boundary value problem for n-symmetric functions.* (Russian) Trudy Sem. Kraev. Zadacam Vyp. 8 (1971), 55–58. For a review of this item see Zbl 236 #30023; MR 58 #17126.
263. Gaiduk, V. N. *The univalence of the solutions of inverse boundary value problems.* (Russian) Trudy Sem. Kraev. Zadacam Vyp. 9 (1972), 39–48. For a review of this item see Zbl 299 #30013; MR 58 #17127.
264. Gaier, D. *Konforme Abbildung mehrfach zusammenhängender Gebiete.* Jahresber. Deutsch. Math.-Verein. 81 (1978/79), no. 1, 25–44.
265. Gal'perin, I. M. *On the problem of the coefficients of multivalent spiral functions.* (Russian) Ukrain. Mat. Z. 32 (1980), no. 4, 545–548; MR 81k:30019.
266. Ganesan, M. S. *Products of close-to-starlike and close-to-convex functions.* Indian J. Pure Appl. Math. 12 (1981), no. 3, 336–346; MR 82d:30014.
267. Garabedian, P. R. *Univalent functions and the Riemann mapping theorem.* Proc. Amer. Math. Soc. 61 (1976), no. 2, 242–244; MR 54 #13052.
268. *Garnett, J. B. *Bounded analytic functions.* Pure and applied Mathematics, 96. Academic Press, Inc. [Harcourt Brace Jovanovich, Publishers], New York-London (1981), xvi + 467 pp. ISBN

0-12-276150-2.
269. Gehring, F. W. *Some problems in complex analysis.* Proceedings of the First Finnish-Polish Summer School in Complex Analysis (Podlesice, 1977), Part II, pp. 61–64, Univ. Łódź, Łódź, 1978; MR 80b:30013.
270. Gehring, F. W. *Univalent functions and the Schwarzian derivative.* Comment. Math. Helv. 52 (1977), no. 4, 561–572; MR 56 #15905.
271. Gehring, F. W. *Remarks on the universal Teichmüller space.* Enseign. Math. (2) 24 (1978), no. 3–4, 173–178; MR 80d:30012.
272. Gel'fer, S. A. *p-valent functions.* (Russian) Izv. Vyss. Ucebn. Zaved. Matematika (1977), no. 5(180), 15–24; MR 57 #6400.
273. Gel'fer, S. A.; Kresnjakova, L. V. *Mean values of analytic functions.* (Russian) Sibirsk. Mat. Z. 18 (1977), no. 4, 747–754, 955; MR 56 #8836.
274. Gevirtz, J. *An upper bound for the John constant.* Proc. Amer. Math. Soc. 83 (1981), no. 3, 476–478; MR 82j:30026.
275. Gocal, Ryszard. *Une remarque sur la mesure des domaines correspondant aux fonctions extrêmales.* (Polish summary) Bull. Soc. Sci. Lett. Łódź 29 (1979), no. 9, 4 pp.; MR 81i:30043.
276. Goel, R. M. *The radii of convexity for certain classes of analytic functions.* J. Math. Sci. 8 (1973), 17–22 (1974); MR 57 #9957.
277. Goel, R. M. *On a class of analytic functions.* Indian J. Math. 17 (1975), no. 1, 9–20; MR 82i:30011
278. Goel, R. M. *A subclass of α-spiral functions.* Publ. Math. Debrecen 23 (1976), no. 1–2, 79–84; MR 54 #2940.
279. Goel, R. M. *Radius of starlikeness of close-to-spirallike functions.* Indian J. Pure Appl. Math. 8 (1977), no. 4, 394–400; MR 57 #9952.
280. Goel, R. M. *Radius of convexity of convex combination of certain analytic functions.* Tamkang J. Math. 10 (1979), no. 1, 75–79; MR 81d:30014.
281. Goel, R. M.; Mehrok, B. S. *On the coefficients of a subclass of starlike functions.* Indian J. Pure Appl. Math. 12 (1981), no. 5, 634–647; MR 82g:30026.
282. Goel, R. M.; Mehrok, B. S. *On a class of close-to-convex functions.* Indian J. Pure Appl. Math. 12 (1981), no. 5, 648–658; MR 82e:30017.
283. Goel, R. M.; Mehrok, B. S. *Some invariance properties of a subclass of close-to-convex functions.* Indian J. Pure Appl. Math. 12 (1981), no. 10, 1240–1249.
284. Goel, R. M.; Sohi, N. S. *On a subclass of univalent functions.* Tamkang J. Math. 10 (1979), no. 2, 151–164; MR 81i:30019.
285. Goel, R. M.; Sohi, N. S. *A new criterion for p-valent functions.*

Proc. Amer. Math. Soc. 78 (1980), no. 3, ,353-357; MR 81b:30020.
286. Goel, R. M.; Sohi, N. S. *On certain classes of analytic functions.* Indian J. Pure Appl. Math. 11 (1980), no. 10, 1308-1324; MR 81j:30024.
287. Goel, R. M.; Sohi, N. S. *New criteria for p-valence.* Indian J. Pure Appl. Math. 11 (1980), no. 10, 1356-1360; MR 82c:30012.
288. Goel, R. M.; Sohi, N. S. *Subclasses of univalent functions.* Tamkang J. Math. 11 (1980), no. 1, 77-81; MR 82g:30025.
289. Goel, R. M.; Sohi, N. S. *On a class of spiral-like functions.* Indian J. Pure Appl. Math. 12 (1981), no. 5, 628-633.
290. Goel, R. M.; Sohi, N. S. *Multivalent functions with negative coefficients.* Indian J. Pur Appl. Math. 12 (1981), no. 7, 844-853; MR 82i:30012.
291. Goel, R. M.; Sohi, N. S. *A new criterion for univalence and its applications.* (Serbo-Croatian summary) Glas. Mat. Ser. III 16(36) (1981), no. 1, 39-49.
292. Gohberg, I. C.; Lerer, L. E. *Resultant operators of a pair of analytic functions.* Proc. Amer. Math. Soc. 72 (1978), no. 1, 65-73.
293. Goldberg, J. L.; Ullman, J. L. *A note on positive real functions.* SIAM J. Math. Anal. 8 (1977), no. 5, 757-762; MR 56 #8854.
294. Goluzina, E. G. *The mutual growth coefficients of p-valent close-to-convex functions.* (Russian) Some problems in modern function theory (Proc. Conf. Modern Problems of Geometric Theory of Functions, Inst. Math., Acad. Sci. USSR, Novosibirsk, 1976) (Russian), pp. 21-27. Akad. Nauk SSSR Sibirsk. Otdel. Inst. Mat., Novosibirsk (1976). For a review of this item see RZMat (1977) #10 B177. [For the entire collection see MR 57 #15813.]; MR 58 #22507.
295. Goluzina, E. G. *On the ranges of values of certain systems of functionals in classes of functions with positive real part.* (Russian) Analytic number theory and the theory of functions, 3. Zap. Naucn. Sem. Leningrad. Otdel. Mat. Inst. Steklov. (LOMI) 100 (1980), 17-25, 173; MR 82b:30036.
296. Goluzina, E. G. *Ranges of values of some systems of functionals in a class of functions that are convex in one direction.* (Russian) Analytic number theory and the theory of functions, 4. Zap. Naucn. Sem. Leningrad. Otdel. Mat. Inst. Steklov. (LOMIA(112 (1981), 51-58, 199.
297. Goodman, A. W. *On the coefficients of a multivalent function.* Mathematical Structures—Computational Mathematics—Mathematical Modelling. Papers dedicated to Professor Iliev's 60th Anniversary, Sofia (1975), pp. 273-279.
298. Goodman, A. W. *The domain covered by a typically-real function.*

Proc. Amer. Math. Soc. 64 (1977), no. 2, 233-237; MR 56 #3301.
299. Goodman, A. W. *The domain of univalence of certain families of rational functions.* Proc. Amer. Math. Soc. 66 (1977), no. 1, 85-90; MR 58 #28463.
300. Goodman, A. W. *Koebe domains for certain families.* Proc. Amer. Math. Soc. 74 (1979), no. 1, 87-94; MR 80d:30006.
301. Goodman, A. W. *An invitation to the study of univalent and multivalent functions.* Internat. J. Math. Math. Sci. 2 (1979), no. 2, 163-186; MR 80f:30013.
302. Goodman, A. W. *Valence sequences.* Proc. Amer. Math. Soc. 79 (1980), 422-426; MR 81c:30026.
303. Goodman, A. W.; Saff, E. B. *On the definition of a close-to-convex function.* Internat. J. Math. and Math. Sci., vol. 1 (1978), 125-132; MR 58 #1118.
304. Goodman, A. W.; Saff, E. B. *On univalent functions convex in one direction.* Proc. Amer. Math. Soc. 73 (1979), no. 2, 183-187; MR 80e:30006.
305. Gopalakrishna, H. S.; Shetiya, V. S. *On the real part of the derivatives of certain analytic functions.* (Italian summary) Atti Accad. Naz. Lincei Rend. Cl. Sci. Fis. Mat. Natur. (8) 59 (1975), no. 1-2, 22-25 (1976); MR 56 #8829.
306. Gopalakrishna, H. S.; Shetiya, V. S. *Coefficient estimates for spirallike functions.* Ann. Polon. Math. 35 (1977/78), no. 1, 1-9; MR 58 #11356.
307. Gopalakrishna, H. S.; Umarani, P. G. *Coefficient estimates for some classes of spiral-like functions.* Indian J. Pure Appl. Math. 11 (1980), no. 8, 1011-1017; MR 82a:30023.
308. Gorjainov, V. V. *Some properties of the solutions of the Löwner-Kufarev equation.* (Russian, English summary) Dokl. Akad. Nauk Ukrain. SSR Ser. A (1978), no. 3, 207-210, 285; MR 58 #11360.
309. Gorjainov, V. V. *A distortion theorem for a class of univalent functions.* (Russian) Theory of functions and mappings (Russian), pp. 75-85, 177, "Naukova Dumka," Kiev (1979); MR 81i:30020.
310. Gorjainov, V. V. *On parametric representation of univalent functions.* (Russian) Dokl. Akad. Nauk SSSR 245 (1979), no. 5, 1038-1041; MR 81d:30040.
311. Gorjainov, V. V. *On a parametric method of the theory of univalent functions.* (Russian) Mat. Zametki 27 (1980), no. 4, 559-568, 669; MR 81f:30017.
312. Gorjainov, V. V. *Distortion theorem in the class of bounded univalent symmetric functions.* (Russian, English summary) Dokl.

Akad. Nauk Ukrain. SSR Ser. A (1980), no. 8, 11-14, 86; MR 81i:30021.
313. Gorjainov, V. V.; Gutljanskii, V. J. *Extremal problems in the class S_M.* (Russian) Mathematics collection (Russian), pp. 242-246. Izdat. "Naukova Dumka," Kiev (1976); MR 56 #12244.
314. Grassmann, E.; Hengartner, W.; Schober, G. *Support points of the class of close-to-convex functions.* Canad. Math. Bull. 19 (1976), no. 2, 177-179; MR 58 #6202.
315. Grassmann, E.; Rokne, J. *Calculation of some extremal conformal mappings.* SIAM J. Math. Anal. 9 (1978), no. 1, 87-105.
316. Grassmann, E.; Rokne, J. *Calculation of extremum problems for univalent functions.* SIAM J. Math. Anal. 10 (1979), no. 4, 850-874; MR 81c:30024.
317. Grinspan, A. Z. *The sharpening of the difference of the moduli of adjacent coefficients of schlicht functions.* (Russian) Some problems in modern function theory (Proc. Conf. Modern Problems of Geometric Theory of Functions, Inst. Math., Acad. Sci. USSR, Novosibirsk, 1976) (Russian), pp. 41-45. Akad. Nauk SSR Sibirsk. Otdel. Inst. Mat., Novosibirsk (1976). For a review of this item see RZMAT (1977) #11 B141. [For the entire collection see MR 57 #15813.]; MR 58 #22520.
318. Grinšpan, A. Z. *Taylor coefficients of certain classes of univalent functions.* (Russian) Metric questions of the theory of functions (Russian), pp. 28-32, 158, "Naukova Dumka," Kiev (1980); MR 82g:30027.
319. Grinšpan, A. Z. *Coefficients of powers of univalent functions.* (Russian) Sibirsk. Mat. Z. 22 (1981), no. 4, 88-93, 230; MR 82i:30023.
320. Grinšpan, A. Z.; Kolomoiceva, Z. D. *The method of areas for functions without common values.* (Russian, English summary) Vestnik Leningrad. Univ. Mat. Meh. Astronom (1979), vyp. 4, 31-36, 122; MR 81e:30035.
321. Gronau, D. *Eine Bemerkung zu einer partiell Differentialgleichung von Hornich im Zusammenhang mit den schlichten Funktionen.* Collection in honor of Hans Hornich. Osterreich. Akad. Wiss. Math.-Naturwiss. Kl. S.-B. II 185 (1976), no. 1-3, 27-30; MR 56 #15906.
322. Grunsky, H. *Lectures on Theory of Functions in Multiply Connected Domains.* Vandenhoeck and Ruprecht in Gottingen und Zurich (1978), 253 pp.; MR 57 #3365.
323. Gupta, V. P.; Ahmad, Iqbal *Certain classes of functions univalent in the unit disc.* Bull. Inst. Math. Acad. Sinica 5 (1977), no. 2,

379-389; MR 57 #6404.
324. Gupta, V. P.; Ahmad, I. *Certain classes of functions univalent in the unit disc.* II. Bull. Inst. Math. Acad. Sinica 7 (1979), no. 1, 7-13; MR 80c:30016.
325. Gupta, V. P.; Ahmad, I. *On starlike functions.* Bull. Austral. Math. Soc. 22 (1980), no. 2, 241-247; MR 82a:30012.
326. Gupta, V. P.; Jain, P. K. *A note on a class of Bazilevič functions.* Tamkang J. Math. 7 (1976), no. 1, 117-119; MR 54 #7769.
327. Gupta, V. P.; Jain, P. K. *Certain classes of univalent functions with negative coefficients.* Bull. Austral. Math. Soc. 14 (1976), no. 3, 409-416; MR 54 #2941.
328. Gupta, V. P.; Jain, P. K. *Certain classes of univalent functions with negative coefficients.* II. Bull. Austral. Math. Soc. 15 (1976), no. 3, 467-473; MR 55 #8334.
329. Gupta, V. P.; Jain, P. K. *On starlike functions.* (Italian summary) Rend. Mat. (6) 9 (1976), no. 3, 433-437; MR 55 #641.
330. Gupta, V. P.; Jain, P. K.; Ahmad, I. *On the radius of univalence of certain classes of analytic functions with fixed second coefficient.* (French summary) Rend. Mat. (6) 12 (1979), no. 3-4, 423-430 (1980); MR 81j:30025.
331. Gutljanskii, V. J. *The method of variations for univalent analytic functions with a quasiconformal extension.* (Russian) Dokl. Akad. Nauk SSSR 236 (1977), no. 5, 1045-1048; MR 57 #3376.
332. Gutljanskii, V. J. *One some classes of univalent functions.* (Russian) Theory of functions and mappings (Russian), pp 85-97, 178, "Naukova Dumka," Kiev (1979); MR 81i:30022.
333. Gutljanskii, V. J. *The method of variations for univalent analytic functions with a quasiconformal extension.* (Russian) Sibirsk. Mat. Z. 21 (1980), no. 2, 61-78, 237.
334. Gutljanskii, V. J.; Scepetev, V.A. *Exact estimates of the modulus of a univalent analytic function with a quasiconformal extension.* (Russian) Akad. Nauk Ukrain. ssr Inst. Mat. Preprint No. 13 (1979), 28 pp.; MR 81j:30031.
335. Haario, H. *On coefficient bodies or univalent functions.* Ann. Acad. Sci. Fenn. Ser. A I Math. Dissertationes No. 22 (1978), 49 pp.; MR 80b:30011.
336. Hall, R. R. *The length of ray-images under starlike mappings.* Mathematika 23 (1976), no. 2, 147-150; MR 55 #642.
337. Hall, R. R. *A conformal mapping inequality for starlike functions of order ½.* Bull. London Math. Soc. 12 (1980), 119-126; MR 81h:30015.

338. Hall, R. R. *On a conjecture of Clunie and Sheil-Small.* Bull. London Math. Soc. 12 (1980), 25-28; MR 81b:30021.
339. Hallenbeck, D. J.; Livingston, A. E. *Applications of extreme point theory to classes of multivalent, functions.* Trans. Amer. Math. Soc. 221 (1976), no. 2, 339-359.
340. Hamilton, D. *Extremal problems for nonvanishing univalent functions.* Aspects of contemporary complex analysis (Proc. NATO Adv. Study Inst., Univ. Durham, Durham, 1979), pp. 415-420, Academic Press, London, 1980; MR 82g:30035.
341. Hamilton, D. H. *On a conjecture of M. S. Robertson.* J. London Math. Soc. (2) 21 (1980), 265-278; MR 81i:30023.
342. Hamilton, D. H.; Tuan, P. D. *An extremal problem for functions of positive real part with application to a radius of convexity problem.* Proc. Amer. Math. Soc. 72 (1978), no. 2, 313-318: MR 80e:30007.
343. Hamilton, D. H.; Tuan, P. D. *Radius of starlikeness of convex combinations of univalent starlike functions.* Proc. Amer. Math. Soc. 78 (1980), no. 1, 56-58; MR 81f:30007.
344. Harris, L. A. *Coefficient inequalities for L_p-valued analytic functions.* Canad. Math. Bull. 24 (1981), no. 3, 347-350.
345. Hartmann, F. W. *Linear homeomorphisms of some classical families of univalent functions.* Proc. Amer. Math. Soc. 63 (1977), no. 2, 265-272.
346. Havinson, S. J. *Extremal functions in problems on the distortion of lenghts and areas under transformations by bounded analytic functions.* (Russian) Moskov. Inz.-Stroitel. Inst. Sb. Trudov No. 130 (1975), 3-6; MR 56 #15911.
347. Hayman, W. K. **Transfinite diameter and its applications.* Notes by K. R. Unni. Second printing. Matscience Report, No. 45. Institute of Mathematical Sciences, Madras (1966), ii + 61 pp.; MR 52 #14272.
348. Hayman, W. K. *Some mathematical achievements of John Edensor Littlewood.* Bull. Institute of Math. and its Appl. 12, no. 7 (July 1976), 196-198.
349. Hayman, W. K. *On a conjecture of Littlewood.* J. Analyse Math. 36 (1979), 75-95 (1980).
350. Hayman, W. K. *The logarithmic derivative of multivalent functions.* Michigan Math. J. 27 (1980), no. 2, 149-179; MR 82k:30022.
351. Hayman, W. K.; Patterson, S. J.; Pommerenke, C. On the coefficients of certain automorphic functions. Math. Proc. Cambridge Philos. Soc. 82 (1977), no. 3, 357-367.

352. Hayman, W. K.; Storvick, D. A. *On normal functions.* Bull. London Math. Soc. 3 (1971), 193-194.
353. Hayman, W. K.; Wu, J. M. G. *Level sets of univalent functions.* Comment. Math. Helv. 56 (1981), no. 3, 366-403.
354. Hazalija, G. J. *The means of the moduli of analytic functions in doubly connected domains.* (Russian) Dokl. Akad. Nauk SSSR 226 (1976), no. 6, 1287-1290 [English translation: Soviet Math. Dokl 17 (1976), no. 1, 300-303]; MR 53 #5850.
355. Heath L. F.; Suffridge, T. J. *Starlike, convex, close-to-convex, spirallike, and Φ-like maps in a commutative Banach algebra with identity.* Trans. Amer. Math. Soc. 250 (1979), 195-212.
356. Heins, M. *Structure theorems for conformal maps of regions convex in a given direction.* Memorial issue for Marston Morse. Bull. Inst. Math. Acad. Sinica 6 (1978), no. 2, part 1, 379-388; MR 80h:30012.
357. Hengartner, W.; Schober, G. *A remark on level curves for domains convex in one direction.* Appl. Anal. 3 (1973), 101-106; MR 52 #14260.
358. Hengartner, W.; Schober, G. *Some new properties of support points for compact families of univalent functions in the unit disk.* Michigan Math. J. 23 (1976), no. 3, 207-216 (1977); MR 58 #11361.
359. Higginson, C. G. *The asymptotic Bieberbach conjecture for weakly p-valent functions.* Proc. London Math. Soc. (3) 35 (1977), no. 2, 291-312; MR 56 #5859.
360. Hohlov, J. E. *Operators and operations on the class of univalent functions.* (Russian) Izv. Vyss. Ucebn. Zaved. Matematika (1978), no. 10 (197), 83-39; MR 81m:30012.
361. Holland, F.; Twomey, J. B. *Hardy spaces of close-to-convex functions and their derivatives.* Trans. Amer. Math. Soc. 246 (1978), 359-372; MR 80h:30029.
362. Holland, F.; Twomey, J. B. *On Hardy classes and the area functions.* J. London Math. Soc. (2) 17 (1978), no. 2, 275-283; MR 58 #6255.
363. Holland F.; Twomey, J. B. *Integral means of functions with positive real part.* Canad. J. Math. 32 (1980), no. 4, 1008-1020; MR 82b:30037.
364. Hornich, H. *Bemerkungen über schlichte Funktionen.* Osterreich. Akad. Wiss. Math.-Natur. Ke. S.-B. II 182 (1974), 103-106; MR 51 #13211.
365. Hornich, H. *Zur Seltenheit der schlichten Funktionen.* Anz. Osterreich. Akad. Wiss. Math.-Naturwiss. K1. (1975), no. 6, 59-61; MR 53 #3289.

366. Hornich, H. *Zur Verteilung der schlichten Funktionen.* (English summary) Monatsh. Math. 82 (1976), no. 1, 27–29; MR 54 #2942.
367. Horowitz, D. *Coefficient estimates for univalent polynomials.* J. Analyse Math. 31 (1977), 112–124; MR 58 #11357.
368. Horowitz, D. *A further refinement for coefficient estimates of univalent functions.* Proc. Amer. Math. Soc. 71 (1978), no. 2, 217–221; MR 58 #1126.
369. Hou, Ming Shu. *α-starlike functions.* (Chinese) Kexue Tongbao 24 (1979), no. 2, 52–57; MR 80g:30006.
370. Hu, Ke. *On the distortion theorem of schlicht functions.* A translation of Kexue Tongbao (Chinese) 25 (1980), no. 13, 577–579. Kexue Tongbao (English) 25 (1980), no. 12, 977–980; MR 82j:30023.
371. Hu, Ke. *Distortion theorems for univalent functions. (Chinese, English summary)* Chinese Ann. Math. 1 (1980), no. 3-4, 421–427.
372. Hummel, J. A. *A variational method for Gel'fer functions.* J. Analyse Math. 30 (1976), 271–280; MR 55 #12906.
373. Hummel, J. A. *Lagrange multipliers in variational methods for univalent functions.* J. Analyse Math. 32 (1977), 222–234; MR 57 #3377.
374. Hummel, J. A. *The b_1, b_2 coefficient body for Bierberbach-Eilenberg functions.* J. Analyse Math. 33 (1978). 168–190; MR 80g:30014.
375. Hummel, J. A. *The second coefficient of univalent Bieberbach-Eilenberg functions near the identity.* Proc. Amer. Math. Soc. 80 (1980), 237–243; MR 81h:30018.
376. Hummel, J. A.; Pinchuk, B.; Schiffer, M. M. *Bounded univalent functions which cover a fixed disc.* J. Analyse Math. 36 (1979), 118–138 (1980); MR 82a:30031.
377. Hummel, J. A.; Scheinberg, S.; Zalcman, L. *A coefficient problem for bounded nonvanishing functions.* J. Analyse Math. 31 (1977), 169–190; MR 58 #11358.
378. Hummel, J. A.; Schiffer, M. M. *Variational methods for Bieberbach-Eilenberg functions and for pairs.* Ann. Acad. Sci. Fenn. Ser. AI Math. 3 (1977), no. 1, 3–42; MR 58 #28473.
379. Hurwitz, A. *Über die Entwicklung der allgemeinen Theorie der analytischen Funktionen in neuerer Zeit.* Mathematische Werke, vol. 1, Funktionentheorie, pp. 461–480, Birkhauser, Basel (1932).
380. Husbaktov, S. D. *Geometric properties of a function of a discrete complex argument.* (Russian) Taskent. Gos. Univ. Naucn. Trudy Vyp. 460 Voprosy Mat. (1974), 144–148; 180; MR 54 #5475.
381. Hwang, Jun Shung. *On a coefficient estimate of angular functions.* Tamkang J. Math. 11 (1980), no. 1, 141–143; MR 82i:30024.

382. Hwang, J. S. *A problem on a geometric property of lemniscates.* Proc. Amer. Math. Soc. 82 (1981), no. 3, 390–392.
383. Hwang, J. S. *On the radial limits of analytic and meromorphic functions.* Tran. Amer. Math. Soc. 270 (1982), no. 1, 341–348.
384. Iliev, L. G. *Extremal problems for univalent functions.* PLISKA Stud. Math. Bulgar. 4 (1981), 137–141.
385. *Imai, I. *Conformal mappings and their applications.* (Japanese) Iwanami Shoten, Tokyo (1979). xiv + 309 pp.;MR 82j:30013.
386. Jablonski, A. *On a certain class of regular functions.* Demonstratio Math. 10 (1977), no. 1, 33–43; MR 56 #3270.
387. Jacob, M.; Offord, C. *Les fonctions aleatoires dans le disque unité.* (English summary) C. R. Acad. Sci. Paris Ser. A-B 290 (1980), no. 8, A367–369.
388. Jain, P. K.; Ahuja, O. P. *A class of univalent functions with negative coefficients.* (Italian summary) Rend. Mat. (7) 1 (1981), no. 1, 47–54; MR 82g:30028.
389. Jakubowski, Z. *On typically-real functions.* Annales Polonici Mathematici, to appear.
390. Jakubowski, Z. J.; Kaminski, J. *On some properties of multivalent alpha-starlike functions.* Demonstratio Math. 9 (1976), no. 2, 257–265; MR 53 #8405.
391. Jakubowski, Z. J.; Kaminski, J. *On some properties of the coefficients of regular functions with positive real part.* (Polish and Russian summaries) Ann. Univ. Mariae Curie-Sklodowska Sect. A 29 (1975), 79–92 (1977); MR 58 #1161.
392. Jakubowski, Z. J.; Kaminski, J. *On some properties of Mocanu-Janowski functions.* Rev. Roumaine Math. Pures Appl. 23 (1978), no. 10, 1523–1532; MR 80i:30018.
393. Jakubowski, Z. J.; Zielinska, A.; Zyskowska, K. *Sharp estimation of even coefficients of bounded symmetric univalent functions.* Ann. Polonici Math., to appear.
394. Janczar, B. z_0-*quasi-convex functions.* (Polish and Russian summaries) Zeszyty Nauk. Politech. Łódź. No. 186 Mat. No. 5 (1974), 27–42; MR 57 #9953.
395. Janczar, B. *On quasi-convex functions.* (Polish and Russian summaries) Zeszyty Nauk. Politech. Łódź. Mat. No. 7 (1976), 37–58; MR 55 #5850, MR 57 #9953.
396. Janczar, B. *Extremal problems in the class of* z_0-*quasi-convex functions* (Polish and Russian summaries) Zeszyty Nauk. Politech. Łódź. No. 245 Mat. No. 8 (1976), 61–68; MR 54 #10588.
397. Jaremenko, L. A. *The radius of starlikeness and univalence of cer-*

tain classes of functions that are regular in the circle $|z| < 1$. (Russian) Questions in the metric theory of mappings and its application (Proc. Fifth Colloq. Quasi-conformal Mappings, Generalizations and Applications, Donetsk 1976). (Russian), pp. 171-175, 188, "Naukova Dumka," Kiev. 1978; MR 82d:30015.

398. Jaremenko, L. A. *The starlikeness boundary and the convexity boundary for certain classes of regular functions in the unit disc.* (Russian) Ukrain. Mat. Z. 31 (1979), no. 3, 329-333, 336; MR 81c:30022.

399. Jaremenko, L. A. *Extremal properties of certain classes of regular functions.* (Russian) Metric questions of the theory of functions (Russian), pp. 146-151, 164, "Naukova Dumka," Kiev (1980).

400. Jenkins, J. A. *On the sharp form of the three-circles theorem.* Kodai Math. Sem. Rep. 27 (1976), no. 1-2, 155-158; MR 53 #11044.

401. Jenkins, J. A. *On quadratic differentials whose trajectory structure consists of ring domains.* Complex analysis (Proc. S.U.N.Y. Conf., Brockport, N.Y., 1976), pp. 65-70. Lecture Notes in Pure and Appl. Math., Vol. 36, Dekker, New York, 1978.

402. Jenkins, J. A. *On explicit bounds in Landau's theorem.* II. Canad. J. Math. 33 (1981), no. 3, 559-562.

403. Jenkins, J. A.; Oikawa, K. *On Ahlfors' "second fundamental inequality."* Proc. Amer. Math. Soc. 62 (1977), no. 2, 266-270; MR 55 #10655.

404. Jenkins, J. A.; Oikawa, K. *On a theorem of Carathéodory and Fejér.* Bull. London Math. Soc. 9 (1977), no. 2, 165-167.

405. Jensen, E.; Waadeland, H. *A coefficient inequality for bi-univalent functions.* Norske Vid. Selsk. Skr. (Trondheim) (1972), no. 15, 11 pp.; MR 54 #10590.

406. Jevtic, M. *Some properties of new Hadamard products of analytic functions.* (Serbo-Croatian, English summary) Mat. Vesnik 4 (17) (32) (1980), no. 2, 157-162.

407. John, F. *A criterion for univalency brought up to date.* Comm. Pure Appl. Math. 29 (1976), no. 3, 293-295; MR 54 #10592.

408. Jondro, H. *Variations of Faber polynomials and their applications to extremal problems in the family of bounded univalent functions.* (Polish; Russian and English summaries) Zeszyty Nauk. Politech. Slask. Mat.-Fiz. No. 30 (1979), 271-287; MR 80h:30020.

409. Jondro, H. *Variations in a class of univalent functions.* (Polish; Russian and English summaries) Zeszyty Nauk. Politech. Slask. Mat.-Fiz. No. 34 (1979), 85-95; MR 81d:30015.

410. Jondro, H. *Sur une methode variationnelle dans la famille des fonctions de Grunsky-Shah.* (English and Russian summaries) Bull. Acad. Polon. Sci. Ser. Sci. Math. 27 (1979), no. 7-8, 541-547 (1980).
411. Juneja, O. P.; Mogra, M. L. *On starlike functions of order α and type β.* Rev. Roumaine Math. Pures Appl. 23 (1978), no. 5, 751-765; MR 80c:30010.
412. Juneja, O. P.; Mogra, M. L. *Radii of convexity for certain classes of univalent analytic functions. Pacific J. Math. 78 (1978), no. 20, 359-368;* MR 80a:30011.
413. Juneja, O. P.; Mogra, M. L. *A class of univalent functions.* (French summary) Bull. Sci. Math. (2) 103 (1979), no. 4, 435-447; MR 81a:30008.
414. Kac, B. A. *Convexity and starlikeness of the level curves of rational functions.* I. (Russian) Izv. Vyss. Ucebn. Zaved. Matematika (1977), no. 3 (178), 32-42; MR 57 #3373a.
415. Kac, B. A. *Convexity and starlikeness of the level curves of rational functions.* II. (Russian) Izv. Vyss. Ucebn. Zaved. Matematika (1977), no. 4 (179), 44-52; MR 57 #3373b.
416. Kaidan, V. O.; Pohilevič, V. O. *The ranges of values of a certain functional on special classes of starlike functions.* (Russian; English summary) Dokl. Akad. Nauk. Ukrain. SSR Ser. A (1976), no. 1, 5-9, 92; MR 53 #13539.
417. Kaidan, V. O.; Pohilevič, V. A. *Some properties of arcs of level curves outside the circle of the radius of convexity.* (Russian) Mathematical analysis and probability theory (Russian), pp. 74-79; 212, "Naukova Dumka," Kiev (1978); MR 80m:30012.
418. Kaidan, V. O.; Pohilevic, V. O. *Boundaries of α-convexity for subclasses of star-like functions* (Russian) Ukrain. Mat. Z. 31 (1979), no. 5, 551-555, 620; MR 81i:30024.
419. Kaminski, J. *Some growth problems for certain α-convex functions.* Demonstratio Math. 12 (1979), no. 1, 211-230; MR 81f:30008.
420. Kamockii, V. I. *A representation of functions of class S.* (Russian) Theory of functions and functional analysis (Russian), pp. 60-66. Leningrad. Gos. Ped. Inst., Leningrad (1975); MR 58 #28465.
421. Kamockii, V. I. *Application of the area theorem for studying the classes S(a).* (Russian) Mat. Zametki 28 (1980), no. 5, 695-706, 802; MR 82g:30032.
422. Kamockii, V. I. *Estimate of integral means in the class S(a).* (Russian) Sibirsk. Mat. Z. 21 (1980), no. 1, 211-215, 239; MR 82c:30013.
423. Kaplan, W. *On the trajectories of a quadratic differential.* II.

Math. Nachr. 93 (1979), 259-278; MR 81m:30044.
424. Kapoor, G. P.; Gopal, K. *On the coefficients of functions analytic in the unit disc having fast rates of growth.* Ann. Mat. Pura Appl. (4) 121 (1979), 337-349.
425. Karunakaran, V. *Certain classes of regular univalent functions.* Pacific J. Math. 61 (1975), no. 1, 173-182; MR 53 #8406.
426. Karunakaran, V. *A certain class of analytic functions in the unit disc. Appl. Math. 7 (1976), no. 12, 1381-*1399; MR 81b:30022.
427. Karunakaran, V. *On a class of meromorphic functions in the unit disc.* Math. Chronicle 4 (1976), no. 2-3, 112-121; MR 53 #5843.
428. Karunakaran, V. *Radius of starlikeness for a class of meromorphic functions.* Comment. Math. Univ. St. Paul. 25 (1976/77), no. 2, 153-157; MR 55 #643.
429. Karunakaran, V. *Radius of starlikeness for a certain class of analytic functions.* Rev. Roumaine Math. Pures Appl. 22 (1977), no. 7, 927-931; MR 57 #605.
430. Karunakaran, V. *A certain radius of convexity problem. Publ. Math. Debrecen 24 (1977), no. 1-2, 1-3*; MR 56 #3271.
431. Karunakaran V. *A subordination theorem for a class of convex functions.* Presented at the meeting of the Tamil Nadu Academy of Sciences held on 27-1-1977 at Matscience, Madras (India).
432. Karunakaran, V.; Padma, K. *Functions of bounded radius rotation..* Indian J. Pure Appl. Math. 12 (1981), no. 5, 621-627; MR 82j:30018.
433. Karunakaran, V.; Ziegler, M. R. *The radius of starlikeness for a class of regular functions defined by an integral.* Pacific J. Math. 91 (1980), no. 1, 145-151;MR 82g:30029.
434. Kasten, V.; Schmieder, G. *Die Koeffizientenkörper schlichter Trinome.* Math. Z. 171 (1980), no. 3, 269-284; MR 81g:30025.
435. Kasten, V.; Schmieder, G. *Über eine Klasse von Trinomen.* Arch. Math. (Basel) 35 (1980), no. 4, 374-385; MR 82e:30018.
436. Kasymov, S. *Tests for the univalence of analytic functions, and their application.* (Russian) Izv. Vyss. Ucebn. Zaved. Matematika (1977), no. 9(184), 38-42; MR 58 #6212.
437. Kennedy, P. B.; Twomey, J. B. *Some properties of bounded univalent functions and related classes of functions.* Proc. Roy. Irish Acad. Sect. A65, 43-49 (1967); MR 35 #4388.
438. Keogh, F. R. *On a theorem of Fejér and Riesz.* Proc. Amer. Math. Soc. 20 (1969), 45-50.
439. Keogh, F. R. *A characterisation of convex domains in the plane.* Bull. London Math. Soc. 8 (1976), no. 2, 183-185; MR 57 #6392.
440. Keogh, F. R. *Some recent developments in the theory of univalent*

functions. Complex analysis (Proc. Conf. Univ. Kentucky, Lexington, Ky., 1976), pp. 76-92. Lecture Notes in Math., Vol. 599, Springer, Berlin (1977); MR 56 #12245.

441. Keogh, F. R.; Merkes, E. P. *Preservation of subordination.* J. D'analyse Math. 36 (179), 1979-183; MR 82a:30013.
442. Kesel'man, G. M. *Extremal properties of functions of the Bazilevič class* (Russian) Mat. Zametki 25 (1979), no. 3, 341-350, 475; MR 82k:30014.
443. Kesel'man, G. M. *Extremal properities of certain spiral-like functions.* (Russian) Sibirsk. Mat. Z. 20 (1979), no. 5, 1050-1059, 1166; MR 81i:30025.
444. Kir'jackii, E. G. (Kirjackis, E.) *A family of functions connected with a linear-fractional transformation of the unit circle.* (Russian, Lithuanian and English summaries) Litovsk. Mat. Sb. 16 (1976), no. 1, 103-110, 247; MR 54 #778.
445. Kir'jackii, E. G. (Kirjackis, E.) *Some operators connected with a linear-fractional transformation of the unit circle.* (Russian, Lithuanian and English summaries) Litovsk. Mat. Sb. 16 (1976), no. 1, 111-122, 247; MR 54 #5453.
446. Kir'jackii, E. G. *On a family of univalent functions.* (Russian, Lithuanian and English summaries) Litovsk. Mat. Sb. 16 (1976), no. 2, 111-116, 242; MR 55 #5851.
447. Kir'jackii, E. G. (Kirjackis, E.) *Some properties of functions with nonzero divided difference. (Russian, Lithuanian and Enlgish summaries)* Litovsk. Mat. Sb. 19 (1979), no. 1, 97-113, 230; MR 80j:30016.
448. Kirwan, W. E. *Extremal properties of slit conformal mappings. Aspects of contemporary complex analysis* (Proc. NATO Adv. Study Inst., Univ. Durham, Durham, 1979), pp. 439-449, Academic Press, London, 1980; MR 82g:30036.
449. Kirwan, W. E.; Pell, R. W. *A note on a class of slit conformal mappings*. Canad. J. Math. 30(1978), no. 6, 1166-1173; MR 80d:30013.
450. Kirwan, W. E.; Pell, R. *Extremal properties of a class of slit conformal mappings.* Michigan Math. J. 25 (1978), no. 2, 223-232; MR 58 #6213.
451. Kirwan, W. E.; Schober, G. *Extremal problems for meromorphic univalent functions.* J. Analyse Math. 30 (1976), 330-348; MR 56 #3283.
452. Kirwan, W. E.; Schober, G. *Inverse coefficients for functions of bounded boundary rotation.* J. Analyse Math. 36 (1979), 167-178 (1980); MR 81j:30026.

453. Kirwan, W. E.; Zalcman, L. *Advances in complex function theory (Proc. Sem., Univ. Maryland, College Park, Md., 1973-1974). Advances in complex function theory. Proceedings of seminars held at Maryland University, College Park, Md., 1973-1974. Edited by W. E. Kirwan and L. Zalcman. Lecture Notes in Mathematics, Vol. 505. Springer-Verlag, Berlin-New York, 1976, viii + 203 pp.; MR 53 #776.
454. Klouth, R. *Abschätzungen im Bereich der Schwarzschen Derivierten und gewisser Verallgemeinerungen. Inaugural-Dissertation zur Erlangung des Koktorgrades der Hohen Mathematisch-Naturwissenschaftlichen Fakultat der Rheinischen Friedrich-Wilhelm-Universitat Bonn, Bonn, 1975. Bonner Mathemtische Schriften, Nr. 82. Mathematisches Institut der Universitat Bonn, Bonn (1976), v + 126 PP.; MR 58 #11346.
455. Klouth, R. Abschätzungen fur verallgemeinerte Schwarzsche Derivierten in linear-invarianten Funktionenfamilien. [Estimations for generalized Schwarz derivatives in linear-invariant families of functions] Arch. Math. (Basel) 36 (1981), no. 5, 455-462; MR 82k:30015.
456. Klouth, R.; Wirths, K-J. Two new extremal properties of the Koebe-function. Proc. Amer. Math. Soc. 80 (1980), 594-596; MR 82d:30017.
457. Kocak, C. An extremum problem in conformal mapping. Istanbul Tek. Univ. Bul. 28 (1975), no. 1, 105-119; MR 58 #28474.
458. Kocetkov, V. K. Extremal properties of certain new special classes of univalent functions. (Russian) Theory of functions. Differential equations and their applications, No. 1 (Russian), pp. 73-89, 216. Kalmyck. Gos. Univ., Elista (1976); MR 58 #28466.
459. Kocetkov, V. K. Higher order parabolic univalent functions. (Russian) Theory of functions. Differential equations and their applications, No. 1 (Russian), pp. 90-95, 216-217. Kalmyck. Gos. Univ., Elista (1976); MR 58 #28469.
460. Kocetkov, V. K. A construction method for some classes of univalent functions. (Russian) Izv. Severo-Kavkaz. Naucn. Centra Vyss. Skoly Ser. Estestv. Nauk. (1980), no. 3, 25-27, 109.
461. Kocur, M. F. Bounds of starlikeness and convexity for certain special classes of analytic functions in the disc. (Russian) Izv. Vyss. Ucebn. Zaved. Matematika (1977), no. 7 (182), 57-60; MR 56 #8830.
462. Kocur, M. F. Curvature bounds and the radius of convexity of the image of the circle $|z| = r$, $0 < r < 1$, in the class $Q(a)$. (Ukrainian, English and Russian summaries) Visnik Kuv. Univ. Ser. Mat. Meh.

No. 19 (1977), 127-129, 153; MR 80g:30007.
463. Kocur, M. F. *Radii of starlikeness and of convexity in some classes of analytic functins in the disc.* (Russian) Mat. Zametki 25 (1979), no. 5, 675-679, 798; MR 81e:30022.
464. Komatu, Yusaku. *A one-parameter family of operators defined on analytic functions in a circle.* Analytic functions, Kozubnik (197) (Proc. Seventh Conf. Kozubnik, 1979), pp. 292-300, Lecture Notes in Math., 798, Springer, Berlin, 1980. [Bull. Fac. Sci. Engrg. Chvo Univ. 22 (1979), 1-22]; MR 81i:30026.
465. Korobeinik, J. F. *The basis property of a certain system of functions.* (Russian, English summary) Anal. Math. 1 (1975), no. 2, 103-113; MR 52 #8388.
466. Kortram, R.; Tammi, O. *On the range of validity of certain inequalities for the fourth coefficient of a bounded univalent functions.* (Polish summary) Bull. Soc. Sci. Lettres Łódź 24 (1974), no. 1, 6 pp. (1975); MR 55 #10657.
467. Kortram, R. A.; Tammi, O. Nonhomogeneous combinations of coefficients of univalent functions. Ann. Acad. Sci. Fenn. Ser. A I Math. 5 (1980), no. 1, 131-144; MR 81m:30017.
468. Krjuckov, B. J.; Popova, G. A. *The reduction of an optimal control problem in Löwner's theory to a Cauchy problem for a system of ordinary differential equations.* (Russian) Theory of optimal processes (Russian), pp. 47-52. Akad. Nauk Ukrain. SSR Inst. Kibernet, Kiev (1974); MR 58 #6217.
469. Kronstadt, E. P. *Compact families of univalent functions.* Michigan Math. J. 23 (1976), no. 4, 367-374 (1977); MR 56 #5861.
470. Krzyz, J. *A counterexample concerning univalent functions.* Folia Societatis Scient. Lubliensis, vol. 2 (1962), pp. 57-58.
471. Krzyz, J. G. *Problems in complex variable theory.* Translation of the 1962 Polish original. Modern Analytic and Computational Methods in Science and Mathematics, No. 36. American Elsevier Publishing Co., Inc., New York; PWN—Polish Scientific Publishers, Warsaw (1971), xvii + 283 pp.; MR 56#5844.
472. Krzyz, J. G. Convolution and quasiconformal extension. Comment. Math. Helv. 51 (1976), no. 1, 99-104; MR 53 #3298.
473. Krzyz, J.; Stankiewicz, J. Quasisubordination and quasimajorization. (Polish and Russian summaries) Ann. Univ. Mariae Curie-Sklodowska Sct. A 31 (1977), 71-74 (1979); MR 81d:30042.
474. Krzyz, J. G.; Zlotkiewicz, E. *Two remarks on typically-real functions.* (Polish and Russian summaries) Ann. Univ. Mariae Curie-Sklodowska Sect. A 30 (1976), 57-61 (1978); MR 58 #17101.

475. Kubo, T.; Owa, Sh. *On some class of analytic functions in an annulus.* Tech. Rep. Osaka Univ. 29 (1979), no. 1459-1491, 1-8; MR 81m:30018.
476. Kubota, Y. *Coefficients of meromorphic univalent functions.* Kodai Math. Sem. Rep. 28 (1976/77), no. 2-3, 253-261; MR 55 #10658.
477. Kubota, Y. *A remark on the third coefficient of meromorphic univalent functions.* Kodai Math. Sem. Rep. 29 (1977), no. 1-2, 197-206; MR 57 #6401.
478. Kühnau, R. *Zur quasikonformen Fortsetzbarkeit schlichter konformer Abbildungen.* (Polish summary) Bull. Soc. Sci. Lettres Lodz 24 (1974), no. 6, 4 pp. (1975); MR 55 #12910.
479. Kühnau, R. *Schlichte konforme Abbildungen auf nichtuberlappende Gebiete mit gemeinsamer quasikonformer Fortsetzung.* Math. Nachr. 86 (1978), 175-180.
480. Kühnau, R. *Zu den Grunskyschen Koeffizienten-Bedingungen.* [On coefficient conditions of Grunsky type] Ann. Acad. Sci. Fenn. Ser. A I Math. 6 (1981), no. 1, 125-130.
481. Kühnau, R. *Der Flächensatz in einer Klasse schlichter Abbildungen.* [The area theorem in a class of univalent mappings] Rev. Roumaine Math. Pures Appl. 26 (1981), no. 8, 1119-1121.
482. Kühnau, R.; Niske, W. *Abschätzung des dritten Koeffizienten bei den quasikonform fortsetzbaren schlichten Funktionen der Klasse S. Math. Nachr. 78 (1977), 185-*192.
483. Kühnau, R.; Renelt, H. *Ein Existenzbeweis fur schlichte Lösungen linearer elliptischer Differentialgleichungs-systeme durch eine Integralgleichung.* Math. Nachr. 79 (1977), 225-232; MR 57 #12840.
484. Kühnau, R.; Thuring, B. *Berechnung einer quasikonformen Extremalfunktion.* Math. Nachr. 79 (1977), 99-113.
485. Kulkarni, V. N.; Shivamurthy, C. G. *The radius of univalence and starlikeness of some classes of regular functions.* J. Karnatak Univ. Sci. 23 (1978), 12-17; MR 82c:30014.
486. Kulkarni, V. N.; Swamy, S. R. *On the univalence of some analytic functions.* Indian J. Pure Appl. Math. 9 (1978), no. 5, 467-480; MR 57 #16562.
487. Kulkarni, V. N.; Swamy, S. R. *On the univalence of some analytic functions.* II. Indian J. Pure Appl. Math. 9 (1978), no. 11, 1131-1137; MR 80d:30007.
488. Kulshrestha, P. K. *Bounded Robertson functions.* Rend. Mat. (6) 9 (1976), no. 1, 137-150. (French summary)
489. Kumbi, V. K. *A note on starlike functions.* J. Karnatak Univ. Sci.

22 (1977), 50-55; MR 80h:30013.
490. Kumbi, V. K. *The radius of univalence and starlikeness of order β of some classes of regular functions.* J. Karnatak Univ. Sci. 22 (1977), 5-10; MR 81a:30009.
491. Kuz'mina, G. V. *Moduli of families of curves and quadratic differentials.* (Russian) Trudy. Mat. Inst. Steklov. 139 (1980), 241 pp.
492. Kuz'mina, G. V. *Covering theorems in classes of Bieberbach-Eilenberg functions.* (Russian) Analytic number theory and the theory of functions, 4. Zap. Naucn. Sem. Leningrad. Otdel. Mat. Inst. Steklov. (LOMI) 112 (1981), 143-158, 201.
493. Kyoto University. **Extremal problems in function theory.* (Japanese) Proceedings of a Symposium held at the Research Institute for Mathematical Sciences, Kyoto University, Kyoto (February 8-10, 1978). Surikaisekikenkyusho Kokyuroku No. 323 (1978). Research Institute for Mathematical Sciences, Kyoto University, Kyoto (1978), pp. i-ii and 1-172; MR 80e:30002.
494. Labelle, G.; Rahman, Q. I. *Remarque sur la moyeene géométrique de fonctions univalentes convexes.* C. R. Acad. Sci. Paris, vol. 266 (1968), pp. 209-210.
495. Lachance, M. *Remark on functions with all derivatives univalent.* Internat. J. Math. Math. Sci. 3 (1980), no. 1, 193-196; MR 81g:30018.
496. Lahi, L. *Some classes of functions generated by an integral.* Ann. Fac. Sci. Univ. Nat. Zaire (Kinshasa) Sect. Math.-Phys. 4 (1978), no. 1, 25-35; MR 81g:30019.
497. Lahi, L.; Merkes, E. P. *Extreme points for the logarithm of functions in certain classes of analytic functions.* Ann. Fac. Sci. Univ. Nat. Zaire (Kinshasa) Sect. Math. Phys. 2 (1976), no. 1, 31-42; MR 55 #8335.
498. Lakshminarasimhan, T. V. *On subclasses of functions starlike in the unit disc.* J. Indian Math. Soc. (N. S.) 41 (1977), no. 3-4, 233-243; MR 80a:30012.
499. Landau, E. *Über einen Satz von Herrn Dieudonne.* Math. Zeit., vol. 24 (1933), pp. 22-27.
500. Landau, E. *Darstellung und Begrundung einiger neuerer Ergebnisse der Funktionentheorie.* Chelsea Publishing Co., New York, N. Y. (1946).
501. Lau, K.; Liu, M. *On a criterion for the univalence of holomorphic functions.* Proc. Amer. Math. Soc. 80 (1980), 651-652; MR 81j:30028.
502. Launonen, E. *On exponentiated Grunsky inequalities for bounded*

univalent functions. Ann. Acad. Sci. Fenn. Ser. A I Math. Dissertationes No. 1 (1975), 34 pp.; MR 57 #611.
503. Laura, P. A. A. *A survey of modern applications of the method of conformal mapping.* Rev. Un. Mat. Argentina 27 (1974/75), no. 3, 167–179 (1976); MR 54 #2937.
504. Lawrynowicz, J.; Tammi, O. *On fourth order Grunsky functionals.* (Polish summary) Bull. Soc. Sci. Lettres Lodz 26 (1976), no. 7, 6 pp.; MR 56 #15901.
505. Leach, R. J. *Multivalent Bazilevič functions.* Rev. Roumaine Math. Pures Appl. 21 (1976), no. 5, 523–527.
506. Leach, R. J. *Strongly starlike functions of higher order.* (Polish and Russian summaries) Ann. Univ. Mariae Curie-Sklodowska Sect. A 30 (1976), 63–67 (1978); MR 80c:30011.
507. Leach, R. J. *Coefficient estimates for certain multivalent functions.* Pacific J. Math. 74 (1978), no. 1, 133–142; MR 58 #22521.
508. Leach, R. J. *The coefficient problem for Bazilevič functions.* Houston J. Math. 6 (1980), 543–547.
509. Lebedev, N. A. *On Hayman's regularity theorem.* (Russian Zap. Naucn. Sem. Leningrad. Otdel. Mat. Inst. Steklov. (LOMI) 44 (1974), 93–99, 187; MR 51 #13209.
510. Lebedev, N. A. *The area principle in the theory of univalent functions.* Izdat. "Nauka," Moscow (1975), 336 pp. 2.19r; MR 56 #8834.
511. Lebedev, N. A. *Univalence of a class of functions.* (Russian, English summary) Vestnik Leningrad. Univ. Mat. Meh. Astronom (1981), vyp. 1, 113–115, 122–123; MR 82e:30019.
512. Lebedev, N. A. *Range of values of a functional in a class of bounded univalent functions.* (Russian, English summary), Vestnik Leningrad. Univ. Mat. Meh. Astronom. (1981), vyp. 3, 33–37, 124–125.
513. Lebedev. N. A.; Milin, I. M. *An inequality.* Vestnik Leningrad University, vol. 20 (1965), no. 19, pp. 157–158 (in Russian).
514. Lebedev, N. A.; Starkov, V. V. *Some extremal problems in Basilevič's class.* (Russian, English summary) Vestnik Leningrad. Univ. Mat. Meh. Astronom. (1979), vyp. 4, 47–49, 122; MR 81e:30034.
515. Lee, S. Y. *On a subclass of close-to-convex functions.* J. Korean Math. Soc. 14 (1977), no. 1, 1–8; MR 56 #12246.
516. Leeman, G. B., Jr. *The seventh coefficient of odd symmetric univalent functions.* Duke Math. J. 43 (1976), no. 2, 301–307; MR 53 #13545.
517. Lehtinen, M. *On the inner radius of univalency for non-circular domains.* Ann. Acad. Sci. Fenn. Ser. A I Math. 5 (1980), no. 1,

45-47; MR 82c:30020.
518. Lehto, O. *On univalent functions with quasiconformal extensions over the boundary.* J. Analyse Math. 30 (1976), 349-354; MR 57 #6422.
519. Lehto, O. *Domain constants associated with Schwarzian derivative.* Comment. Math. Helv. 52 (1977), no. 4, 603-610; MR 56 #15907.
520. Lehto, O. *Univalent functions and Teichmüller theory.* Proceedings of the First Finnish-Polish Summer School in Complex Analysis (Podlesice, 1977); Part I, pp. 11-33. Univ. Lodz, Lodz, 1978; MR 58#6229.
521. Lehto, O. *Univalent functions, Schwarzian derivatives and quasiconformal mappings.* Enseign. Math. (2) 24 (1978), no. 3-4, 203-214; MR 80c:30020.
522. Lehto, O. *Remarks on Nehari's theorem about the Schwarzian derivative and Schlicht functions.* J. D'analyse Math. 36 (1979), 184-190; MR 81i:30034.
523. Lehto, O.; Tammi, O. *Area method and univalent functions with quasiconformal extensions.* Ann. Acad. Sci. Fenn. Ser. A I Math. 2 (1976), 307-313; MR 57 #6421.
524. Lehto, O.; Tammi, O. *Schwarzian derivative in domains of bounded rotation.* Ann. Acad. Sci. Fenn. Ser. A I Math. 4 (1979), no. 2, 253-257; MR 81b:30014.
525. Leung, Y. *Successive coefficients of starlike functions.* Bull. London Math. Soc. 10 (1978), no. 2, 193-196; MR 58 #1127.
526. Leung, Y. J. *Integral means of the derivatives of some univalent functions.* Bull. London Math. Soc. 11 (1979), no. 3, 289-294; MR 80m:30013.
527. Leung, Y. *Robertson's conjecture on the coefficients of close-to-convex functions.* Proc. Amer. Math. Soc. 76 (1979), no. 1, 89-94; MR 80i:30019.
528. Lewandowski, Z. *On a univalence criterion.* (Russian summary) Bull. Acad. Polon. Sci. Ser. Sci. Math. 29 (1981), no. 3-4, 123-126; MR 82k:30023.
529. Lewandowski, Z.; Libera, R. J.; Złotkiewicz, E. J. *Values assumed by Gel'fer functions.* (Polish and Russian summaries) Ann. Univ. Mariae Curie-Sklodowska Sect. A 31 (1977), 75-84 (1979); MR 81f:30023.
530. Lewandowski, Z.; Miazga, J. *The Koebe domain for the class of typically-real functions under Montel's normalization.* (Russian summary) Bull. Acad. Polon. Sci. Math. 28 (1980), no. 9-10, 465-470 (1981); MR 82i:30013.
531. Lewandowski, Z.; Miller, S.; Złotkiewicz, E. *Gamma-starlike*

functions. (Polish and Russian summaries) Ann. Univ. Mariae Curie-Sklodowska Sect. A 28 (1974), 53–58 (1976); MR 54 #530.

532. Lewandowski, Z.; Miller, S.; Złotkiewicz, E. *Generating functions for some classes of univalent functions.* Proc. Amer. Math. Soc. 56 (1976), 111–117).

533. Lewandoswki, Z.; Stankiewicz, J. *On angularly accessible univalent functions.* (Russian summary) Bull. Acad. Polon. Sci. Ser. Sci. Math. Astronom. Phys. 24 (1976), no. 4, 217–221; MR 54 #2973.

534. Lewandowski, Z.; Wajler, S. *Sur les fonctions typiquement-reelles bornées.* (Polish and Russian summaries) Ann. Univ. Mariae Curie-Sklodowska Sect. A 28 (1974), 59–64 (1976); MR 53 #13578.

535. Lewandowski, Z.; Wajler, S. *L'equation de Löwner généralisée pour certaines sous-classes de fonctions univalentes.* (Russian summary) Bull. Acad. Polon. Sci. Ser. Sci. Math. Astronom. Phys. 24 (1976), no. 4, 223–229; MR 54 #531.

536. Lewis, J. L. *Convolutions of starlike functions.* Indiana Univ. Math. J. 27 (1978), 671–688; MR 58 #1124.

537. Lewis, J. L. *Applications of a convolution theorem to Jacobi polynomials.* SIAM J. Math. Anal. 10 (1979), no. 6, 1110–1120.

538. Libera, R. J.; Złotkiewicz, E. J. *Loewner-type equations for some classes of functions.* Complex analysis (Proc. S.U.N.Y.) Conf., Brockport, N. Y. (1976), pp. 75–85. Lecture Notes in Pure and Appl. Math., Vol. 36, Dekker, New York, 1978; MR 58 #1119.

539. Libera, R. J.; Złotkiewicz, E. J. *Loewner-type approximations for convex functions.* Colloq. Math. 36 (1976), no. 1, 143–151; MR 54 #13057.

540. Lin, C. *On some classes of univalent functions.* Soochow J. Math. Natur. Sci. 3 (1977), 101–107; MR 58 #11350.

541. Liu, L. Z. *The rotation theorem for meromorphic starlike functions.* (Chinese, English summary) Kexue Tongbao 24 (1979), no. 16, 724–726; MR 80j:30017.

542. Livingston, A. E. *Weakly starlike meromorphic univalent functions.* II. Proc. Amer. Math. Soc. 62 (1976), no. 1, 47–53; MR 55 #8336.

543. Livingston, A. E. *On the integral means of univalent, meromorphic functions.* Pacific J. Math. 72 (1977), no. 1, 167–180; MR 56 #12247.

544. Livingston, A. E.; Pfaltzgraff, J. A. *Structure and extremal problems for classes of functions analytic in an annulus.* Colloq. Math. 43 (1980), no. 1, 161–181 (1981); MR 82g:30030.

545. Lohwater, A. J. *Some function-theoretic results involving Baire category.* Topics in analysis (Colloq. Math. Anal., Jyvaskyla, 1970), pp. 253–259. Lectures Notes in Math., Vol. 419, Springer, Berlin, 1974; MR 51 #13234.
546. London, R. R. *A note on Hadamard's three circles theorem.* Bull. London Math. Soc. 9 (1977), no. 2, 182–185.
547. London, R. R. *On derivatives of the maximum modulus of a starlike function.* Bull. London Math. Soc. 13 (1981), no. 3, 207–213; MR 82i:30014.
548. Lupu, M.; Pascu, N.; Podaru, V. *Calculation of the order of starlikeness for α-starlike functions.* (Romanian, English summary) Bull. Univ. Brasov Ser. C 20 (1978), 73–77; MR 81i:30027.
549. Lyzzaik, A. *Multivalent linearly accessible functions and close-to-convex functions.* Proc. London Math. Soc. (3) 44 (1982), no. 1, 178–192.
550. Lyzzaik, A.; Styer, D. *Goodman's conjecture and the coneffficients of univalent functions.* Proc. Amer. Math. Soc. 69 (1978), no. 1, 111–114; MR 57 #612.
551. Lyzzaik, A.; Styer, D. *A covering surface conjecture. Brannan and Kirwan.* Bull. London Math. Soc. 14 (1982), no. 1, 39–42.
552. Macaev, V. I.; Mogul'skii, E. Z. *A division theorem for analytic functions with a given majorant, and some of its applications.* (Russian, English summary) Investigations on linear operators and theory of functions, VI. Zap. Naucn. Sem. Leningrad. Otdel. Mat. Inst. Steklov. (LOMI) 56 (1976), 73–89, 196.
553. MacFarlane, A. G. J.; Postlethwaite, I. *Extended principle of the argument.* Internat. J. Control 27 (1978), no. 1, 49–55.
554. Majchrzak, W. *On univalent p-symmetric functions in the unit disc.* Ann. Polon. Math. 34 (1977), no. 2, 135–163; MR 57 #6396.
555. Majchrzak, W. *An extremal arclength problem in some classes of univalent and p-symmetric functions.* Ann. Polon. Math. 36 (1979), no. 3, 287–297; MR 80j:30018.
556. Makowka, B. *On some subclasses of univalent functions.* (Polish and Russian summaries) Zeszyty Nauk. Politech. Łódź. Mat. No. 9 (1977), 71–76; MR 58 #28467.
557. Makowka, B. *Convex functions of order-α and type-β.* (Polish and Russian summaries) Zeszyty Nauk. Politech. Łódź. Mat. No. 10 (1978), 47–57; MR 80j:30019.
558. Mandik, V. P. *Pairs of functions that are simultaneously p-valent in the mean with respect to the circle.* (Russian, English summary) Some problems in modern function theory (Proc. Conf. Modern

Problems of Geometric Theory of Functions, Inst. Math., Acad. Sci. USSR, Novosibirsk, 1976) (Russian), pp. 83-87. Akad. Nauk SSSR Sibirsk. Otdel. Inst. Mat., Novosibirsk (1976). For a review of this item see RZMat 1977 #11 B153. [For the entire collection see MR 57 #15813.] MR 58 #22525.

559. Mandik, V. P. *Area theorems for systems of multivalent functions.* (Russian) Sibirsk. Mat. Z. 20 (1979), no. 6, 1275-1281, 1408; MR 81a:30013.

560. Marcus, M. *Some geometric properties of the image of the unit disk by conformal maps.* J. London Math. Soc. (2) 13 (1976), no. 1, 177-182; MR 54 #527.

561. Marden, M. *Much Ado About Nothing.* Amer. Math. Monthly, vol. 83, no. 10, December 1976, 788-798.

562. Martio, O. *New methods in injectivity theorems connected with Schwarzian derivative.* 18th Scandinavian Congress of Mathematicians (Aurhus, 1980), pp. 411-415, Progress in Math., 11, Birkhauser, Boston, Mass., 1981; MR 82k:30025.

563. Marusciac, I. *Some estimations for strongly αm-convex functions.* (Romanian summary) Studia Univ. Babes-Bolyai Math. 25 (1980), no. 2, 39-44.

564. Mateljevic, M. *The isoperimetric inequality in the Hardy class H^1.* Mat. Vesnik 3(16)(31)(1979), no. 2, 169-178; MR 82i:30052.

565. Mazur, Ryszard. *Estimation of the modulus and argument of a quasi-β-starlike function.* Demonstratio Math. 9 (1976), no. 4, 639-645; MR 55 #3233.

566. Mazur, R. *On some extremal problems in the class of functions with bounded rotation.* (Polish and Russian summaries) Zeszyty Nauk. Politech. Lódź. Mat. No. 9 (1977), 77-88; MR 58 #22508.

567. Mazur, R. *On a subclass of convex functions.* (Polish and Russian summaries) Zeszyty Nauk. Politech. Lódź. Mat. No. 13 (1981), 15-20.

568. Mazur, R.; Treska, A. *On some classes of β-starlike meromorphic k-symmetric functions.* Demonstratio Math. 11 (1978), no. 3, 735-750; MR 81a:30010.

569. McCall, J. D.; Fricke, G. H.; Beyer, W. A. *Remainders of power series.* Internat. J. Math. Math. Sci. 2 (1979), no. 2, 239-250; MR 80i:30003.

570. McCoy, T. L. *A partial surface variation for extremal schlicht functions.* Trans. Amer. Math. Soc. 234 (1977), no. 1, 119-138; MR 57 #12841a.

571. McCoy, T. L. Erratum: *"A Partial surface variation for extremal*

schlicht functions." Trans. Amer. Math. Soc. 240 (1978), 393; MR 57 #12841b.

572. McLaughlin, R. *Two inequalities for starlike functions.* Colloq. Math. 34 (1975/76), no. 2, 287–292; MR 53 #11039.

573. Merkes, E. P. *A covering theorem for odd typically-real functions.* Internat. J. Math. Math. Sci. 3 (1980), no. 1, 189–192; MR 82c:30008.

574. Merkes, E. P. *Partial sum subordination of univalent functions.* Houston J. Math. 6 (1980), 77–83; MR 81f:30009.

575. Mikka, V. P. *The univalent solvability of certain inverse boundary value problems.* (Russian) Trudy Sem. Kraev. Zadacam Vyp. 9 (1972), 200–207. For a review of this item see Zbl 299 #30037; MR 58 #17131.

576. Mikka, V. P. *Two sufficient conditions for the univalence of analytic functions.* (Russian) Mat. Zametki 19 (1976), no. 3, 331–346; MR 53 #13547.

577. Milcetich, J. G. *A general extremal problem for the class of close-to-convex functions.* Trans. Amer. Math. Soc. 225 (1977), 307–323; MR 58 #22509.

578. Milcetich, J. G. *Analytic mappings of the unit disk on a convex domain.* American Math. Monthly 84, no. 8 (October 1977) (Advanced Problems and Solutions), p. 663.

579. Milcetich, J. G. *Inequality of L^p norms of a derivative of a function.* American Mathematical Monthly 87, no. 9, (November 1980) (Solutions of Advanced Problems), p. 739.

580. Milin, I. M. **Univalent functions and orthonormal systems.* Trans. Math. Monographs, Vol. 49, Amer. Math. Soc., Providence, R. I. (1977), iv + 202 pp.; MR 55 #651.

581. Milin, I. M. *Loose errata:* Univalent functions and orthonormal systems, p. 194. Amer. Math. Soc., Providence, R. I. (1977); MR 55 #651, MR 56 #8835.

582. Milin, I. M. *A property of logarithmic coefficients of univalent functions.* (Russian) Metric questions of the theory of functions (Russian), pp. 86–90, 161, "Naukova Dumka," Kiev, (1980); MR 82i:30026.

583. Milin, V. N. [V. I.] *A certain family of non-univalent functions that satisfy the Prawitz area theorem.* (Russian) Sibirsk. Mat. Z. 17 (1976), no. 2, 304–317, 478; MR 53 #13548.

584. Milin, V. I. *Estimate of the coefficients of odd univalent functions.* (Russian) Metric questions of the theory of functions (Russian), pp. 78–86, 160, "Naukova Dumka," Kiev, (1980); MR 82i:30025.

585. Milin, V. I. *Neighboring coefficients of odd univalent functions.* (Russian) Sibirsk. Mat. Z. 22(1981), no. 2, 149–157, 237; MR 82i:30027.
586. Miller, J. E. *Convex and starlike meromorphic functions.* Proc. Amer. Math. Soc. 80 (1980), 607–613; MR 81k:30015.
587. Miller, S. S. *Arclength and subordination properties of alpha-convex functions.* Mathematica (Cluj) 16(39) (1974), no. 1, 93–98; MR 53 #13540.
588. Miller, S. S. *Problems in complex function theory.* Edited by Sanford S. Miller. Complex analysis (Proc. S.U.N.Y. Conf., State Univ. New York, Brockport, N.Y., (1976), pp. 167–177. Lecture Notes in Pure and Appl. Math., Vol. 36, Dekker, New York, 1978; MR 58 #11363.
589. Miller, S. S. *On a class of starlike functions.* Ann. Polon. Math. 32 (1976), no. 1, 77–81. (errata insert); MR 53 #3283.
590. Miller, S. S. *A class of differential inequalities implying boundedness.* Illinois J. Math. 20 (1976), no. 4, 647–649.
591. Miller, S. S. *Complex analysis (Proc. S.U.N.Y. Conf., Brockport, N. Y., 1976) Complex analysis.* Proceedings of the S.U.N.Y Brockport Conference on Complex Function Theory held at the State University College, Brockport, N. Y., June 7–9, 1976. Edited by Sanford S. Miller. Lecture Notes in Pure and Applied Mathematics, Vol. 36. Marcel Dekker, Inc., New York-Basel, 1978. xii + 177 pp. ISBN 0-8247-6725-X.
592. Miller, S. S. *Condition for* $|f(z)| < 1$, $|z| < 1$. American Math. Monthly, Vol. 84, no. 3 (March 1977) (Advanced Problems Department), p. 223.
593. Miller, S. S. *"A problem" for functions of positive real part.* Amer. Math. Monthly 85, no. 3 (1978), p. 203. (Solution: same journal, Vol. 86, no. 8, October 1979, p. 711).
594. Miller, S. S.; Mocanu, P. T. Alpha-convex functions and derivatives in the Nevanlinna class. Studia Univ. Babes-Bolyai Math. 20 (1975), 35–40; MR 53 #820.
595. Miller, S. S.; Mocanu, P. T. *Second order differential inequalities in the complex plane.* J. Math. Anal. Appl. 65 (1978), no. 2, 289–305; MR 80f:30009.
596. Miller, S. S.; Mocanu, P. T. *Differential subordinations and univalent functions.* Michigan Math. J. 28 (1981), no. 2, 157–172.
597. Miller, S. S.; Mocanu, P. T. *On classes of functions subordinate to the Koebe function.* Rev. Roumaine Math. Pures Appl. 26(1981), no. 1, 95–99; MR 82i:30033.
598. Miller, S. S.; Mocanu, P.; Reade, M. O. *Janowski alpha-convex*

functions. (Polish and Russian summaries) Ann. Univ. Mariae CUrie-Sklodowska Sect. A 29 (1975), 93-98 (1977); MR 56 #8831.
599. Miller, S. S.; Mocanu, P. T.; Reade, M. O. *On generalized convexity in conformal mappings.* II. Rev. Roumaine Math. Pures. Appl. 21 (1976), no. 2, 219-225; MR 54 #2938.
600. Miller, S. S.; Mocanu, P. T.; Reade, M. O. *On the radius of alpha-convexity. (Romanian summary) Studia Univ. Babes-Bolyai Math. 22 (1977), 35-39*; MR 55 #8337.
601. Miller, S. S.; Mocanu, P. T.; Reade, M. O. *A particular starlike integral operator.* (Romanian summary) Studia Univ. Babes-Bolyai Math. 22 (1977), no. 2, 44-47; MR 58 #17067.
602. Miller, S. S.; Mocanu, P. T.; Reade, M. O. *Starlike integral operators.* Pacific J. Math. 79 (1978), no. 1, 157-168; MR 80g:30008.
603. Miller, S. S.; Mocanu, P. T.; Reade, M. O. *The order of starlikeness of alpha-convex functions.* Mathematica (Cluj) 20(43) (1978), no. 1, 25-30; MR 81d:30016.
604. Mil'man, D. A.; Musaev, B. I. *The approximate solution of a problem of conformal mapping of nonstarlike domains.* (Russian, Azerbaijani summary) Azerbaidzan. Gos. Univ. Ucen. Zap. (1974), no. 1, Voprosy Prikl. Mat. i Kibernet, 60-68; MR 54 #7767.
605. Minda, C. David. *The image of the Ahlfors function.* Proc. Amer. Math. Soc. 83 (1981), no. 4, 751-756.
606. Minda, C. D.; Rodin, B. *Extremal length, extremal regions and quadratic differentials.* Comment. Math. Helv. 50 (1975), no. 4, 455-475.
607. Miniowitz, R. *On almost bounded functions.* Trans. Amer. Math. Soc. 223 (1976), 93-102; MR 55 #644.
608. Mista, K. *Estimates of the second and third derivative in a class of functions a multivalent in the mean.* (Polish, Russian and English summaries) Zeszyty Nauk. Politech. Slask. No. 441 Mat.-Fiz. Zeszyt 26 (1976), 169-174; MR 56 #598.
609. Mista, K. *Sur les inégalités billineaires de Grunsky pour les couples de fonctions qui n'empletent pas l'une sur l'autre.* I. Demonstratio Math. 12 (1979), no. 1, 71-103; MR 80i:30031.
610. Mista, K. *Interior pairs of Aharonov.* (Polish, Russian and English summaries) Zeszyty Nauk. Politech. Slask. Mat.-Fiz. No. 34 (1979), 137-145; MR 81d:30017.
611. Mista, K. *Sur les inégalités bilineaires de Grunsky pour les couples de fonctions qui n'empietent pas l'une sur l'autre.* II. Demonstratio Math. 14 (1981), no. 1, 33-58; MR 82i:30015.
612. Mitjuk, I. P. *Symmetrization methods, and their application in*

geometric function theory. (Russian) Izv. Severo-Kavkaz. Naucn. Centra Vyss. Skoly Ser. Estestv. Nauk. (1976), no. 2, 3–10, 116; MR 55 #5853.
613. Mitrinovic, D. S. *On the univalence of rational functions.* Univ. Beograd. Publ. Elektrotehn. Fak. Ser. Mat. Fiz. (1979), no. 634–677, 221–227; MR 81i:30028.
614. Mocanu, P. T. *Starlikeness and convexity for nonanalytic functions in the unit disc.* (Mathematic (Cluj) 22(45) (1980), no. 1, 77–83; MR 82i:30016.
615. Mocanu, P. T.; Reade, M. O.; Ripeanu, D. *The order of starlikeness of a Libera integral operator.* Mathematica (Cluj) 19(42) (1977), no. 1, 67–73 (1978); MR 58 #22511.
616. Mocanu, P. T.; Ripeanu, D. (Ripianu, D.) *An extremal problem for the transfinite diameter of a continuum.* Romanian-Finnish Seminar on Complex Analysis (Proc., Bucharest, 1976), pp. 323–330, Lecture Notes in Math., 743, Springer, Berlin, 1979; MR 81c:30043.
617. Mogra, M. L. *On a class of starlike functions in the unit disc.* I. J. Indian Math. Soc. (N. S.) 40 (1976), no. 1–4, 159–161 (1977); MR 58 #6203a.
618. Mogra, M. L. *On a class of starlike functions in the unit disc.* II. Indian J. Pure Appl. Math. 8 (1977), no. 2, 157–165; MR 58 #6203b.
619. Mogra, M. L. *On a class of functions with positive real part.* Riv. Mat. Univ. Parma (4) 4 (1978), 101–108 (1979); MR 80i:30020.
620. Mogra, M. L.; Juneja, O. P. *Coefficient estimates for starlike functions.* Bull, Austral. Math. Soc. 16 (1977), no. 3, 415–425; MR 56 #5860.
621. Montel, P. *Sur une formula de Darboux et les polynômes d'interpolation.* Annali della R. Scuola Normale superiore di Pisa II, vol. 1 (1932), pp. 371–384.
622. Moulis, E. J., Jr. *The univalence of a class of analytic functions.* Complex analysis (Proc. S.U.N.Y. Conf., Brockport, N. Y., 1976), pp. 89–94. Lecture Notes in Pure and Appl. Math., Vol. 36, Dekker, New York, 1978; MR 58 #6204.
623. Moulis, E. J. *Generalizations of the Robertson functions.* Pacific J. Math. 81 (1979), no. 1, 167–174; MR 81c:30023.
624. Murai, T. *On area integrals and radial variations of analytic functions in the unit disk.* Nagoya Math. J. (1976), 135–159; MR 54 #10603.
625. Mustafa, C. *The extension of starshaped bounded Lipschitz functions.* Anal. Numer. Theor. Approx. 9 (1980), no. 1, 93–99.

626. Nasr, M. A. *On convex combination of certain analytic functions.* Indian J. Pure Appl. Math. 6 (1975), no. 4, 337-346; MR 56 #3272.
627. Nasr, M. A. *Convex hull and extreme points of two families of univalent functions.* Proc. Indian Acad. Sci. Sect. A. 83 (1976), no. 4, 138-144; MR 54 #532.
628. Nasr, M. A. *On a class of functions with bounded boundary rotation.* Bull. Inst. Math. Acad. Sinica 5(1977), no. 1, 27-36; MR 58 #6214.
629. Nasr, M. A. *On the Hardy class of certain subclass of Bazilevič function.* Ann. Polon. Math. 34(1977), no. 3, 265-268; MR 57 #640.
630. Nasr, M. A. *On the radius of α-convexity of certain classes of starlike functions.* Proc. Indian Acad. Sci. Sect. A 85(1977), no. 5, 367-378; MR 56 #596.
631. Nasr, M. A. *On Bazilevič functions and generalized convexity.* Rev. Roumaine Math. Pures Appl. 22(1977), no. 9, 1279-1281; MR 58 #22512.
632. Nehari, Zeev. *A property of convex conformal maps.* J. Analyse Math. 30(1976), 390-393; MR 55 #12901.
633. Nehari, Zeev. *Univalence criteria depending on the Schwarzian derivative.* Illinois J. Math 23(1979), no. 3, 345-351; MR 80i:30033.
634. Netanyahu, E. *Extremal problems for schlicht functions in the exterior of the unit circle.* Canadian Journ. Math., vol. 17(1965), pp. 335-341.
635. Netanyahu, E. *Extemal properties of some classes of univalent functions.* Ann. Acad. Sci. Fenn. Ser. A I Math. 2 (1976), 345-360; MR 57 #16569.
636. Netanyahu, E. *On the local maxima of the coefficients of univalent functions.* J. Analyse Math. 30(1976), 394-403; MR 56 #3280.
637. Netanyahu, E.; Pinchuk, B. *Symmetrization and extremal bounded univalent functions.* J. Analyse Math. 36(1979), 139-144 (1980); MR 82a:30032.
638. Netanyahu, E.; Schiffer, M. *On the monotonicity of some functionals in the family of univalent functions.* Israel J. Math. 32 (1979), no. 1, 14-26; MR 80m:30014.
639. Nevanlinna, R. **Proceedings of the Rolf Nevanlinna Symposium on Complex Analysis.* Held at the Mathematical Research Institute of the University of Istanbul in Silivri (September 20-25, 1976). Publication of the Mathematical Research Institute, Istanbul, 7. Mathematical Research Institute, University of Istanbul, Istanbul, 1978. iv + 142 pp.; MR 80e:30004.

640. Nishimiya, H. *On coefficient regions of Laurent series with positive real part.* Kodai Math. Seminar Reports, vol. 11(1959), pp. 25-39.
641. Noonan, J. W. *Powers of p-valent functions.* J. Austral. Math. Soc. Ser. A 25(1978), no. 1, 66-70; MR 57 #6402.
642. Noonan, J. W.; Thomas, D. K. *On the second Hankel determinant of areally mean p-valent functions.* Trans. Amer. Math. Soc. 223(1976), 337-346; MR 54 #10593.
643. Noor, K. I. *A note on close-to-convex functions.* Punjab Univ. J. Math. (Lahore) 8(1975), 11-14; MR 57 #9954.
644. Noor, K. I. *On Hankel determinants of close-to-convex functions.* Porceedings of the Seventh National Mathematics Conference (Dept. Math., Azarabadegan Univ., Tabriz, 1976), pp. 137-151. Azarabadegan Univ., Tabriz, 1977.
645. Noor, K. I. *On a subclass of close-to-convex functions.* Comment. Math. Univ. St. Paul. 29(1980), no. 1, 25-28; MR 81k:30016.
646. Noor, K. I. *On the Hankel determinants of close-to-convex univalent functions.* Internat. J. Math. Math. Sci. 3(1980), no. 3, 447-481; MR 81k:30020.
647. Noor, K. I. *On analytic functions related with functions of bounded boundary rotation.* Comment. Math. Univ. St. Paul 30 (1981), no. 2, 113-118.
648. Noor, K. I. *Bieberbach conjecture for univalent functions.* Comment. Math. Univ. St. Paul. 30(1981), no. 2,, 131-133.
649. Noor, K. I.; Thomas, D. K. *Quasi-convex univalent functions.* Internat. J. Math. Math. Sci. 3 (1980), no. 2, 255-266; MR 81f:30010
650. Nosenko, A. S. (Nosenko, O. S.) *Classes of functions with increasing argument of the circular derivative of order p.* (Russian) Ukrain. Mat. Z. 27(1975), no. 6, 827-830, 863; MR 53 #13541.
651. Nunokawa, M. *On the starlike boundary of univalent functions.* (Japanese) Sugaku 31(1979), no. 3, 255-256; MR 81e:30023.
652. Obrock, A. E. *Teichmüller inequalities without coefficient normalization.* Trans. Amer. Math. Soc. 159(1971), 391-416; MR 52 #14269.
653. Oikawa, K. *On angular derivatives of univalent functions.* Kodai Math. Sem. Rep. 27(1976), no. 1-2, 193-210; MR 55 #639.
654. Osgood, B. G. *Univalence criteria in multiply-connected domains.* Trans. Amer. Math. Soc. 260(1980), no. 2, 459-473; MR 81h:30021.
655. Osserman, R. *The isoperimetric inequality.* Bull. Amer. Math. Soc. 84, no. 6(1978), 1182-1238.
656. Owa, S. *On the univalent functions with a quasiconformal extension.* Math. Japon. 23(1978/79), no. 2, 259-261; MR 80b:30017.

657. Owa, S. *On the distortion theorems.* I. Kyungpook Math. J. 18(1978), no. 1, 53–59; MR 58 #22513.
658. Owa, S. *A remark on new criteria for univalent functions.* Kyungpook Math. J. 21(1981), no. 1, 15–23; MR 82j:30019.
659. Owa, S. *On applications of the fractional calculus.* Math. Japon. 25(1980), no. 2, 195–206; MR 81i:30004.
660. Paatero, V. *Über die Konforme Abbildung mehrblättriger Gebiete von beschränkter Randdrehung.* Ann. Acad. Sci. Fenn., A. I. 128 (1952), pp. 1–7.
661. Paatero, V. *Über die Verzerrung bei der Abbildung mehrblättriger Gebiete von beschränkter Randdrehung.* Ann. Acad. Sci. Fenn., A. I. 147(1953), pp. 1–7.
662. Paatero, V. *Über die Randdrehung der mehrblättrigen einfach zusammenhängenden Gebiete.* Ann. Acad. Sci. Fenn., A. I. 194 (1955), pp. 1–7.
663. Pacevic, E. L. *Sufficient conditions for univalence of certain integral representations.* (Russian) Mat. Zametki 27(1980), no. 3, 399–410, 493; MR 81h:30016.
664. Pachulski, Z. *On the radius of convexity for certain family of functions meromorphic in unit disc.* Demonstratio Math. 12(1979), no. 1, 281–287; MR 80h:30014.
665. Padmanabhan, K. S.; Parvatham, R. *On the univalence and convexity of partial sums of a certain class of analytic functions whose derivatives have positive real part.* Indian J. Math. 16(1974), no. 2, 67–77.
666. Padmanabhan, K. S.; Parvatham, R. *Radius of convexity of partial sums of a certain power series.* Indian J. Math. 17(1975), no. 3, 133–138; MR 82h:30011.
667. Padmanabhan, K. S.; Parvatham, R. *On functions with bounded boundary rotation.* Indian J. Pure Appl. Math. 6(1975), no. 11, 1236–1247; MR 57 #607, #12832.
668. Padmanabhan, K. S.; Parvatham, R. *On a certain class of functions with bounded boundary rotation.* Rev. Roumaine Math. Pures Appl. 21(1976), no. 8, 1077–1084; MR 57 #12832.
669. Padmanabhan, K. S.; Parvatham, R. *Radii of starlikeness of certain class of analytic functions.* Rev. Roumaine Math. Pures Appl. 23(1978), no. 10, 1545–1551; MR 81d:30018.
670. Padmanabhan, K. S.; Parvatham, R. *On certain generalized close-to-star functions in the unit disc.* Ann. Polon. Math. 37 (1980), no. 1, 1–11; MR 81i:30029.
671. Parvatham, R.; Padmanabhan, K. S. *Radii of starlikeness of cer-*

tain classes of analytic functions. Mathematica (Cluj) 16(39) (1974), no. 1, 143–157; MR 53 #13542.
672. Pascu, N. N. *Alpha-close-to-convex functions.* Romanian-Finnish Seminar on Complex Analysis (Proc., Bucharest, 1976), pp. 331–335, Lecture Notes in Math., 743, Springer, Berlin, 1979; MR 80j:30021.
673. Pascu, N. N. *Janowski alpha-starlike-convex functions.* Studia Univ. Babes-Bolyai Math. 21(1976), 23–27; MR 53 #789.
674. Pascu, N. α-*starlike functions.* (Romanian, English summary) Bul. Univ. Brasov Ser. C 19 (1977), 37–39(1978); MR 80j:30020.
675. Pascu, N.; Podaru, V.; Rotaru, P. *Calculus of the alpha-starlike radius for starlike functions of the beta order.* (Romanian, German summary) Bul. Univ. Brasov Ser. C 19 (1977), 41–46(1978); MR 80j:30022.
676. Paskuleva, D. Z. *The radius of spiral convexity of a class of spirallike functions.* (Russian) C. R. Acad. Bulgare Sci. 30(1977), no. 12, 1675–1677; MR 58 #11351.
677. Paskuleva, D. Z. *Univalent functions that are convex in one direction.* (Russian) C. R. Acad. Bulgare Sci. 31(1978), no. 7, 795–798; MR 81f:30011.
678. Paskuleva, D. Z.; Dimkov, G. M. *Proof of a conjecture for α-convex functions.* (Russian) C. R. Acad. Bulgare Sci. 30(1977), no. 12, 1671–1673; MR 58 #22514.
679. Patil, D. A.; Thakare, N. K. *On extreme points of functions of order α with positive real part and related results.* Indian J. Pure Appl. Math. 9(1978), no. 12, 1353–1358; MR 80e:30013.
680. Patil, D. A.; Thakare, N. K. *On coefficient bounds of p-valent λ-spiral functions of order α.* Indian J. Pure Appl. Math. 10(1979), no. 7, 842–853; MR 80k:30017.
681. Patil, D. A.; Thakare, W. K. *Radii of convexity of certain classes of p-valent analytic functions.* Bull. Math. R. S. Roumaine (N.S.) 23(71)(1979), no. 1, 71–84; MR 80i:30021.
682. Patil, D. A.; Thakare, N. K. *On univalence of certain integrals.* Indian J. Pure Appl. Math. 11(1980), no. 12, 1626–1642; MR 82f:30011.
683. Paul, S. *An extremal problem for starlike functions.* Indian J. Pure Appl. Math. 7(1976), no. 10, 1199–1202; MR 81f:30012.
684. Pearce, K. *New support points of S and extreme points of HS.* Proc. Amer. Math. Soc. 81(1981), no. 3, 425–428; MR 82b:30012.
685. Pearce, K. *A product theorem for F_p classes and an application.* Proc. Amer. Math. Soc. 84(1982), no. 4, 509–515.

686. Pell, R. *Support point functions and the Loewner variation.* Pacific J. Math. 86(1980), no. 2, 561-564; MR 82a:30014.
687. Pengra, R. W. *On entire conformal mappings of simply connected regions.* Proc. Amer. Math. Soc. 50(1975), 249-254; MR 52 #5950.
688. Peschl, E. *Über unverzweigte konforme Abbildungen.* Collection in honor of Hans Hornich. Osterreich. Akad. Wiss. Math.-Naturwiss. Kl. S.-B. II 185(1976), no. 1-3, 55-78; MR 58 #1158.
689. Pethe, K. *An estimate of the coefficients A_4 and A_5 as a function of A_2 in a subclass of the univalent functions.* (Polish; Russian and English summaries) Zeszyty Nauk. Politech. Slask. Mat.-Fiz. Zeszyt. 25(1974), 53-64; MR 55 #5852.
690. Pethe, K. *An estimate of the coefficient b_n of an areally mean p-valent function outside the unit disc.* (Polish; Russian and English summaries) Zeszyty Nauk. Politech. Slask. Mat.-Fiz. Zeszyt. 25(1974), 137-157; MR 55 #10659.
691. Pethe, K. *The coefficients of areally p-valent functions.* (Polish; Russian and English summaries) Zeszyty Nauk. Politech. Slask. Mat.-Fiz. Zeszyt. 25(1974), 223-246; MR 55 #10660.
692. Pethe, K. *The domain of the first coefficients of p-valent functions in the unit disc.* (Polish; Russian and English summaries) Zeszyty Nauk. Politech. Slask. Mat.-Fiz. No. 35(1979), 39-47; MR 81d:30023.
693. Pethe, K. *Maximum of the functional $|c_3| - |c_2|$ in the class of functions S_M and S_M^{-1}.* (Polish; Russian and English summaries) Zeszyty Nauk. Politech. Slask. Mat.-Fiz. No. 35(1979), 49-58; MR 81d:30024.
694. Pethe, K. *Estimation of the functionals $|c_{p+2} - \alpha c_{p+1}^2|$, $|c_{p+2}| - |c_{p+1}|$ of p-valent functions in the unit disc.* (Polish; Russian and English summaries) Zeszyty Nauk. Politech. Slask. Mat.-Fiz. No. 35(1979), 59-70; MR 81d:30025.
695. Pethe, K. *Maximum and minimum of the functional $Re(c_{p+2} + \alpha c_{p+1})$ in the class of p-valent functions.* (Polish; Russian and English summaries) Zeszyty Nauk. Politech. Slask. Mat.-Fiz. No. 35(1979), 71-78; MR 81d:30026.
696. Pfaltzgraff, J. A.; Reade, M. O.; Umezawa, T. *Sufficient conditions for univalence.* Ann. Fac. Sci. Univ. Nat. Zaire (Kinshasa) Sect. Math.-Phys. 2(1976), No. 2, 211-218; MR 55 #8338.
697. Pfluger, A. *On a coefficient problem for Schlicht functions.* Advances in complex function theory (Proc. Sem., Univ. Maryland, College Park, Md., 1973-1974), pp. 79-91. Lecture Notes in Math., Vol 505, Springer, Berlin, 1976; MR 54 #5455.

698. Pfluger, A. *Functions of bounded boundary rotation and convexity.* J. Analyse Math. 30(1976), 437–451; MR 56 #599.
699. Pfluger, A. *Some coefficient problems for starlike functions.* Ann. Acad. Sci. Fenn. Ser. A I Math. 2(1976), 383–396; MR 58 #11359.
700. Pfluger, A. *On a uniqueness theorem on conformal mapping.* Michigan Math. J. 23(1976), no. 4, 363–365(1977); MR 56 #593.
701. Pfluger, A. *On a coefficient inequality for schlicht functions.* Romanian-Fennish Seminar on Complex Analysis (Proc., Bucharest, 1976), pp. 336–343. Lecture Notes in Math., 743. Springer, Berlin, 1979; MR 81d:30027.
702. Pfluger, A. **Contributions to analysis (Internat. Sympos. Analysis, Eidgenoss. Tech. Hochsch., Zurich, 1978).* Contributions to analysis. International Symposium on Analysis held at the Eidgenossische Technische Hochschule, Zurich, April 10–15, 1978. In honour of Professor Albert Pfluger. Monographies de l'Enseignement Mathematique, 27. L'Enseignement Mathematique, Universite de Geneve, Geneva, 1979. 106 pp. (1 plate).
703. Pfluger, A. *Nonlinear extremal problems for starlike functions.* J. Analyse. Math. 36(1979), 217–226(1980); MR 82a:30033.
704. Pfluger, A. *Über die Koeffizienten schlichter Funktionen.* Lectures from the Colloquium on the Occasion of Ernst Peschl's 70th birthday (German), pp. 41–61, Bonner Math. Schriften, 121, Univ. Bonn, Bonn, 1980; MR 82f:30014.
705. Piotrowska, J. *Extremal problems in some classes of univalent functions in a half-plane.* Ann. Polon. Math. 34(1977), no. 2, 201–220; MR 57 #6405.
706. Piranian, G. *The points of maximum modulus of a univalent function.* Complex analysis (Proc. Conf., Univ. Kentucky, Lexington, Ky., 1976), pp. 96–100. Lecture Notes in Math., Vol. 599, Springer, Berlin, 1977; MR 56 #12256.
707. Piranian, G.; Weitsman, A. *Level sets of infinite length.* Comment. Math. Helv. 53(1978), no. 2, 161–164; MR 81j:30029.
708. Platynowicz, B. *On some extremal problems in special classes of holomorphic functions at certain additional condition.* Demonstratio Math. 12(1979), no. 3, 753–768; MR 81b:30043.
709. Platynowicz, B. *On some extremal problems in classes of holomorphic functions $S_p^*(\varrho)$, $S_p^0(\varrho)$, $\Sigma_p^*(\varrho)$.* Mat. Vesnik 4(17)(32)(1980), no. 2, 203–219; MR 82f:30012.
710. Pohilevič, V. A. (Pohilevič, V. O.) *Ranges of values of certain functionals on special classes of analytic functions.* (Russian) Mathematics collection (Russian), pp. 13–17. Izdat. "Naukova

Dumka," Kiev (1976); MR 57 #12833.
711. Pohilevič, V. A. (Pohilevič, V. O.); Kaidan, V. A. (Kaidan, V. O.) *The ranges of values of certain functionals on special classes of analytic functions.* (Russian) Ukrain. Mat. Z. 30(1978), no. 2, 192–200, 282; MR 80m:30050.
712. Pommerenke, C. *Probleme und Methoden für schlichte Funktionen.* Jber. Deutsche. Math.-Verein. 76(1974/75), no. 4, 183–190.
713. Pommerenke, C. *Univalent functions.* With a chapter on quadratic differentials by Gerd Jensen. Studia Mathematica/Mathematische Lehrbucher, Band XXV. Vandenhoeck & Ruprecht, Gottingen (1975), 376. pp. DM 88.00; MR 58 #22526.
714. Pommerenke, C. *On the angular derivative and univalence.* (Russian summary) Anal. Math. 3(1977), no. 4, 291–297; MR 57 #3367.
715. Pommerenke, C. *Schlichte Funktionen und analytische Funktionen von beschränkter mittlerer Oszillation.* Comment. Math. Helv. 52 (1977), no. 4, 591–602; MR 56 #12268.
716. Pommerenke, C. *On univalent functions, Bloch functions and VMOA.* Math. Ann. 236(1978), no. 3, 199–208; MR 58 #11352.
717. Pommerenke, C. *On the growth of normal analytic functions.* J. Analyse Math. 36(1979), 227–232(1980)
718. Pommerenke, C. *Boundary behaviour of conformal mappings.* Aspects of contemporary complex analysis (Proc. NATO Adv. Study Inst., Univ. Durham, Durham, 1979), pp. 313–331, Academic Press, London, 1980; MR 82i:30008.
719. Popova, G. A. *A certain optimal control problem.* (Russian) Theory of optimal processes (Russian), pp. 72–81. Akad. Nauk. Ukrain. SSR Inst. Kibernet., Kiev, 1972; MR 53 #8411.
720. Popova, G. A. *A study of the extremal properties of a class of univalent functions on the basis of L. S. Pontrjagin's maximum principle.* (Russian) Theory of optimal processes (Russian), pp. 35–41. Akad. Nauk. Ukrain. SSR Inst. Kibernet., Kiev (1973); MR 57 #16571.
721. Poreda, T. *On structure of coefficients of quasi-starlike functions.* (Polish and Russian summaries) Zeszyty Nauk. Politech. Lodz. No. 245 Mat. No. 8(1976), 69–79; MR 54 #13058.
722. Prohorov, D. V. *A certain generalization of the Christoffel-Schwarz formula.* (Russian) Mat. Zametki 17(1975), no. 5, 749–756; MR 56 #592.
723. Prohorov, D. V. *The geometric characterization of functions of Bazilevič subclasses.* (Russian) Differencial'nye Uravnenija i Vycisl. Mat. Vyp. 6(1976), part 2, 124–135, 171; MR 57 #16563.
724. Prohorov, D. V. *Integrals of univalent functions.* (Russian) Mat.

Zametki 24(1978), no. 5, 671-678, 734; MR 80h:30015.
725. Prohorov, D. V. *Linearly invariant extensions of families of analytic functions.* (Russian) Izv. Vyss. Ucebn. Zaved. Matematika (1979), no. 9, 41-47; MR 81i;30030.
726. Prohorov, D. V. *Integral transformations in some classes of univalent functions.* (Russian) Izv. Vyss. Ucebn. Zaved. Matematika (1980), no. 12, 45-49; MR 82h:30012.
727. Prohorov, D. V.; Rahmanov, B. N. *The integral representation of a certain class of univalent functions.* (Russian) Mat. Zametki 19(1976), no. 1, 41-48; MR 58 #1172.
728. Prohorov, D. V.; Rahmanov, N. B. *The radius of close-to-convexity of order α in the class of close-to-convex functions of order β.* (Russian) Differencial'nye Uravnenija i Vycisl. Mat. Vyp. 6(1976), part 2, 135-140, 171; MR 57 #9955.
729. Prohorov, D. V.; Szynal, J. *Coefficient estimates for bounded nonvanishing functions.* (Russian summary) Bull. Acad. Polon. Sci. Ser. Sci. Math. 29(1981), no. 5-6, 223-230.
730. Quine, J. R. *The geometry of $p(S_1^1)$.* Pacific J. Math. 64(1976), no. 2, 551-557; MR 56 #589.
731. Quine, J. R. *On univalent polynomials.* Proc. Amer. Math. Soc. 57 (1976), no. 1, 75-78.
732. Quine, J. R. *Some topological theorems relating to close-to-convex functions.* J. London Math. Soc. (2) 14(1976), no. 1, 39-42; MR 58 #22515.
733. Quine, J. *Criterion for a subclass of univalent polynomials.* Complex analysis (Proc. S.U.N.Y. Conf., Brockport, N.Y., 1976), pp. 99-104. Lecture Notes in Pure and Appl. Math., Vol. 36, Dekker, New York, 1978; MR 57 #16549.
734. Rahman, Q. I. *Some inequalities for polynomials.* Proc. Amer. Math. Soc. 56(1976), 225-230; MR 57 #3358.
735. Rahman, Q. I.; Schmeisser, G. *Inequalities for polynomials on the unit interval.* Trans. Amer. Math. Soc. 231(1977), no. 1, 93-100; MR 57 #3358.
736. Rahman, Q. I.; Szynal, J. *On some classes of univalent polynomials.* Canad. J. Math. 30(1978), no. 2, 332-349; MR 57 #3359.
737. Rahman, Q. I.; Waniurski, J. *Coefficient regions for univalent trinomials.* Canad. J. Math. 32(1980), no. 1, 1-20; MR 81b:30027.
738. Rahmanov, B. N. *Certain classes of univalent functions.* (Russian) Studies in differential equations and the theory of functions, No. 3 (Russian), pp. 63-65. Izdat. Saratov. Univ., Saratov (1971); MR 53 #8400.
739. Rahmanov, B. N. *The radii of convexity and starlikeness of*

univalent functions. (Russian) Studies in differential equations and the theory of functions, No. 3 (Russian), pp. 58-62. Izdat. Saratov. Univ., Saratov (1971); MR 53 #8407.
740. Rahmanov, B. N. *Star-like functions.* (Russian) Differencial'nye Uravnenija i Vycisl. Mat. Vyp. 2 (1975), 137-142, 181; MR 55 #8339.
741. Rahmanov, B. N. *On series of univalent functions.* (Russian) Mat. Zametki 23(1978), no. 4, 593-600; MR 58 #11353.
742. Rahmanov, B. N. *The product of univalent functions.* (Russian) Mat. Zametki 23(1978), no. 5, 697-708; MR 58 #11354.
743. Rangarajan, M. R. *On the radius of univalence and starlikeness of a class of analytic functions.* Indian J. Pure Appl. Math. 11 (1980), no. 2, 245-251; MR 81g:30020.
744. Range, R. M. *On a Lipschitz estimate for conformal maps in the plane.* Proc. Amer. Math. Soc. 58(1976), 375-376; MR 53 #11033.
745. Ratti, J. S. *The radius of convexity of certain analytic functions.* II. Internat. J. Math. Math. Sci. 3(1980), no. 3, 483-489; MR 82b:30013.
746. Rauch, J. *Illumination of bounded domains.* Amer. Math. Monthly 85, no. 5, May (1978), p. 359.
747. Reade, M. O.; Todorov, P. G. *The radii of starlikeness and convexity of certain analytic functions.* Proc. Amer. Math. Soc. 83 (1981), no. 2, 289-295.
748. Red'kov, M. I. *The domain of values of a certain functional on certain classes of bounded univalent functions.* (Russian) Tomsk. Gos. Univ. Ucen. Zap. No. 36(1960), 33-50; MR 52 #5951.
749. Red'kov, M. I. *A correction to the paper: "The domain of values of a certain functional on certain classes of bounded univalent functions"* (Tomsk. Gos. Univ. Ucen. Zap. No. 36 [1960], 33-50). (Russian) Trudy Tomsk. Gos. Univ. 163(1963), 152-154; MR 52 #5952.
750. Ren, F. Y. *Coefficients of inverses of meromorphic univalent functions.* A translation of Kexue Tongbao (Chinese) 25(1980), no. 5, 193-195. Kexue Tongbao (English) 25(1980), no. 4, 277-280; MR 82f:30015.
751. Renelt, H. *Über Extremalprobleme für schlichte Lösungen elliptischer Differentialgleichungssysteme.* Comment. Math. Helv. 54(1979), no. 1, 17-41.
752. Reza, F. M. *Restrictions of the derivatives of positive real functions.* J. Franklin Inst. 312(1981), no. 5, 327-334.
753. Ripeanu, D. *Sur la fonctionnelle $f(z_1)/f'(z_2)$ au cas des fonctions*

typiquement-reelles. I. Mathematica (Cluj) 21(44)(1979), no. 2, 163-188; MR 81m:30013.

754. Ripeanu, D. *Sur la fonctionnelle $f(z_1)/f'(z_2)$ au cas des fonctions typiquement-reelles*. II. Mathematica (Cluj) 22(45)(1980), no. 1, 131-165; MR 82j:30020.

755. Robertson, M. S. *Complex powers of p-valent functions and subordination*. Complex analysis (Proc. S.U.N.Y. Conf., Brockport, N.Y., 1976), pp. 1-33. Lecture Notes in Pure and Appl. Math., Vol. 36, Dekker, New York, 1978; MR 58 #6215.

756. Robertson, M. S. *Coefficient theorems for analytic functions*. Rev. Roumaine Math. Pures Appl. 23(1978), no. 3, 479-486; MR 57 #12838.

757. Robertson, M. S. *A characterization of the class of starlike univalent functions*. Michigan Math. J. 26(1979), no. 1, 65-69; MR 80f:30010.

758. Robertson, M. S. *Univalent functions starlike with respect to a boundary point*. J. Math. Anal. Appl. 81(1981), 327-345; MR 82i:30017.

759. Rodin, B.; Warschawski, S. E. *Extremal length and univalent functions*. I. The angular derivative. Math. Z. 153(1977), no. 1, 1- 17; MR 58 #28461.

760. Rodin, B.; Warschawski, S. E. *Extremal length and univalent functions*. III. Consequences of the Ahlfors distortion property. Bull. Inst. Math. Acad. Sinica 6(1978), no. 2, part 2, 583-597; MR 81k:30013.

761. Rodin, B.; Warschawski, S. E. *Extremal length and univalent functions*. II. Integral estimates of strip mappings. J. Math. Soc. Japan 31(1979), no. 1, 87-99; MR 81k:30012.

762. Royster, W. C.; Ziegler, M. *Univalent functions convex in one direction*. Publ. Math. Debrecen 23(1976), no. 3-4, 339-345; MR 54 #13059.

763. Rubel, L. A.; Yang, C. C. *Values shared by an entire function and its derivative*. Complex analysis (Proc. Conf., Univ. Kentucky, Lexington, Ky., 1976), pp. 101-103. Lecture Notes in Math., Vol. 599, Springer, Berlin, 1977; MR 57 #633.

764. Rung, D. C. *A local form of Lappan's five-point theorem for normal functions*. Michigan Math. J. 23(1976), no. 2, 141-145; MR 53 #13575.

765. Ruscheweyh, S. *On starlike functions*. (Polish and Russian summaries) Ann. Univ. Mariae Curie-Sklodowska Sect. A 28(1974), 65-70 (1976); MR 55 #645.

766. Ruscheweyh, S. *A subordination theorem for Φ-like functions*. J.

London Math. Soc. (2)13(1976), no. 2, 275-280; MR 54#533.
767. Ruscheweyh, S. *On functions with non-vanishing divided differences.* Math. Nachr. 73(1976), 143-146; MR 55 #646.
768. Ruscheweyh, S. *An extension of Becker's univalence condition.* Math. Ann. 220(1976), no. 3, 285-290; MR 53 #3284.
769. Ruscheweyh, S. *Linear operators between classes of prestarlike functions.* Comment. Math. Helvetici 52(1977), 497-509; MR 58 #1123.
770. Ruscheweyh, S. *Some convexity and convolution theorems for analytic functions.* Math. Ann. 238(1978), no. 3, 217-228; MR 80i:30023.
771. Ruscheweyh, S. *On the Kakeya-Eneström theorem and Gegenbauer polynomial sums.* SIAM J. Math. Anal. 9(1978), no. 4, 682-686; MR 57 #16551.
772. Ruscheweyh, S. *On the partial sums of prestarlike and related functions.* Indian J. Pure Appl. Math. 11(1980), no. 12, 1587-1589; MR 82i:30018.
773. Ruscheweyh, S. *Neighborhoods of univalent functions.* Proc. Amer. Math. Soc. 81(1981), no. 4, 521-527; MR 82c:30016.
774. Ruscheweyh, S.; Singh, V. *On certain extremal problems for functions with positive real part.* Proc. Amer. Soc. 61(1976), no. 2, 329-334; MR 54 #13060.
775. Ruscheweyh, S.; Singh, V. *On a Briot-Bouquet equation related to univalent functions.* Rev. Roumaine Math. Pures Appl. 24(1979), no. 2, 285-290; MR 80i:30022.
776. Ruscheweyh, S.; Wirths, K. J. *Riemann's mapping theorem for n-analytic functions.* Math. Z 149(1976), no. 3, 287-297; MR 53 #13543.
777. Ruscheweyh, S.; Wirths, K. J. *Convex sums of convex univalent functions.* Indian J. Pure Appl. Math. 7(1976), no. 1, 49- 52; MR 56 #8832.
778. Sabalin, P. L. *The univalence of the general solution of an interior inverse boundary value problem.* (Russian) Izv. Vyss. Ucebn. Zaved. Matematika 1975, no. 12(153), 92-95(1976). (English translation: Soviet Math. [Iz. Vuz.] 19 [1975], no. 12, 77-80 [1976]); MR 55 #8371.
779. Sabalin, P. L. *Univalence classes and V. I. Smirnov's domains.* (Russian) Trudy Sem. Kraev. Zadacam No. 16(1979), 218-226, 246; MR 82c:30022.
780. Saff, E. B.; Sheil-Small, T. *Coefficient and integral mean estiamtes for algebraic and trigonometric polynomials with restricted zeroes.*

J. London Math. Soc. (2)9(1974/75), 16-22; MR 52 #3491.
781. Sakaguchi, K. *On the coefficients of p-fold symmetric and univalent functions.* Bull. Nara Univ. Ed. Natur. Sci. 30(1981), no. 2, 11-15
782. Sakaguchi, K.; Fukui, S. *On alpha-starlike functions and related functions.* Bull. of Nara Univ. of Education, 28(1979), pp. 5-12; MR 82a:30015.
783. Sakai, M. *On extremal sets of parallel slits.* Hiroshima Math. J. 5 (1975), no. 3, 499-516; MR 55 #12907.
784. Salagean, G. S. *Properties of starlikeness and convexity preserved by some integral operators.* Romanian-Finnish Seminar on Complex Analysis (Proc., Bucharest, 1976), pp. 367-372. Lecture Notes in Math., 743. Springer, Berlin, 1979; MR 81e:30024.
785. Salmassi, M.; Shah, S. M. *Functions with univalent derivatives and of irregular growth.* Indian J. Pure Appl. Math. 12 (1981), no. 6, 749-752; MR 82i:30019.
786. Sarangi, S. M.; Uralegaddi, B. A. *Functions whose real parts are greater than α.* J. Karnatak Univ. Sci. 22(1977), 37-43; MR 81a:30011.
787. Sarangi, S. M.; Uralegaddi, B. A. *The radius of convexity and starlikeness for certain classes of analytic functions with negative coefficients.* I. Atti Accad. Naz. Lincei Rend. Cl. Sci. Fis. Mat. Natur. (8)65(1978), no. 1-2, 38-42(1979); MR 81d:30019.
788. Sarangi, S. M.; Uralegaddi, B. A. *On the radius of univalence of certain analytic functions.* J. Karnatak Univ. Sci. 23(1978), 18-24; MR 82c:30015.
789. Sarangi, S. M.; Uralegaddi, B. A. *The radius of convexity and starlikeness for certain classes of analytic functions with negative coefficients.* II. (Italian summary) Atti. Accad. Naz. Lincei Rend. Cl. Sci. Fis. Mat. Natur. (8)67(1979), no. 1-2, 16-20 (1980); MR 82k:30016.
790. Sarason, D. *Function theory on the unit circle.* Notes for lectures given at a Conference at Virginia Polytechnic Institute and State University, Blacksburg, Va., June 19-23, 1978. Department of Mathematics,Virginia Polytechnic Institute and State University, Blacksburg, Va. (1978), iv + 138 pp. $6.00; MR 80d:30035.
791. Schiffer, M.; Schober, G. *A distortion theorem for quasiconformal mappings.* Advances in complex function theory (Proc. Sem., Univ. Maryland, College Park, Md., 1973-1974), pp. 138-147. Lecture Notes in Math., Vol. 505, Springer, Berlin, 1976; MR 54 #10598.
792. Schild, A.; Silverman, H. *Convolutions of univalent functions with negative coefficients.* (Polish and Russian summaries) Ann. Univ.

Mariae Curie-Sklodowska Sect. A 29(1975), 99–107 (1977); MR 56 #15902.
793. Schild, A.; Silverman, H. *Some properties of schlicht functions with "small" coefficients.* Complex analysis (Proc. S.U.N.Y. Conf., Brockport, N.Y., 1976), pp. 105–114. Lecture Notes in Pure and Appl. Math., Vol. 36, Dekker, New York, 1978; MR 58 #17069.
794. Schober, G. *Univalent functions—selected topics.* Lecture Notes in Mathematics, Vol. 478. Springer-Verlag, Berlin-New York (1975). v + 200 pp. $9.90; MR 58 #22527.
795. Schober, G. *Coefficients of inverses of meromorphic univalent functions.* Proc. Amer. Math. Soc. 67(1977), no. 1, 111–116; MR 56 #12251.
796. Schober, G. *Coefficients of inverses of univalent functions with quasiconformal extensions.* Kodai Math. J. 2(1979), no. 3, 411–419; MR 81b:30039.
797. Schober, G. *Coefficient estimates for inverses of Schlicht functions.* Aspects of contemporary complex analysis (Proc. NATO Adv. Study nst., Univ. Durham, Durham, 1979), pp. 503–513, Academic Press, London (1980); MR 82j:30021.
798. Schwarz, G. *Zur partiellen Differentialgleichung der schlichten Funktionen.* Collection in honor of Hans Hornich. Osterreich. Akad. Wiss. Math.-Naturwiss. Kl. S.-B. II 185(1976), no. 1–3, 89–93; MR 56 #15908.
799. Seiler, A. *Inegalites de Grunsky du type Garabedian-Schiffer pour des paires de fonctions univalentes.* (English summary) C. R. Acad. Sci. Paris Ser. A-B 283(1976), no. 10, Aii, A755–A757; MR 54 #13064.
800. Sekigawa, H. *Schwarzian derivatives of some conformal mappings.* Tohoku Math. J. (2)31(1979), no. 3, 309–318; MR 80j:30029.
801. Selberg, H. L. *Uber die Potenzreihen mit positiven Koeffizienten.* Norske Vid. Selsk. Skr. (Trontheim) (1975), no. 5, 6 pp.; MR 54 #10578.
802. Selljahova, T. N.; Sobolev, V. V. *The study of extremal properties of a class of univalent conformal mappings of the half-plane onto itself.* (Russian) Some problems in modern function theory (Proc. Conf. Modern Problems of Geometric Theory of Functions, Inst. Math., Acad. Sci. USSR, Novosibirsk, 1976) (Russian), pp. 142–145. Akad. Nauk SSSR Sibirsk. Otdel. Inst. Mat., Novosibirsk, 1976. For a review of this item see RZMat 1977 #10 B178. [For the entire collection see MR 57 #15813.]; MR 58 #22501.

803. Seretov, V. G. *Harmonic mappings, and univalent functions.* (Russian) Mathematical analysis, No. 2. Kuban. Gos. Univ. Naucn. Trudy Vyp. 180 (1974), 143–153, 166–167; MR 55 #12911.
804. Shaffer, D. B. *On the order of convexity and starlikeness of a special class of analytic and meromorphic functions.* J. Math. Anal. Appl. 64 (1978), no. 1, 216–222; MR 57 #16564.
805. Shah, S. M. *Analytic functions with some derivatives univalent and a related conjecture.* (Italian summary) Atti. Accad. Naz. Lincei Rend. Cl. Sci. Fis. Mat. Natur. (8) 61 (1976), no. 5, 344–353 (1977); MR 58 #6236.
806. Shah, S. M.; Trimble, S. Y. *Univalence of derivatives of functions defined by gap power series.* II. J. Math. Anal. Appl. 56 (1976), no. 1, 28–40.
807. Shah, S. M.; Trimble, S. Y. *Analytic functions with univalent derivatives.* Indian J. Math. 20 (1978), 265–299; MR 82a:30016.
808. Shah, S. M.; Trimble, S. Y. *On the zeros of univalent functions with univalent derivatives.* Ann. Mat. Pura Appl. (4) 121 (1979), 309–317; MR 81c:30027.
809. Sharma, A.; Tzimbalario, J. *Classes of functions defined by differential inequalities.* J. Math. Anal. Appl. 61 (1977), no. 1, 122–135; MR 80j:30036.
810. Sheil-Small, T. *The Hadamard product and linear transformations of classes of analytic functions.* J. Analyse Math. 34 (1978), 204–239 (1979); MR 81b:30023.
811. Sheil-Small, T. *Applications of the Hadamard product.* Aspects of contemporary complex analysis (Proc. NATO Adv. Study Inst., Univ. Durham, Durham, (1979), pp. 515–523, Academic Press, London, 1980; MR 82g:30042.
812. Sheng, G. *A simple proof of Bieberbach conjecture for sixth coefficient.* Sci. Sinica 23 (1980), no. 1, 1–15; MR 81g:30026.
813. Sheng, G. *On FitzGerald inequalities.* J. London Math. Soc. (2) 24 (1981), no. 2, 227–242; MR 82k:30021.
814. Siewierski, L. *Sharp estimation of the coefficients of bounded univalent functions close to identity.* Dissertationes Math. Rozprawy Mat. 86 (1971), 149 pp.; MR 55 #10661.
815. Siewierski, L. *The maximum of the functional $a_3 - \alpha a_2^2$ in classes of quasi-convex functions.* (Polish; Russian and English summaries) Zeszyty Nauk. Politech. Łódź. No. 134 Mechanika No. 28 (1971), 5–21; MR 52 #733.
816. Siewierski, L.; Śmiałkówna, H. *Zeszyty Naukowe Politech-niki Lodzkiej, Mat.* (1973), pp. 105–124.

817. Silverman, H. *Products of starlike and convex functions.* (Polish and Russian summaries) Ann. Univ. Mariae Curie-Sklodowska Sect. A 29 (1975), 109–116; MR 57 #608.
818. Silverman, H. *Applications of a theory of Zmorovic.* Complex analysis (Proc. S.U.N.Y. Conf., Brockport, N.Y., 1976), pp. 141–157. Lecture Notes in Pure and Appl. Math., Vol. 36, Dekker, New York, 1978; MR 57 #16565.
819. Silverman, H. *Extreme points of univalent functions with two fixed points.* Trans. Amer. Math. Soc. 219 (1976), 387–395.
820. Silverman, H. *Convexity theorems for subclasses of univalent functions.* Pacific J. Math. 64 (1976), no. 1, 253–263; MR 54 #2944.
821. Silverman, H. *Subclasses of starlike functions.* Rev. Roumaine Math. Pures Appl. 23 (1978), no. 7, 1093–1099; MR 80g:30009.
822. Silverman, H. *Univalent functions with varying arguments.* Houston J. Math. 7 (1981), no. 2, 283–287.
823. Silverman, H.; Silvia, E. *On linear combinations of convex functions of order β.* Rev. Roumaine Pures Appl. 22 (1977), no. 6, 851–855; MR 57 #609.
824. Silverman, H.; Silvia, E. *Convex families of starlike functions.* Houston J. Math. 4 (1978), no. 2, 263–268; MR 58 #1120.
825. Silverman, H.; Silvia, E. M. *Prestarlike functions with negative coefficients.* Internat. J. Math. Math. Sci. 2 (1979), no. 3, 427–439; MR 80i:30024.
826. Silverman, H.; Silvia, E. *The influence of the second coefficient on prestarlike functions.* Rocky Mountain J. Math. 10 (1980), 469–474; MR 81m:30014.
827. Silverman, H.; Silvia, E. M. *Fixed coefficients for subclasses of starlike functions.* Houston J. Math. 7 (1981), 129–136.
828. Silverman, H.; Silvia, E. M.; Telage, D. N. *Locally univalent functions and coefficient distortions.* Pacific J. Math. 77 (1978), no. 2, 533–539; MR 80a:30013.
829. Silverman, H.; Silvia, E. M.; Telage, D. *Convolution conditions for convexity, starlikeness and spiral-likeness.* Math. Z. 162 (1978), no. 2, 125–130; MR 80i:30025.
830. Silverman, H.; Telage, D. N. *Spiral functions and related classes with fixed second coefficient.* Rocky Mountain J. Math. 7 (1977), no. 1, 111–116; MR 56 #3273.
831. Silverman, H.; Telage, D. N. *Extremal properties of a subclass of close-to-convex functions.* Rocky Mountain J. Math. 7 (1977), no. 2, 371–376; MR 56 #3274.
832. Silverman, H.; Telage, D. N. *Extreme points of subclasses of close-*

to-convex functions. Proc. Amer. Math. Soc. 74 (1979), no. 1, 59–65; MR 80c:30012.
833. Silverman, H.; Ziegler, M. *Functions of positive real part with negative coefficients.* Houston J. Math. 4 (1978), no. 2, 269–275; MR 58 #1121.
834. Silvia, E. M. *A variational method on certain classes of functions.* Rev. Roumaine Math. Pures Appl. 21(1976), no. 5, 549–557; MR 54 #534.
835. Silvia, E. M. *p-valent classes related to functions of bounded boundary rotation.* Rocky Mountain J. Math. 7 (1977), no. 2, 265–274; MR 55 #8340.
836. Silvia, E. M. *A note on special classes of p-valent functions.* Rocky Mountain J. Math. 9 (1979), no. 2, 365–370; MR 80c:30013.
837. Sindalovskii, G. H. *Differentiability and analyticity of univalent mappings.* (Russian) Dokl. Akad. Nauk SSSR 249 (1979), no. 6, 1325–1327; MR 81d:30020.
838. Singh, D. B. *On the radius of univalence of some functions analytic in the unit disc.* Rev. Roumaine Math. Pures Appl. 23 (1978), no. 7, 1107–1114; MR 80f:30011.
839. Singh, D. B.; Srivastava, R. S. L. *Some classes of regular univalent functions.* Ganita 30 (1979), no. 1–2, 146–154.
840. Singh, P.; Tygel, M. *On some univalent functions in the unit disc.* Indian J. Pure Appl. Math. 12 (1981), no. 4, 513–520; MR 82k:30024.
841. Singh, Prithvipal; Singh, Prem *Integral representation of functions in certain classes of univalent functions.* Indian J. Pure Appl. Math. 12 (1981), no. 4, 459–471; MR 82i:30020.
842. Singh, R. *A sufficient condition for univalence.* Publ. Math. Debrecen 25 (1978), no. 1–2, 1–3; MR 58 #6205.
843. Singh, R.; Singh, S. *Integrals of certain univalent functions.* Proc. Amer. Math. Soc. 77 (1979), no. 3, 336–340; MR 81e:30025.
844. Singh, R.; Singh, V. *On a class of bounded starlike functions.* Indian J. Pure Appl. Math. 5 (1974), no. 8, 733–754; MR 55 #647.
845. Singh Sohi, N. *A class of p-valent analytic functions.* Indian J. Pure Appl. Math. 10 (1979), no. 7, 826–834; MR 80i:30026.
846. Singh, S.; Singh, R. *On a subclass of α-spiral-like functions.* Indian J. Pure Appl. Math. 11 (1980), 160–164; MR 82a:30017.
847. Singh, S.; Singh, R. *Subordination by univalent functions.* Proc. Amer. Math. Soc. 82 (1981), no. 1, 39–47; MR 82f:30019.
848. Singh, S.; Singh, R. *On new subclasses of close-to-convex functions.* Indian J. Pure Appl. Math. 12 (1981), no. 6, 743–748.
849. Singh, V. *Univalent functions with bounded derivative in the unit*

disc. Indian J. Pure Appl. Math. 8 (1977), no. 11, 1370–1377; MR 80k:30018.
850. Singh, V. *Bounds on the curvature of level lines under certain classes of univalent and locally univalent mappings.* Indian J. Pure Appl. Math. 10 (1979), no. 2, 129–144; MR 80c:30014.
851. Singh, V.; Chichra, P. N. *Univalent functions $f(z)$ for which $zf'(z)$ is α-spiral-like.* Indian J. Pure Appl. Math. 8 (1977), no. 2, 253–259; MR 57 #6397.
852. Singh, V.; Chichra, P. N. *An extension of Becker's criterion of univalence.* J. Indian Math. Soc. (N.S.) 41 (1977), no. 3–4, 353–361 (1978); MR 80j:30024.
853. Singh, V.; Gupta, R. S. *Radius of convexity of α-spirallike functions.* Indian J. Pure Appl. Math. 6 (1975), no. 3, 322–336; MR 56 #12248.
854. Singh, V.; Gupta, R. S. *An extremal problem for functions with positive real part.* Indian J. Pure Appl. Math. 8 (1977), no. 11, 1279–1297; MR 81d:30054.
855. Singh, V.; Paul, S. *Constrained extremal problems for symmetric functions with real part positive.* Indian J. Pure Appl. Math. 12 (1981), no. 4, 492–512; MR 82g:30039.
856. Singh, V.; Paul, S. *Constrained extremal problems for functions with positive real part.* Indian J. Pure Appl. Math. 8 (1977), no. 4, 454–462; MR 58 #28511.
857. Singh, V.; Singh, R. *On a class of functions schlicht in the unit disc.* Indian J. Pure Appl. Math. 7 (1976), no. 1, 116–120; MR 57 #610.
858. Singh, V.; Singh, R. *The radii of starlikeness of certain classes of close-to-convex functions.* Indian J. Pure Appl. Math. 8 (1977), no. 12, 1497–1504; MR 80h:30016.
859. Sirokov, N. A. *Approximation of continuous analytic functions in domains with bounded boundary rotation.* (Russian) Dokl. Akad. Nauk SSSR 228 (1976), no. 4, 809–812.
860. Sirokova, E. A. *The univalence of certain integrals.* (Russian) Izv. Vyss. Ucebn. Zaved. Mathematika (1977), no. 9 (184), 107–114; MR 58 #6206.
861. Sitarski, R. *On some application of a parametric method.* Polish and Russian summaries) Zeszyty Nauk. Lodz. Mat. No. 9 (1977), 63–69; MR 58 #22530.
862. Sitarski, R. *Extremal problems for certain classes of holomorphic and bounded functions.* (Polish and Russian summaries) Zeszyty Nauk. Politech. Lodz. Mat. No. 12 (1979), 45–61.

863. Sižuk, P. I. *The radius of almost convexity of order α in a class of schlicht functions.* (Russian) Mat. Zametki 20 (1976), no. 1, 105-112; MR 54 #13061.
864. Sižuk, P. I.; Cernikov, V. V. *The coefficients of α-convex functions of order β.* (Russian) Mat. Zametki 24 (1978), no. 5, 679-686, 734; MR 80g:30010.
865. Sižuk, G. I.; Sižuk, P. I. *On the theory of special classes of analytic functions.* (Russian) Ukrain. Mat. Z. 32 (1980), no. 2, 256-262, 286; MR 81c:30044.
866. Skalska, K. *Certain subclasses of the class of typically real functions.* Ann. Polon. Math. 38 (1980), no. 2, 141-152; MR 82a:30018.
867. Sladkowska, J. *Coefficient inequalities for Bieberbach-Eilenberg functions.* (Polish; Russian and English summaries) Zeszyty Nauk. Politech. Slask. Mat.-Fiz. Zeszyt 25 (1974), 11-46; MR 55 #12903.
868. Sladkowska, J. *Théorème des aires dans la theorie des fonctions univalentes bornées.* I. Demonstratio Math. 10 (1977, no. 2, 287-316; MR 58 #28471a. Théorème des aires dans la théorie des fonctions univalentes bornées. II. Demonstratio Math. 10 (1977), no. 3-4, 547-569. For reviews of these items see RZMat 1978 #7 B211 and RZMat 1978 #9 B115; MR 58 #28471b.
869. Sladkowska, J. *Les polynômes de Faber dans les familles compactes de fonctions univalentes bornées.* Comment. Math. Prace Mat. 20 (1977), no. 1, 205-214; MR 58 #22528. For a review of this item see Zbl 371 #30012.
870. Sladkowska, J. *Sur les fonctions univalentes, bornées, satisfaisant deux au moins D_n-equations.* Demonstratio Math. 11 (1978), no. 2, 351-378; MR 82c:30017.
871. Slyk, V. A. *Distortion theorems for a family of weakly univalent functions in a circle.* (Russian) Mat. Zametki 27 (1980), no. 6, 927-933, 990; MR 81m:30015.
872. Slyk, V. A. *On the theory of nonunivalent mappings of multiply connected domains.* (Russian) Analytic number theory and the theory of functions, 4. Zap. Naucn. Sem. Leningrad. Otdel. Math. Inst. Steklov. (LOMI) 112 (1981), 184-197, 203.
873. Slyk, V. A. *Some estimates in an annulus for weakly univalent functions that omit values on a circle.* (Russian) Izv. Vyss. Ucebn. Zaved. Matematika (1981), no. 8, 85-86.
874. Smigielska, J. *The Grunsky-Nehari inequality for univalent, symmetric and bounded functions.* (Polish; Russian and English summaries) Zeszyty Nauk. Politech. Slask. Mat.-Fiz. No. 34 (1979), 97-105; MR 81e:30027.

875. Smigielska, J. *Variational formulas in a class of univalent, symmetric and bounded functions.* (Polish; Russian and English summaries) Zeszyty Nauk. Politech. Slask. Mat.-Fiz. 34 (1979), 107–116; MR 81d:30021.
876. Smith, H. V. *Bi-univalent polynomials.* Simon Stevin 50 (1976/77), no. 2, 115–122; MR 54 #13065.
877. Sohi, N. S. *On a subclass of p-valent functions.* Indian J. Pure Appl. Math. 11 (1980), no. 11, 1504–1508; MR 81m:30016.
878. Špak, G. S. *The range of bounded functions that have prescribed values at two points.* (Russian) Mat. Zametki 23 (1978), no. 2, 231–235; MR 58 #1113.
879. Srivastava, G. S.; Juneja, O. P. *The maximum term of a power series.* J. Math. Anal. Appl. 81 (1981), no. 1, 1–7.
880. Srivastava, J. K. *Semigroups of classes of univalent functions.* Math. Balkanica 6 (1976), 237–244 (1978); MR 80g:30012.
881. Srivastava, R. S. L.; Bajpai, S. K. *Some distortion theorems for univalent α-spiral functions.* J. Math. Sci. 7 (1972), 101–108 (1974); MR 57 #12834.
882. Stankiewicz, J. *Quasisubordination and quasimajorization of analytic functions.* (Polish and Russian summaries) Ann. Univ. Mariae Curie-Sklodowska Sect. A 31 (1977), 127–135 (1979); MR 81e:30036.
883. Stankiewicz, J.; Stankiewicz, Z. *Some remarks on subordination and majorization of functions.* (Polish and Russian summaries) Ann. Univ. Mariae Curie-Sklodowska Sect. A 31 (1977), 119–126 (1979); MR 81d:30044.
884. Stankiewicz, J.; Waniurski, J. *Some classes of functions subordinate to linear transformation and their applications.* (Polish and Russian summaries) Ann. Univ. Mariae Curie-Sklodowska Sect. A 28 (1974), 85–94 (1976); MR 56 #5858.
885. Stankiewicz, Z. *Sur la subordination en domaine de certains operateurs.* (Polish and Russian summaries) Ann. Univ. Mariae Curie-Sklodowska Sect. A 28 (1974), 95–103 (1976); MR 56 #8837.
886. Starkov, V. V. *The subclasses S_α of univalent functions.* Vestnik Leningrad. Univ. (1976), no. 7 Mat. Meh. Astronom. vyp. 2, 82–87, 163; MR 56 #12254.
887. Starkov, V. V. *A variational formula in the class of Bazilevič functions.* (Russian; English summary) Vestnik Leningrad. Univ. Mat. Meh. Astronom. No. 19 Mat. Meh. Astronom. Byp. 4 (1978), 81–88, 151; MR 80d:30009.
888. Starkov, V. V. *Certain properties of functions of the Basilevič*

class. (Russian; English summary) Vestnik Leningrad. Univ. Mat. Meh. Astronom. No. 13 Mat. Meh. Astronom. no. 3 (1978), 148-149; 160; MR 80d:30008.

889. Starkov, V. V. *Letter to the editors: "Certain properties of functions of the Bazilevič class"* (Vestnik Leningrad. Univ. Mat. Meh. Astronom. 1978, vyp. 3, 148-160). (Russian) Vestnik Leningrad. Univ. Mat. Meh. Astronom. 1979, vyp. 1, 132.

890. Starkov, V. V . *A property of the Bazilevič class functions.* (Russian; English summary) Vestnik Leningrad. Univ. Mat. Meh. Astronom. (1979), vyp. 1, 84-87, 135.

891. Stegbuchner, H. *Einige Bemerkungen uber die Fixpunkte der schlichten Funktionen.* Collection in honor of Hans Hornich. Osterreich. Akad. Wiss. Math.-Naturwiss. Kl. S.-B. II 185 (1976), no. 1-3, 105-113; MR 56 #12255.

892. Stegbuchner, H. *Carleson-sets and fixed-points of schlicht functions.* Romanian-Finnish Seminar on Complex Analysis (Proc., Bucharest, 1976), pp. 373-388. Ledture Notes in Math., 743. Springer, Berlin, 1979; MR 81c:30028.

893. Stephenson, K. *Weak subordination and stable classes of meromorphic functions.* Trans. Amer. Math. Soc. 262 (1980), 565-577.

894. Styer, D. *An analytic characterization of geometrically starlike functions.* Michigan Math. J. 23 (1976), no. 2, 137-139; MR 53 #13544.

895. Styer, D.; Wright, D. J. Typically-real functions and their associated functions of positive real part. Proc. London Math. Soc. (3) 34 (1977), no. 2, 193-212; MR 56 #3275.

896. Styer, D.; Wright, D. J. *Results on bi-univalent functions.* Proc. Amer. Math. Soc. 82 (1981), no. 2, 243-248; MR 82c:30018.

897. Suffridge, T. J. *Starlike functions as limits of polynomials.* Advances in complex function theory (Proc. Sem., Univ. Maryland, College Park, Md., 1973-1974), pp. 164-203. Lecture Notes in Math., Vol. 505, Springer, Berlin, 1976; MR 58 #1122.

898. Suffridge, T. J. *A new criterion for starlike functions.* Indiana Univ. Math. J. 28 (1979), no. 3, 429-443; MR 81b:30024.

899. Suvorov, G. D. *Theory of functions and mappings.* (Russian) Edited by G. D. Suvorov. "Naukova Dumka," Kiev (1979), 180 pp.; MR 81b:30002.

900. Suvorov, G. D. *Metric questions of the theory of functions. (Russian) "Naukova Dumka," Kiev (1980), 164 pp.; MR 81j:30002.

901. Swamy, S. R. *(A) The radius of univalence of certain analytic functions. I. (B) The radius of univalence of certain analytic functions. II.* J. Karnatak Univ. Sci. 22 (1977), (A) 11-15; MR 81f:30013a, (B)

16-23; MR 81f:30013b.
902. Szapiel, M. *Sur la subordination dans la classe de fonctions typiquement-reelles.* Demonstratio Math. 13 (1980), no. 2, 461–467; MR 82a:30019.
903. Szász, O. *Über Potenzreihen die im Einheitskreise beschränkte Funktionen darstellen.* Math. Zeit., vol. 8 (1920), pp. 222–236.
904. Szilárd, K. *Über die Abschätzung der Koeffizienten der Potenzreihe einer analytischen Funktionen, die einch schlichte Abbildung des Einheitskreises auf ein Sterngebiet verwirklicht.* Proccedings of the Conference on the Constructive Theory of Functions (Approximation Theory), Budapest, 1969, pp. 475–476. Akademiai Kiado, Budapest (1972); MR 52 #5954.
905. Szynal, J. *Regions of variability of functionals represented by Stieltjes integral with side conditions.* (Russian summary) Bull. Acad. Polon. Sci. Ser. Sci. Math. Astronom. Phys. 22 (1974), 897–901; MR 53 #794.
906. Szynal, A.; Szynal, J. *On some problems concerning subordination and majorization of functions.* Demonstratio Math. 11 (1978), no. 2, 331–350; MR 80j:30023.
907. Szynal, A.; Szynal, J.; Wajler, S. *On the Gronwall's problem for some classes of univalent functions.* Analytic functions, Kozubnik (1979) (Proc. Seventh Conf., Kozubnik, 1979), pp. 429–434, Lecture Notes in Math., 798, Springer, Berlin, 1980; MR 82a:30020.
908. Szynal, J.; Wajler, S. *Sur les fonctions H-quasi-étoilées.* Demonstratio Math. 8 (1975), no. 2, 205–214; MR 57 #3374.
909. Szynal, J.; Waniurski, J. *Some problems for linearly invariant families.* (Polish and Russian summaries) Ann. Univ. Mariae Curie-Sklodowska Sect. A 30 (1976), 91–102 (1978); MR 80c:30017.
910. Taladaj, H. *Quasi-convex functions.* Proc. of the conference on analytic functions, Lublin, pp. 23–29, August (1970).
911. Tammi, O. *Use of the Power inequality in connection of coefficient regions for bounded univalent functions.* Proceedings of the Sixth Conference on Analytic Functions (Krakow, 1974). Ann. Polon. Math. 33 (1976), no. 1–2, 117–123; MR 57 #16566.
912. Tammi, O. *Inequalities for the fifth coefficient for bounded real univalent functions.* J. Analyse Math. 30 (1976), 481–497; MR 55 #8342.
913. Tammi, O. *On generalizing the power inequality for the first coefficient region in the class of bounded univalent functions.* Proceedings of the Rolf Nevanlinna Symposium on Complex Analysis (Math. Res. Inst., Univ. Instanbul, Silivri, 1976), pp. 103–129.

Publ. Math. Res. Inst. Instanbul, 7. Math. Res. Inst., Univ. Instanbul, Istanbul, 1978; MR 81f:30014.
914. Tammi, O. *Extremum problems for bounded univalent functions.* Lecture Notes in Mathematics, Vol. 646. Springer-Verlag, Berlin-New York (1978). vii + 313 pp. $13.50. ISBN 3-540-08756-7; MR 58 #11337.
915. Tammi, O. *On optimized inequalities in connection with coefficient bodies of bounded univalent functions.* Ann. Acad. Sci. Fenn. Ser. A I Math. 4 (1979), no. 1, 45–52; MR 80f:30012.
916. Tammi, O. *A survey of some classes of univalent functions and methods for studying them.* (Finnish; English summary) Arkhimedes 33 (1981), no. 4, 215–231.
917. Targosz, J.; Targosz, R. *The variational method applied to Gel'fer bounded functions.* (Polish; Russian and English summaries) Zeszyty Nauk. Politech. Slask. Mat.-Fiz. No. 34 (1979), 127–135; MR 81i:30042.
918. Targosz, R. *Formulas for the Nehari coefficients for Aharonov pairs of univalent functions.* (Polish; Russian and English summaries) Zeszyty Nauk. Politech. Slask. Mat.-Fiz. No. 34 (1979), 117–126; MR 81i:30032.
919. Thangamani, J. *The radius of univalence of certain analytic functions.* Indian J. Pure Appl. Math. 10 (1979), no. 11, 1369–1373; MR 81a:30012.
920. Thangamani, J. *On starlike functions with respect to symmetric points.* Indian J. Pure Appl. Math. 11 (1980), no. 3, 392–405; MR 81i:30031.
921. Thangamani, J. *On a generalization of the class of functions with bounded Mocanu variation.* Proc. Indian Acad. Sci. Math. Sci. 90 (1981), no. 3, 213–218.
922. Timoney, R. M. *A necessary and sufficient condition for Bloch functions.* Proc. Amer. Math. Soc. 71 (1978), no. 2, 263–266; MR 58 #1159.
923. Todorov, P. *On certain univalent conformal mappings that are realized by Gauss's hypergeometric function.* (Bulgarian; Russian and French summaries) Plovdiv. Univ. Naucn. Trud. 12 (1974), no. 1, 59–64; MR 53 #11034.
924. Todorov, P. *Maximal univalent mappings with bounded rotation that are realized by the class of integrals of Schwarz-Christoffel type.* (Bulgarian; Russian and French summaries) Plovdiv. Univ. Naucn. Trud. 12 (1974), no. 1, 49–58.
925. Todorov, P. G. *Application of maximal univalence domains of*

classes of rational functions with simple poles and politive residues to the uncovering of the structure of magnetic fields of systems of equally directed direct currants. (Russian; French summary) Plovdiv. Univ. Naucn. Trud. 13(1975), no. 1, 317–321 (1977); MR 58 #22529.

926. Todorov, P. G. *On the unique maximal domain of univalence of the derivative class* $\Phi(v-1)(R_v)$ *of the rational class* $\Phi(R_v)$. (Russian; French summary) Plovdiv. Univ. Naucn. Trud. 16(1978), no. 1, 303–314 (1980).

927. Todorov, P. G. *New explicit formulas for the coefficients of p-symmetric functions*. Proc. Amer. Math. Soc. 77(1979), no. 1, 81–86; MR 81g:30021.

928. Todorov, P. G. *A simple proof of the Kirwan theorem for the radius of starlikeness of the typically real functions and one new result*. Acad. Roy. Belg. Bull. Cl. Sci. (5)66(1980), no. 4, 334–342; MR 82a:30021.

929. Todorov, P. G. *Radius of convexity for a Nevanlinna class of analytic functions*. (Bulgarian; Russian and English summaries) Plovdiv. Univ. Naucn. Trud. 16(1978), no. 1, 315–323 (1980); MR 82f:30013.

930. Todorov, P. G. *Limite de convexité d'une classe de Nevanlinna de fonctions analytiques*. Arch. Math. (Basel) 34(1980), no. 2, 127–121; MR 81k:30017.

931. Todorov, P. G. *New explicit formulas for the nth derivative of composite functions*. Pacific J. Math. 92(1981), no. 1, 217–236.

932. Todorov, P. G. *Explicit formulas for the coefficients of Faber polynomials with respect to univalent functions of the class* Σ. Proc. Amer. Math. Soc. 82(1981), no. 3, 431–438; MR 82f:30016.

933. Todorov, P. G. *Uniqueness of the maximal univalence domain of the derivative class* $\Phi^{(v-1)}(R_v)$ *of the rational class* $\Phi(R_v)$. (Russian; Lithuanian and French summaries) Litovsk. Mat. Sb. 21 (1981), no. 1, 203–208; MR 82j:30024.

934. Townsend, D. W. *Imaginary values of meromorphic functions in the disk*. Pacific J. Math. 96(1981), no. 1, 225–242.

935. Trimble, S. Y.; Wright, D. J. *Close-to-convex functions and their extreme points in Hornich space*. Monatsh. Math. 85(1978), no. 3, 235–244; MR 58 #11355.

936. Tsanov, V. V. (Canov, V. V.) *Extreme points and uniformization*. C. R. Acad. Bulgare Sci. 30(1977), no. 2, 179–181; MR 56 #5868.

937. Tsuji, H. *A generalization of Schwarz lemma*. Math. Ann. 256 (1981), no. 3, 387–390.

938. Tuan, P. D.; Anh, V. V. *Radii of starlikeness and convexity for certain classes of analytic functions.* J. Math. Anal. Appl. 64(1978), no. 1, 146-158; MR 58 #1125.
939. Tuan, P. D.; Anh, V. V. *Radii of convexity of two classes of regular functions.* Bull. Austral. Math. Soc. 21(1980), no. 1, 29-41; MR 81g:30022.
940. Tuan, P. D.; Anh, V. V. *Extremal problems for functions of positive real part with a fixed coefficient and applications.* With a loose Russian summary. Czechoslovak Math. J. 30(105) (1980), no. 1, 302-312; MR 81f:30015.
941. Turzynski, A. *Estimation of coefficients in some classes of holomorphic functions.* (Polish and Russian summaries) Zeszyty Nauk. Politech. Lodz. No. 245 Mat. 8(1976), 81-84; MR 54 #7770.
942. Turzynski, A. *On a subclass of starlike functions.* (Polish and Russian summaries) Zeszyty Nauk. Politech. Lodz. Mat. No. 11 (1978), 73-79; MR 82k:30017.
943. Twomey, J. B. *An integral mean inequality for starlike functions.* Mathematika 28(1981), no. 1, 88-98; MR 82k:30018.
944. Uluçay, C. *On an extremalization method in the theory of analytic functions.* (Turkish summary) Comm. Fac. Sci. Univ. Ankara Ser. A 23(1974), no. 9, 85-93; MR 54 #13062.
945. Umarani, P. G. *On a subclass of spiral-like functions.* Indian J. Pure Appl. Math. 10(1979), no. 10, 1292-1297; MR 80i:30027.
946. Umarani, P. G. *On a generalized class of functions of bounded boundary rotation.* (French summary) Rend. Mat. (7)1(1981), no. 1, 127-138; MR 82i:30021.
947. Umezawa, T.; Reade, M. O. *Functions whose derivative has a positive real part.* Collections of articles on mathematics and mathematical education, No. 1, pp. 103-108, Saitama Univ. Urawa (1980); MR 82k:30019.
948. Umezawa, T.; Takijima, K. *On the univalence and close-to-convexity of a certain integral.* J. Saitama Univ. Fac. Ed. Math. Natur. Sci. 23(1974), 3-8; MR 54 #13063.
949. Umgeher, K. *Zur Frage multivalenter Lösungen von linearen Differential-gleichungen.* Osterreich. Akad. Wiss. Math.-Natur. Kl. S.-B. II 183(1975), no. 8-10, 413-421; MR 54 #535.
950. Umgeher, K. *Lineare Differentialgleichungen mit multivalenten Lösungen.* Collection in honor of Hans Hornich. Osterreich. Akad. Wiss. Math.-Naturwiss. Kl. S.-B. II 185(1976), no. 1-3, 115-120; MR 56 #15909.
951. Umgeher, K. *Die Verteilung der schlichten Funktionen in einem*

Funktionenraum. (English summary) Monatsh. Math. 81(1976), no. 4, 311–314; MR 54 #2947.
952. Umgeher, K. *Über eine Produktdarstellung holomorpher Funktionen.* (Italian summary) Atti Accad. Sci. Instit. Bologna Cl. Sci. Fis. Rend. (13)5(1977/78), no. 1, 149–156; MR 80f:30003.
953. Vazdaev, V. P. *Extremal problems in classes of functions that are regular in the half-plane.* (Russian) Izv. Vyss. Ucebn. Zaved. Matematika (1975), no. 11(162), 33–40; MR 53 #13550.
954. Viswanath, G. R. *On convex univalent functions.* Portugal. Math. 35(1976), no. 3–4, 221–225; MR 57 #6398.
955. Vo, D. T. *Verhalten schlicht-konformer Abbildungen in Kreisringe eingebetteter Gebiete.* Math. Nachr. 74(1976), 99–134; MR 57 #3369.
956. Vo, D. T. *Über einige Flächeninhaltsformeln bei schlicht-konformer Abbildung von Kreisbogenschlitzgebieten.* Math. Nachr. 74(1976), 253–261.
957. Volkovyskii, L. I. **Questions in Mathematics.* (Russian) Partial Proceedings of the Seminar-Conference on Geometric Theory of Functions of a Complex Variable held in Tashkent, May 10–19 (1975). Edited by L. I. Volkovyskii. Taskent. Gos. Univ., Tashkent, 1976, 156 pp.; MR 57 #6377.
958. Volkovyskii, L. I.; Lunc, G. L.; Aramanovic, I. G **Problems in the theory of functions of a complex variable.* Translated from the Russian by Victor Shiffer. Second edition, revised from the 1975 Russian edition published by Izdat. "Nauka." Mir Publishers, Moscow (1977).
959. Volynec, I. A. *Distortion under mappings of class BL.* (Russian) Sibirsk. Mat. Z. 18(1977), no. 6, 1259–1270, 1435.
960. Waadeland, H. *Some univalent compositions of polynomials with univalent functions.* Ann. Polon. Math. 38(1980), no. 1, 65–73; MR 82e:30021.
961. Walczak, S. (A) *Investigations of conditional extrema in the class of Hardy functions and in the family of univalent functions.* I. (Polish summary) (B) Same title as in (A). II. (Polish summary) Bull. Soc. Sci. Lett. Lodz, (A) 27(1977), no. 7, 15 pp.; MR 80f:30026a; (B) 28(1978), no. 1, 14 pp.; MR 80f:30026b.
962. Wald, J. K. *On starlike functions.* Ph.D. thesis (1978), University of Delaware, Newark, Delaware.
963. Wang, S. L. *The behavior of functions regular in the unit circle.* (Chinese) Acta Math. Sinica 21(1978), no. 1, 91–93; MR 80d:30036.

964. Waniurski, J. *A note on extremal properties for certain family of convex mappings.* (Polish and Russian summaries) Ann. Univ. Mariae Curie-Sklodowska Sect. A 29(1975), 151–158(1977); MR 56 #15910.
965. Waniurski, J. *Some relations between starlike and convex functions.* Porceedings of the Sixth Conference on Analytic Funcitons (Krakow, 1974). Ann. Polon. Math. 33(1976), no. 1-2, 131–135; MR 55 #648.
966. Wellstein, H. *Schlichte Funktionen im Einheitskreis, die einen Randbogen in einen Randbogen abbilden.* J. Reine Angew. Math. 285(1976), 24–27; MR 54 #538.
967. Wesolowski, A. *Sur certaines estimations dans la classe $\Sigma^*(\alpha, \gamma)$.* (Polish and Russian summaries) Ann. Univ. Mariae Curie-Sklodowska Sect. A 29(1975), 159–165(1977); MR 56 #8833.
968. Wilken, D. R.; Feng, J. *A remark on convex and starlike functions.* J. London Math. Soc. (2), 21(1980), 287–290; MR 81f:30016.
969. Williams, R. K. *On the linearity of one-to-one entire functions.* Delta (Waukesha) 6(1976), no. 2, 72–77; MR 55 #3256.
970. Wirths, K.-J. *Über Drehungssätze fur gewisse Klassen holomorpher Funktionen.* Festband anlablich des 65 Geburtstages von Ernst Peschl, pp. 71–76. Gesellsch. Math. Datenverarbeitung Bonn, Ber. No. 57, Gesellsch. Math. Datenverarbeitung, Bonn (1972); MR 51 #13207.
971. Wirths, K.-J. *Bemerkungen zu einem Satz von Fejér.* (Russian summary) Anal. Math. 1(1975), no. 4, 313–318; MR 53 #781.
972. Wirths, K.-J. *Über totalmonotone Zahlenfolgen.* Arch. Math., vol. 26(1975), pp. 508–517.
973. Wirths, K.-J. *Über holomorphe Funktionen, die einer Wachstumsbeschränkung unterliegen.* Arch. Math. (Basel) 30(1978), no. 6, 606–612; MR 58 #11369.
974. Wlodarczyk, K. *Inequalities of Grunsky-Nehari type for pairs of vector functions.* Ann. Polon. Math 37(1980), no. 2, 179–198; MR 81h:30019.
975. Wlodarczyk, K. *Analogues of the Garabedian-Schiffer theorem for pairs of vector functions.* Demonstratio Math. 14(1981), no. 2, 321–341; MR 82k:30020.
976. Wrzesien, A. *On meromorphic quasi-β-spiral-starlike functions.* (Polish and Russian summaries) Zeszyty Nauk. Politech. Lodz. Mat. No. 12(1979), 63–81.
977. Wu, Z. R. *On a class of analytic functions related to starlike functions.* (Chinese) Acta Math. Sinica 24(1981), no. 2, 283–290.

978. Yamashita, S. *The derivative of a bounded holomorphic function in the disk.* Proc. Amer. Math. Soc. 53(1975), no. 1, 60–64; MR 51 #13235.
979. Yamashita, S. *Almost locally univalent functions.* Monatsh. Math. 81(1976), no. 3, 235–240; MR 53 #11042.
980. Yamashita, S. *Local schlichtness of a function meromorphic in the disk.* Math. Nachr. 77(1977), 163–166; MR 56 #12243.
981. Yamashita, S. *Schlicht holomorphic functions and the Riccati differential equation.* Math. Z. 157(1977), no. 1, 19–22; MR 58 #6216.
982. Yamashita, S. *Conformality and semiconformality of a function holomorphic in the disk.* Trans. Amer. Math. Soc. 245(1978/79), 119–138; MR 80h:30018.
983. Yamashita, S. *On the John constant.* Math. Z. 161(1978), no. 2, 185–188; MR 58 #22516.
984. Yamashita, S. *A univalent function nowhere semiconformal on the unit circle.* Proc. Amer. Math. Soc. 69(1978), no. 1, 85–86; MR 57 #6403.
985. Yamashita, S. *The normality of the logarithmic derivative.* Comment. Math. Univ. St. Paul. 27(1978/79), 25–28.
986. Yamashita, S. *Inequalities for the Schwarzian derivative.* Indiana Univ. Math. J. 28(1979), no. 9, 131–135; MR 81j:30030.
987. Yamashita, S. *On a theorem of Duren, Shapiro and Schields.* Proc. Amer. Math. Soc. 73(1979), no. 2, 180–182; MR 80b:30012.
988. Yamashita, S. *Criteria for functions to be of Hardy class H^p.* Proc. Amer. Math. Soc. 75(1979), no. 1, 69–72.
989. Yamashita, S. *Criteria for functions to be Bloch.* Bull. Austral. Math. Soc. 21(1980), no. 2, 223–227; MR 81g:30041.
990. Yamashita, S. *Criteria for functions to be Bloch.* Bull. Austral. Math. Soc. 21(1980), no. 2, 223–227; MR 81g:30041.
991. Yamashita, S. *The normality of the derivative.* Rev. Roumaine Math. Pures Appl. 25(1980), no. 3, 481–484; MR 81g:30023.
992. Yamashita, S. *Gap series and α-Bloch functions.* Yokohama Math. J. 28(1980), no. 1–2, 31–36.
993. Yamashita, S. *Length estimates for holomorphic functions.* Proc. Amer. Math. Soc. 81(1981), no. 2, 250–252.
994. Yamashita, S. *The order of normality and meromorphic univalent functions.* Colloq. Math. 44(1981), no. 1, 159–163; MR 82j:30044.
995. Yoshikawa, H.; Yoshikai, T. *Some notes on Bazilevič functions.* J. London Math. Soc. (2)20(1979), no. 1, 79–85; MR 80i:30028.
996. Zderkiewicz, J. *Sur la courbure des lignes de niveau dans la classe des fonctions convexes d'ordre α.* (Polish and Russian summaries)

Ann. Univ. Mariae Curie-Sklodowska Sect. A 27(1973), 131–138 (1975); MR 53 #3285.

997. Zderkiewicz, J. *Sur la courbure des lignes de niveau dans la classe* Σ_a^c. (Polish and Russian summaries) Ann. Univ. Mariae Curie-Sklodowska Sect. A 28(1974), 121–126(1976); MR 54 #5454.

998. Zderkiewicz, J. *Sur la subordination dans la classe $S^*(p)$ des fonctions etoilees d'ordre* ϱ. Comment. Math. Prace Mat. 19(1976/77), no. 2, 409–414; MR 57 #12835.

999. Zderkiewicz, J. *Sur la courbure de l'image du rayon dans la classe S_α^c des fonctions convexes d'ordre* α. Comment. Math. Prace Mat. 19(1976/77), no. 2, 415–419; MR 57 #12836.

1000. Zderkiewicz, J. *Sur le rayon de β-convexité de la famille* Σ_a^*. Ann. Polon. Math. 34(1977), no. 1, 39–42; MR 55 #649.

1001. Zeheb, E. *Root locus techniques applied to proving equivalence between positive realness conditions.* Israel J. Tech. 13(1975), no. 4, 276–278; MR 54 #10577.

1002. Zemyan, S. M. *A minimal outer area problem in conformal mapping.* J. Analyse Math. 39(1981), 11–23; MR 82i:30022.

1003. Ziegler, M. R. *An extremal problem for functions of positive real part with vanishing coefficients.* (Polish and Russian summaries) Ann. Univ. Mariae Curie-Sklodowska Sect. A 30(1976), 85–89 (1978); MR 58 #22518.

1004. Zielinska, A.; Zyskowska, K. *On estimation of the eighth coefficient of bounded univalent functions with real coefficients.* Demonstratio Math. 12(1979), no. 1, 231–246; MR 80i:30032.

1005. Zmorovič, V. A.; Gudz, L. A. *On A. Marx's conjecture for the convex hull of the class of functions that are star-shaped in the disc.* (Russian) Ukrain. Mat. Z 28(1976), no. 3, 390–393, 430; MR 55 #650.

1006. Zmorovič, V. A.; Gudz, L. A. *Some extremal estimates for classes of normalized univalent functions with the second coefficient of their Taylor expansion fixed.* (Russian) Mathematics collection (Russian), pp. 68–72. Izdat. "Naukova Dumka," Kiev (1976); MR 56 #12249.

1007. Zmorovič, V. A.; Gudz, L. A. *The conjecture of A. Marx for the class of k-tuply symmetric starlike functions in the disc* $|z| < 1$. (Russian; English summary). Dokl. Akad. Nauk Ukrain. SSR Ser. A (1976), no. 7, 587–590, 670; MR 55 #3234.

1008. Zmorovič, V. A.; Gudz, L. A. *A parametric method for the solution of extremal problems of the theory of special classes of analytic functions.* (Russian) Ukrain. Mat. Z. 30(1978), no. 3,

362–367; MR 58 #17070.
1009. Zmorovič, V. A.; Gudz, L. A. *Some classes of univalent functions in the disc* $|x| < 1$ *and in the domain* $|x| < 1$. (Russian) Mathematical analysis and probability theory (Russian), pp. 59–63, 210, "Naukova Dumka," Kiev (1978); MR 81b:30025.
1010. Zmorovič, V. A.; Gudz, L. A. *On a method of the theory of extremal problems.* (Russian; English summary) Dokl. Akad. Nauk Ukrain. SSR Ser. A (1981), no. 8, 20–22, 96.
1011. Zmorovič, V. A.; Jakubenko, A. A. *A certain generalization of a class of α-convex Mocanu functions.* (Russian) Mathematics collection (Russian), pp. 254–257. Izdat. "Naukova Dumka," Kiev (1976); MR 58 #6208.
1012. Zmorovič, V. A.; Jakubenko, A. A. (Jakubenko, O. A.) *Some sharp estimates in the class* $B_0(b)$. (Russian; English summary) Dokl. Adac. Nauk Ukrain. SSR Ser. A (1977), no. 8, 686–689, 763; MR 57 #9956.
1013. Zmorovič, V. A.; Jakubenko, A. A. *A generalization of the Mocanu-Reade theorem on the boundary of α-convexity of a class of star functions.* (Russian) Mathematical analysis and probability theory (Russian), pp. 70–74, 211, "Naukova Dumka," Kiev (1978); MR 81b:30026.
1014. Zmorovič, V. A.; Jakubenko, A. A. (Jakubenko, O. A.) *The boundary of α-convexity of order* γ *of class* $S^*(m)$. (Russian; English summary) Dokl. Akad. Nauk Ukrain. SSR Ser. A No. 3 (1979), 165–168, 237; MR 80i:30029.
1015. Zmorovič, V. A.; Jakubenko, O. A. *On the boundary of* α*-convexity of order* γ *of class* $S_\beta^*(m)$. (Russian; English summary) Dokl. Akad. Nauk Ukrain. SSR Ser. A (1981), no. 10, 6–9, 93.
1016. Zmorovič, V. A.; Korobkova, I. K. *On the order of starlikeness of convex functions of order* α. (Russian; English summary) Dokl. Akad. Nauk Ukrain. SSR Ser. A (1977), no. 7, 584–587, 669; MR 58 #6209.
1017. Zmorovič, V. A.; Korobkova, I. K. *The order of starlikeness of a class of α-convex functions for* $\alpha = 2$. (Russian) Mathematical analysis and probability theory (Russian), pp. 63–66, 211, "Naukova Dumka," Kiev (1978); MR 81j:30027.
1018. Zmorovič, V. A.; Korobkova, I. K. *The order of starlikeness of certain subclasses of α-convex functions for* $\alpha > 0$. (Russian) Ukrain. Mat. Z. 30(1978), no. 4, 536–539, 573; MR 80a:30014.
1019. Zmorovič, V. A.; Nikolaeva, R. V. *Certain integral transforms in classes of univalent functions.* (Ukrainian; English and Russian

summaries) Dopovidi Akad. Nauk Ukrain. RSR Ser. A (1975), no. 11, 976–978, 1050; MR 58 #6207.
1020. Zmorovič, V. A.; Nikolaeva, R. V. *Certain integral transforms in classes of univalent functions.* (Ukrainian; English and Russian summaries) Dopovidi Akad. Nauk Ukrain. RSR Ser. A (1975), no. 11, 976–978, 1050; MR 58 #6207.
1021. Zmorovič, V. A.; Nikolaeva, R. V. *Integral transformation with logarithmic kernels in some classes of univalent functions.* (Russian) Mathematical analysis and probability theory (Russian), pp. 66–70, 211 "Naukova Dumka," Kiev (1978; MR 82d:30016.
1022. Zmorovič, V. A.; Pohilevič, V. A. (Pohilevic, V. O.] *α-convex functions for $\alpha < 0$.* (Russian) Ukrain. Mat. Z. 31(1979), no. 4, 431–433, 478; MR 81e:30026.
1023. Zmorovič, V. A.; Pohilevič, V. O. *On α-almost convex functions.* (Russian) Ukrain. Mat. Z. 33(1981), no. 5, 670–673, 718.
1024. Zuravlev, I. V. *Some problems for Bloch functions.* (Russian) Dokl. Akad. Nauk SSSR 236(1977), no. 1, 21–22; MR 56 #15903.
1025. Zuravlev, I. V. *Univalent functions and Teichmüller spaces.* (Russian) Dokl. Akad. Nauk SSSR 250 (1980), no. 5, 1047–1050; MR 81b:30085.

Table 1

Following is a list of those references in this Bibliography (Part III) which were published prior to the year 1976 and which were not included in either Bibliography (Part I) or Bibliography (Part II).

Year Reference

Year	Reference
1975	19, 69, 88, 89, 90, 92, 125, 133, 145, 146, 152, 181, 183, 184, 185, 190, 238, 277, 297, 305, 346, 365, 420, 425, 457, 465, 502, 510, 572, 594, 598, 606, 626, 643, 650, 666, 667, 687, 713, 722, 740, 783, 792, 794, 801, 817, 853, 908, 925, 949, 953, 957, 964, 967, 971, 972, 978, 1001, 1019, 1020.
1974	12, 17, 51, 67, 81, 166, 201, 211, 215, 247, 364, 380, 394, 466, 468, 478, 503, 509, 531, 534, 587, 604, 665, 671, 689, 690, 691, 712, 765, 780, 803, 844, 867, 884, 885, 905, 911, 923, 924, 944, 948, 997.
1973	1, 49, 218, 276, 357, 453, 697, 720, 791, 816, 897, 996.

Year	References
1972	263, 405, 575, 719, 881, 904, 970.
1971	11, 214, 234, 262, 352, 471, 652, 738, 739, 814, 815.
1970	50, 261, 545, 910.
1969	20, 260, 438.
1968	494.
1967	437.
1966	347.
1965	513, 634.
1963	749.
1962	470.
1960	748.
1959	640.
1955	662.
1953	661.
1952	660.
1946	500.
1934	114.
1933	499.
1932	379, 621.
1928	244.
1920	903.
1914	243.
1913	242.

Table 2

Following is a list of those references in this Bibliography (Part III) which were published during the year *1982:*

References: 52, 167, 383, 549, 551, 685.

Table 3

Following is a listing of the total number of research papers in this Bibliography (Part III) which were published in each of the given years.

Year	Total Number of Papers
1976	199
1977	148
1978	132
1979	143
1980	130
1981	107

Notes For the year 1976, Bibliography Part II contains 38 references, while Bebliography Part III contains 199 references,—for a total of 237 references.

Corrections

Following is a list of references in this Bibliography (Part III) indicating duplications of papers:

Reference	same as	*Reference*
990	......	989
1020	......	1019